Joint Communications and Sensing

IEEE Press
445 Hoes Lane
Piscataway, NJ 08854

Joint Communications and Sensing

From Fundamentals to Advanced Techniques

Kai Wu
J. Andrew Zhang
Y. Jay Guo

Global Big Data Technologies Centre
University of Technology Sydney
Sydney, Australia

IEEE PRESS
WILEY

Published by John Wiley & Sons, Inc., Hoboken, New Jersey.
Published simultaneously in Canada.

Library of Congress Cataloging-in-Publication Data applied for:

Hardback: 9781119982913

Cover Design: Wiley
Cover Image: © RCW.studio/Shutterstock

Set in 9.5/12.5pt STIXTwoText by Straive, Chennai, India

Contents

About the Authors

Kai Wu received a BE from Xidian University, Xi'an, China, in 2012, a PhD from Xidian University in 2019, and a PhD from the University of Technology Sydney (UTS), Sydney, Australia, in 2020. His Xidian-PhD won the "Excellent Thesis Award 2019" (a highest national award for PhD in electronic engineering fields) from the Chinese Institute of Electronics. His UTS-PhD was included in the "Chancellor's list 2020" of UTS.

Dr Kai Wu is now a research fellow at the Global Big Data Technologies Centre, UTS. His research focus has been on joint communications and sensing post-PhD. In particular, he has advanced waveform and system designs of the frequency-hopping MIMO radar-based communications, with four IEEE journal papers plus a book chapter published on this topic. Dr Wu has also been developing efficient algorithms to perform high-quality sensing using mainstream and future communication waveforms, such as OFDM and OTFS. One of his recent works established a unified sensing framework employing most, if not all, communication waveforms, which has aroused great interest and attention from academia and industry. Moreover, Dr Wu was also a tutorial speaker on joint communications and sensing in IEEE ICC'21 and WCNC'21.

Dr. J. Andrew Zhang received the BSc degree from Xi'an JiaoTong University, China, in 1996, the MSc degree from Nanjing University of Posts and Telecommunications, China, in 1999, and the PhD degree from the Australian National University, in 2004.

Currently, Dr. Zhang is an associate professor in the School of Electrical and Data Engineering, University of Technology Sydney, Australia. He was a researcher with Data61, CSIRO, Australia, from 2010 to 2016, the Networked Systems, NICTA, Australia, from 2004 to 2010, and ZTE Corp., Nanjing, China, from 1999 to 2001. Dr. Zhang's research interests are in the area of signal processing for wireless communications and sensing. He has published more than 250 papers in leading international journals and conference proceedings, and has won five best paper awards. He is a recipient of CSIRO Chairman's Medal and the Australian Engineering Innovation Award in 2012 for exceptional research achievements in multigigabit wireless communications.

Prof. Zhang is one of the leading researchers in joint communication and radar/radio sensing (JCAS) technologies. He initiated the concept of perceptive mobile network, by defining its system framework and demonstrating its feasibility in a set of papers back to 2017. Prof. Zhang co-organized multiple IEEE workshops on JCAS and served as a guest editor for multiple JCAS special issues on leading journals. He has also delivered JCAS conference tutorials in WCNC 2021, ICC 2021, and ICC2022, and offered numerous keynotes and invited talks. He is serving as the publication cochair of ISAC-ETI and the editor-in-chief of its official publications, ISAC-Focus. In addition to academic research, Prof. Zhang is also working with the industry to make JCAS a reality and has developed two proof-of-concept demonstrators. He has led the completion of four industrial projects of applying JCAS technologies in cellular, Wi-Fi, and UAV communication networks and has attracted more than 4 million dollars of research fund in this area.

Y. Jay Guo is a distinguished professor and the founding director of Global Big Data Technologies Centre (GBDTC) at the University of Technology Sydney (UTS), Australia. He is also the founding technical director of the New South Wales Connectivity Innovation Network. Prior to joining UTS in 2014, he served as a research director in CSIRO for over nine years. Before joining CSIRO, he held various senior technology leadership positions in Fujitsu, Siemens, and NEC in the United Kingdom. His research interests include antennas, mm-wave, and THz communications and sensing systems as well as big data technologies. He has published six books and over 600 research papers including over 300 IEEE journal papers, and he holds 26 international patents. His main technical contributions are in the fields of Fresnel antennas, reconfigurable antennas, hybrid antenna arrays and, most recently, analogue multibeam antennas, and joint communications and sensing (JCAS) systems for 6G.

Prof Guo is a fellow of the Australian Academy of Engineering and Technology and a fellow of IEEE. He was a member of the College of Experts of Australian Research Council (ARC, 2016–2018). He has won a number of the most prestigious Australian national awards including the Engineering Excellence Awards (2007, 2012) and CSIRO Chairman's Medal (2007, 2012). He was named one of the most influential engineers in Australia in 2014 and 2015, and one of the top researchers across all fields in Australia in 2020 and 2021, respectively. He and his students have won numerous best paper awards.

Acknowledgments

We would like to thank our families, colleagues, students, and collaborators for their sacrifices and contributions made to our research. We are indebted to our industrial and government sponsors, especially the Australian Research Council (ARC), for the generous funding we have received.

Preface

Wireless communications and radar are traditionally two different industries, academic disciplines, and research communities. Motivated by spectrum and hardware sharing, and energy saving, there has been a rapidly increasing market pull and technology push to integrate the two systems into one for many practical applications. This makes a lot of sense as both of these systems require radio spectrum and employ similar hardware such as antennas, radio frequency (RF), and digital circuits. From the point of signal processing, they both need filtering, Fourier transforms, and, most recently, machine learning.

Early efforts in this endeavor were focused on cognitive radio, with the aim to share resources and reconfigure the system to deliver the functionality required on the fly and manage mutual interference between the two coexisting systems when they arise. However, the current trends in research and industrial development are to design new systems with both functions optimized jointly, thereby maximizing the spectrum and hardware sharing and energy savings while delivering the best performance trade-off. As radar technologies and applications have been expanded and diversified to realize more general sensing, the new dual-function systems are now widely referred to as joint communications and sensing (JCAS) systems, a.k.a., integrated sensing and communications (ISAC). Great examples of JCAS systems are beyond 5G and 6G communication networks that will extend the communication-only designs in previous generation systems to include wireless sensing capabilities. Just imagine: if we can leverage the expected universal wireless communication infrastructure to conduct wide-area and pervasive sensing, we would have unprecedented availability of real-time data to manage our environment, our cities, and industrial activities much more effectively yet with marginal extra cost. Even for this reason alone, we can easily see the huge potential of JCAS and be excited to participate in this technology innovation.

This book has two objectives. First, we intend to provide the fundamentals of basic architectures, applications, and signal processing techniques of JCAS. Second, we aim to present a range of advanced techniques for designing high-performance JCAS systems, including both communication-centric and radar-centric systems. The two systems have communications and sensing as primary functions, respectively. We sincerely hope that the book will prove valuable to both engineers working in wireless communications and sensing industries and graduate students and academic researchers.

Kai Wu, J. Andrew Zhang, and Y. Jay Guo

Acronyms

AoA	angle of arrival
AoD	angle of departure
ANM	atomic norm minimization
BBU	baseband unit
BS	base-station
CACC	cross-antenna cross-correlation
CPI	coherent processing interval
CRAN	cloud radio access network
CS	compressive sensing
CSI	channel state information
C&S	wireless communications and radar/radio sensing
DFRC	dual-function(al) radar communications
DFT	discrete Fourier transform
DMRS	demodulation reference signals
FDD	frequency division duplexing
FMCW	frequency-modulated continuous-wave
GMM	Gaussian mixture model
IFFT	inverse fast Fourier transform
IoT	Internet-of-things
JCAS	joint communications and radio/radar sensing
LFM	linear frequency modulation
LFM-CPM	LFM-continuous phase modulation
LOS	line of sight
MAC	medium access
MI	mutual information
MIMO	multiple-input and multiple-output
MISO	multiple-input and single-output
MMSE	minimum mean-square error
MMV	multi-measurement vector

mmWave	millimeter wave
NLOS	none line of sight
NR	new radio
OFDM	orthogonal frequency-division multiplexing
OFDMA	orthogonal frequency-division multiple access
PHY	physical
PAPR	peak-to-average power ratio
PMN	perceptive mobile network
PDSCH	physical downlink shared channel
PUSCH	physical uplink shared channel
PRB	physical resource-block
RFID	radio frequency identification
RIP	restricted isometry property
RIS	reconfigurable intelligent surface
RMA	recursive moving averaging
RMSE	root mean square error
RRU	remote radio unit
Rx	receiver
SC	single carrier
SDMA	spatial division multiple access
SISO	single input single output
SRS	sounding reference signals
SSB	synchronization signal and broadcast blocks
TDD	time-division duplexing
Tx	transmitter
UE	user equipment
ULA	uniform linear array
V2V	vehicle to vehicle

Part I

Fundamentals of Joint Communications and Sensing (JCAS)

1

Introduction to Joint Communications and Sensing (JCAS)

1.1 Background

Wireless communications and radar sensing have been advancing in parallel for decades. Numerous new system architectures and algorithms have been developed in the names of new generation wireless communications systems and modern radar, respectively. However, despite the fact that they share many commonalities in terms of signal processing algorithms, devices and, to certain extent, system architecture, there have been very limited intersections between the designs and the deployment of the two systems. Chiefly motivated by spectrum and cost sharing and energy saving, we are witnessing a rapidly growing interest in the coexistence, cooperation, and, most importantly, joint design of the two systems recently [1–5].

The coexistence of communication and radar systems is not new, and the issue has been extensively studied in the past decade. The focus was on developing efficient interference management techniques in order for the two individually deployed systems to operate simultaneously without interfering with each other [6]. In this setup, radar and communication systems may be colocated and or spatially separated, and they may transmit two different signals overlapped in time and/or frequency domains. They can operate simultaneously by sharing the same resources cooperatively, with a goal of minimizing interference to each other. Great efforts have been devoted to mutual interference cancelation in this case, using, for example, beamforming design, cooperative spectrum sharing, opportunistic primary–secondary spectrum sharing, and dynamic coexistence. However, effective interference cancelation typically has stringent requirements on the mobility of nodes and information exchange between them. The spectral efficiency improvement is hence limited in such schemes.

Since the interference in coexisting systems is caused by transmitting two separate signals, it is natural to ask whether it is possible to use one single transmitted

Joint Communications and Sensing: From Fundamentals to Advanced Techniques, First Edition.
Kai Wu, J. Andrew Zhang, and Y. Jay Guo.

signal for both communications and radar sensing. Radar systems typically use specially designed waveforms such as short pulses and chirps, which enable high-power radiation and simple receiver processing. However, these waveforms are not necessarily required for radar sensing. *Passive radar* or *passive sensing* is a good example of exploring diverse radio signals for sensing [7, 8]. In principle, the objects to be sensed or detected can be illuminated by any radio signals of sufficient power, such as TV signals, Wi-Fi signals, and mobile (cellular) signals. This is because the propagation of radio signals is always affected by the static and dynamic environments such as transceiver movement, surrounding objects' movement and profile variation, and even weather changes. Hence, the environmental information is embedded in the received radio signals and can be extracted by using passive radar techniques. However, there are two major limitations with passive sensing. First, the clock phases between transmitter and receiver are not synchronized in passive sensing, and there are always unknown and possibly time-varying timing, frequency, and phase offsets between the transmitted and received signals. This leads to timing and therefore ranging ambiguity in the sensing results, as well as causing difficulties in aggregating multiple measurements for joint processing. Second, the sensing receiver may not know the signal structure. As a result, passive sensing lacks the capability of interference suppression, and it cannot separate multiuser signals from different transmitters. Admittedly, the radio signals are usually not optimized for sensing in any way.

The most recent trend is that radar systems are evolving toward more general *radio sensing*. We prefer the term "radio sensing" to radar due to its generality and comprehensiveness. Radio sensing here refers to retrieving information from received radio signals; this is in contrast to extracting information from the communication data modulated to the signal at the transmitter. It can be achieved through the measurement of *sensing parameters* related to location and movement, such as time delay, angle-of-arrival (AoA), angle-of-departure (AoD), Doppler frequency, and magnitude of multipath signals, and *physical feature parameters* such as inherent "radio signature" of devices/objects/activities. The two corresponding processing activities are called *sensing parameter estimation* and *pattern recognition* in this book. In this sense, radio sensing refers more to general sensing techniques and applications using radio signals, just like video sensing using video signals. Radio sensing has a diverse range of applications such as object, activity, and event recognition in Internet of Things (IoT), Wi-Fi, and 5G networks. These radio signals are transmitted by an existing infrastructure and are not specifically designed for sensing purpose. The paper [9] presents numerous Wi-Fi sensing applications where, for instance, Wi-Fi signals have been used for people and behavior recognition in indoor environments. In [10], it is shown that other radio signals, such as radio-frequency identification (RFID) and ZigBee, can also be used for activity recognition. These publications demonstrate the strong potential of using low-bandwidth communication signals for radio-sensing applications.

Joint communications and radar/radio sensing (JCAS) [11, 12] is emerging as an attractive solution to integrating communications and sensing into one system. It has also been known under different terms, such as radar-communications (RadCom) [1], joint radar (and) communications (JRC) [3, 13, 14], joint communications (and) radar (JCR) [15], dual-function(al) radar communications (DFRC) [16, 17], and more recently, integrated sensing and communications (ISAC). In a JCAS system, a single transmitted signal for both communications and sensing is jointly designed and employed. The objective for JCAS is that the majority of transmitter modules can be shared by communications and sensing. In such a system, most of the receiver hardware can also be shared, but some receiver baseband signal processing would be different for communications and sensing. By virtue of joint design, JCAS can also potentially overcome the many limitations in passive sensing. These properties make JCAS significantly different from existing spectrum sharing concepts such as cognitive radio, the aforementioned coexisting communication-radar systems, and "integrated" systems using separated waveforms [18] where communications and sensing signals are separated in such resources as time, frequency, and code, despite the two functions may physically be combined in one system. In Table 1.1, we briefly compare the signal formats and key features, advantages, and disadvantages of five types of systems: communications and sensing with separated waveforms, coexisting communications and sensing, passive sensing, cognitive radio, and JCAS.

1.2 Three Categories of JCAS Systems

The initial concept of integrated communications and sensing may be traced back to the 1960s [3] and had been primarily investigated for developing multimode or multifunction military radars. In early days, most of such systems belonged to the type in which communications and sensing use separated waveforms, as detailed in Table 1.1. There has been very limited research on JCAS for civil systems before 2010. In the past 10 years, JCAS has been receiving rapidly growing interest and is being considered as a candidate for next generations of communications, radar, and sensing systems.

Based on the design priority and the underlying signal formats, the current JCAS systems may be classified into the following three categories:

- *Communication-centric design*: In this class, radio sensing is an add-on to a communication system, where the design priority is on communications.

Table 1.1 Comparison of communications and sensing (C&S) with separated waveforms, coexisting communications and sensing, passive sensing, cognitive radio, and JCAS.

Systems	Signal formats and key features	Advantages	Disadvantages
C&S with separated waveforms (e.g. [18])	– C&S signals are separated in time, frequency, code and/or polarization – C&S hardware and software are partially shared	– Low mutual interference – Almost independent design of C&S waveforms	– Low-spectrum efficiency – Low order of integration – Complex transmitter hardware
Coexisting C&S (e.g. [6, 19])	C&S use separated signals but share the same resource	Higher-spectrum efficiency	– Interference is a major issue – Nodes cooperation and complicated signal processing are typically required
Passive sensing (e.g. [7, 8, 20, 21])	– Received radio signals are used for sensing at a specifically designed sensing receiver, external to the communication system – No joint signal design at transmitter	– Without requiring any change to existing infrastructure – Higher-spectrum efficiency	– Require dedicated sensing receiver – Timing ambiguity – No waveform optimization – Noncoherent sensing and limited-sensing capability when signal structure is complicated and unknown, e.g. incapable of separating multiuser signals from different transmitters
Cognitive radio (e.g. [22])	Secondary systems coexist with primary ones by sensing spectrum holes or via interference mitigation	– Improved spectrum efficiency – Negligible impact on the operation of primary systems	Performance of secondary systems cannot be guaranteed. They also have higher complexities due to requirement for spectrum sensing and potential interference suppression
JCAS (e.g. [3, 13, 17, 23, 24])	A common transmitted signal is jointly designed and used for communications and sensing	– Highest-spectral efficiency – Fully shared transmitter and largely shared receiver – Joint design and optimization on waveform, system, and network – "Coherent sensing"	– Requirement for full-duplex or equivalent capability of a receiver colocating with the transmitter – Sensing ambiguity when transmitter and receiver are separated without clock synchronization

The aim of such a design is to exploit communication waveform to extract sensing information through target echoes. Enhancements to hardware and algorithms are required to support radio sensing. Possible enhancements to communication standards may be introduced to enable better reuse of the communication waveform for radio sensing. In this design, the communication performance can be largely unaffected; however, the sensing performance may be scenario-dependent and difficult-to-optimize.

- *Radar-centric design*: Conversely, such approaches aim at modulating or introducing information signaling in known radar waveforms. Since the radar signaling remains largely unaltered, a near optimal radar performance can be achieved. The main drawback of such approaches is the limited achievable data rates. If some performance loss can be tolerated by the radar system, better communication data rates could be obtained. Given the high-transmission power of typical radar systems, very long range communications can generally be achieved.
- *Joint design and optimization*: This class encompasses systems that are jointly designed from the start to offer a tunable trade-off between communications and sensing performance. Such systems may not be limited by any of the existing communication or radar standards and can be optimized by jointly and fairly considering the requirements for both communications and sensing.

Owing to the significant differences between traditional communication and sensing systems, the design problems in these three categories are quite different. In the first two categories, the design and research focus are typically on how to realize the other function based on the signal formats of the primary system, with the principle of not significantly affecting the primary system, though slight modifications and optimizations may be applied to the system and signals. The last category considers the design and optimization of the signal waveform, system, and network architecture, without bias to either communications or sensing, aiming at fulfilling the desired applications only.

Next, we first briefly discuss the major differences between traditional communication and radar signals, which are important for understanding the design philosophy of the three categories of JCAS systems. We then provide a brief review on the recent research progress in each of the categories, referring to the classification of the three categories of JCAS systems in terms of their technical scope, as shown in Figure 1.1.

1.2.1 Major Differences Between Communications and Sensing

Communication and radar signals are originally designed for different objectives and are generally not directly applicable to each other. Radar signals are typically

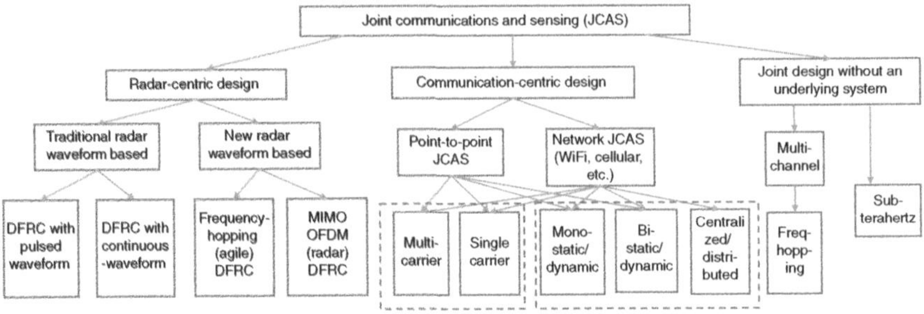

Figure 1.1 A classification of the three categories of JCAS systems.

designed to achieve high localization and tracking accuracy and to enable simple sensing parameter estimation. The following properties of radar signals are desired: low peak-to-average-power ratio (PAPR) to enable high-efficiency power amplifier and long-range operation; and a waveform ambiguity function with steep and narrow mainlobes for high resolution. In contrast, communication signals are designed to maximize the information-carrying capabilities and are typically modulated and packet-based. To support diverse devices and meet various quality-of-services requirements, communication signals can have complicated structures, with advanced modulations applied across time, frequency, and spatial domains, and being discontinuous and fragmented over these domains.

Figure 1.2 presents the simplified transceivers and signal structures of C&S to illustrate their major differences.

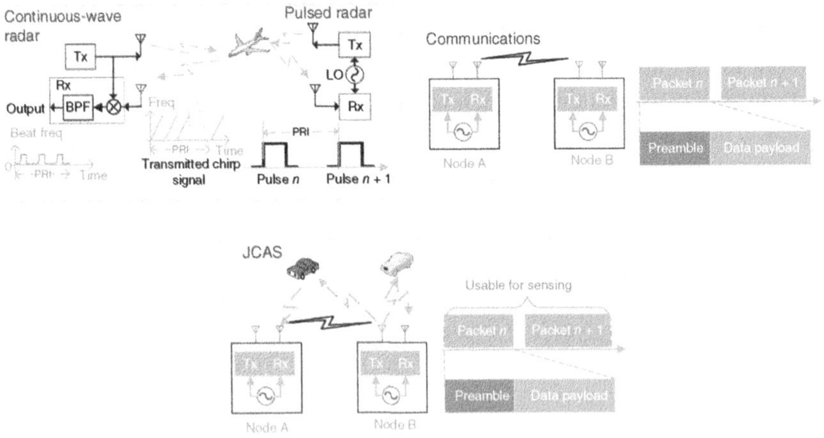

Figure 1.2 Illustration of basic pulse and continuous-wave radar, communication systems, and JCAS systems. Tx stands for transmitter; Rx for receiver; PRI for pulse repetition interval; and BPF for bandpass filter.

Conventional radar systems include pulsed and continuous-wave radars [25], as shown in Figure 1.2. In pulsed radar systems, short pulses of large bandwidth are transmitted either individually or in a group, followed by a silent period for receiving the echoes of the pulses. Continuous-wave radars transmit waveforms, such as chirp, continuously, typically scanning over a large range of frequencies. In either system, the waveforms are typically nonmodulated. These waveforms are used in both single-input and single-output (SISO) and multiple-input and multiple-output (MIMO) radar systems, with orthogonal waveforms used in MIMO radars [25].

In most of radar systems, low PAPR is a desired feature for the transmitting signal, which enables the use of high-efficiency power amplifier and long-range operation. The transmitting waveform is also desired to have an ambiguity function with steep and narrow mainlobes, which are the correlation function of the received echo signals and the local template signal. These waveforms are designed to enable low-complexity hardware and signal processing in radar receivers, for estimating key sensing parameters such as delay, Doppler frequency, and angle of arrival. However, these waveforms are not indispensable for estimating these parameters. A pulsed radar receiver typically samples the signal at a high-sampling rate twice of the transmitted pulse bandwidths, or at a relatively lower-sampling rate at the desired resolution of the delay (ranging); while the receiver of a continuous-wave radar, e.g. frequency modulated continuous wave (FMCW) radar, typically samples signals of "beat" frequency at a rate much smaller than the scanning bandwidth, proportional to the desired detection capability of the maximal delay. Here, the beat frequency refers to the difference between the frequencies of the echo signal and the transmitted signal that is used as the input to the local oscillator of the receiver, and contains the range information. Note that the FMCW signal has a unique time–frequency relationship, typically, the frequency (f) is a linear function of time (t), namely, $f = kt$ with a nonzero coefficient k determined by radar configurations. Since the echo signal is essentially the scale of the transmitted signal with a time delay, the time–frequency relationship can be preserved yet also with a time delay, i.e. $f = k(t - \tau)$ with τ being the delay. Consequently, the frequency conversion at the FMCW receiver cancels kt in the frequency, yielding a signal with a constant frequency $k\tau$ that is solely related to target delay. Due to their special signal form and hardware, radar systems generally cannot support very high-rate communications without significant modifications to the waveforms and/or receiver structure [4].

In contrast, communication signals are designed to maximize the information-carrying capabilities. They are typically modulated, and modulated signals are usually appended with nonmodulated intermittent training signals in a packet, as can be seen from Figure 1.2. To support diverse devices and meet various service requirements, communication signals can be very complicated. For example, they can be discontinuous and fragmented over time and frequency domains, have high PAPR, have complicated signal structures due to advanced modulations applied across time, frequency, and spatial domains. Although being designed without considering the demand for sensing, communication signals can potentially be used for estimating all the key sensing parameters. However, different from conventional channel estimation, which is already implemented in communication receivers, sensing parameter estimation requires extraction of the channel composition rather than channel coefficients only. Such detailed channel composition estimation is largely limited by the hardware capability. The complicated communication signals are very different to conventional radar and demand new sensing algorithms. We note that the detailed information on the signal structure, such as resource allocation for time, frequency, and space, and the transmitted data symbols, can be critical for sensing. For example, the knowledge on signal structure is important for coherent detection. In comparison, most passive radar sensing can only perform noncoherent detection with the unknown signal structure, and hence only limited sensing parameters can be extracted from the received signals with degraded performance [7].

There are also practical limitations in communication systems, such as full-duplex operation and asynchronization between transmitting node and receiving node, which requires new sensing solution to be developed. It is a fundamental challenge to address the potential requirements for full duplex operation of JCAS systems where the transmitter and sensing receivers are colocated. On the one hand, monostatic radar addresses the requirement for full-duplex operation in mainly two approaches, as illustrated in Figure 1.2, which may not be replicable in communication systems. One approach is typically applied in a pulsed radar, by applying a long silent period to receive echo signals, which essentially bypasses full-duplex operation and make the radar work in a time-division duplex mode; the other is typically used in a continuous-wave radar, via using the transmitter signal as the local template signal to the oscillator at the receiver, and detecting only the "beat" signal, the difference between the transmitted and received signals. Such designs enable low-complexity and efficient radar sensing. However, they constrain the options for integrating communications functionalities and limit the achievable communication rates. For example, there is large uncertainty with the availability and bandwidth of the beat signal; hence, information conveying will be unreliable. On the other hand, full-duplex operation is still immature for communications, and there is typically clock asynchronism between spatially separated transmitting and receiving nodes. These impose significant limits on integrating radar sensing into communications.

The differences and benefits of JCAS in comparison with individual radar or communication system are summarized in Table 1.2.

Table 1.2 Comparison among radar, communications, and JCAS.

Specifications	Radar	Communications	JCAS system
Signal waveform	Typically simple, unmodulated single-carrier signals occupying large bandwidth; pulse or continuous-wave frequency modulated; orthogonal if multiple spatial streams and orthogonality can be realized in one or more domains of time, frequency, space, and code; typically low PAPR. Radars with advanced waveforms such as orthogonal frequency-division multiplexing (OFDM) and frequency hopping is also emerging	Mix of unmodulated (pilots/training sequences) and modulated data symbols; Complicated signal structure and resource usage; advanced modulations, e.g. orthogonal frequency-division multiple access (OFDMA) and multiuser-MIMO; High PAPR	JCAS can use both traditional radar and communication signals, with appropriate modifications to support both communications and sensing and optimize their performance jointly
Tx power	Typically high in large-scale and long-range radar; low in short-range radar such as FMCW radar used in vehicular networks	Typically low, supporting linkage distance up to a few kilometers	Communications integrated into the radar can achieve very long link distance. Sensing integrated into a single communication device can only support short range, but overall JCAS can cover very large areas due to the wide coverage of communication networks
Bandwidth	Large signal bandwidth. Range resolution proportional to bandwidth. But the bandwidth of the output signal in FMCW radar may be narrow, depending on the signal propagation distance	Typically much smaller than radar	mmWave signals are very promising for JCAS, due to large signal bandwidth. In addition, sensing applications do not have to rely on large bandwidth, such as known Wi-Fi sensing applications

(Continued)

Table 1.2 (Continued)

Specifications	Radar	Communications	JCAS system
Signal band	X, S, C, and Ku	Sub-6 GHz and mmWave bands	Have an impact on operation distances and resolution capabilities of JCAS
Duplex	Full-duplex (continuous-wave radar) or half-duplex (pulse radar)	Colocated transmitter and receiver typically cannot operate on the same time and frequency block. Communications are in either time division duplex (TDD) or frequency division duplex (FDD)	Full-duplex is a favorite condition, but not essential
Clock synchronization	Transmitter and receiver are clock-locked	Colocated transmitter and receiver share the same clock, but noncolocated nodes typically do not	Clock-level synchronization removes ambiguity in sensing parameter estimation but is not essential for some sensing applications

1.2.2 Communications-Centric Design

In communications-centric (CC) JCR systems, radio sensing is integrated into existing communication systems as a secondary function. Revision and enhancement to communication infrastructure and systems may be required, but the primary communication signals and protocols largely remain unchanged.

Two fundamental problems in integrating sensing into communications are the following: (i) how to realize full-duplex operation in a monostatic setup where the sensing receiver and transmitter are colocated, and (ii) how to remove the clock asynchronization impact in a bistatic or multistatic setup due to typically unsynchronized clocks between spatially separated transmitters and (sensing) receivers. Full-duplex here means that the receiver and transmitter work at the same time over the same frequency band. For a monostatic radar, full-duplex operation is avoided in pulsed radar via temporally separating the transmitting and receiving timeslots, leading to blind spots in near-field sensing; for FMCW radar, it is realized via using the transmitted signal as the input to the local oscillator to suppress the leakage signal from the transmitter, which leads to the output of the beat-frequency signal with little information on the transmitted signal. Modern communication systems primarily transmit continuous waveform and have unmodulated sinusoidal signals as the input to the oscillator. Hence, both radar methods are not practical in communication systems, unless a dedicated sensing receiver hardware similar to FMCW radar is integrated. In the long term, full-duplex technologies, as have been widely investigated for communications, would be a desired solution for monostatic sensing. However, the technology is still largely immature for practical applications. For bistatic and multistatic radars, clock synchronization is typically realized via wired connections or locking to the GPS signals [26]. These methods are feasible for some communication setups, but lack the generality. In the presence of clock asynchronization, it is also possible to apply signal-processing techniques to overcome it, which will be elaborated in Chapter 4.

Considering the topology of communication networks, CC JCAS systems can be classified into two subcategories, namely, those realizing sensing in point-to-point communication systems particularly for applications in vehicular networks, and those realizing sensing in networks such as mobile/cellular networks. Depending on how the transmitter and sensing receiver are spatially distributed, in terms of sensing, these systems are analog to traditional monostatic, bistatic, and multistatic radars.

There have been numerous reports on sensing in vehicular networks using IEEE 802.11 signals. In [27], the authors implemented active radar sensing functions into a communication system with OFDM signals for vehicular applications. The presented radar sensing functions involve Fourier transform algorithms that estimate the velocity of multiple reflecting objects in IEEE 802.11.p based JCAS system. In [28], automotive radar sensing functions are performed using the single carrier (SC) physical (PHY) frame of IEEE 802.11ad in an IEEE 802.11ad millimeter wave (mmWave) vehicle to vehicle (V2V) communication system. In [29], OFDM communication signals, conforming to IEEE 802.11a/g/p, are used to perform radar functions in vehicular networks. More specifically, a brute-force optimization algorithm is developed based on received mean-normalized channel energy for radar ranging estimation. The processing of delay and Doppler information with IEEE 802.11p OFDM waveform in vehicular networks is shown in [30] by applying the estimation of signal parameters via rotational invariant techniques (ESPRIT) method.

There has been rapidly increasing JCAS work reported for modern mobile networks. In [31], some early work on using OFDM signals for sensing was reported. In [32], sparse array optimization is studied for MIMO JCAS systems. Sparse transmit array design and transmit beampattern synthesis for JCAS are investigated in [33], where antennas are assigned to different functions.

In [34], mutual information for an OFDM JCAS system is studied, and power allocation for subcarriers is investigated based on maximizing the weighted sum of the mutual information for communications and sensing. In [35], waveform optimization is studied for minimizing the difference between the generated signal and the desired sensing waveform. In [36], the multiple access performance bound is derived for a multiple antenna JCAS system. In [37], a multicarrier waveform is proposed for dual-use radar-communications, for which interleaved subcarriers or subsets of subcarriers are assigned to the radar or the communications tasks. These studies involve some key signal formats in modern mobile networks, such as MIMO, multiuser MIMO, and OFDM. In [11, 38–41], the authors systematically studied how JCAS can be realized in mobile networks by considering their specific signal, system and network structures, and how radar sensing can be done based on modern mobile communication signals. Such a mobile network with integrated communications and sensing capability is called "*perceptive mobile network (PMN).*"

In addition to JCAS for mobile networks, sensing using Wi-Fi signals for primarily indoor applications has also received significant research interests in the past 10 years, with various sensing applications being demonstrated. Although mobile and Wi-Fi networks adopt similar modulation technologies, there exist significant differences between them in terms of communication protocols and network topologies, which also lead to different JCAS features. In Table 1.3, we compare the JCAS-versions of the two systems.

Table 1.3 Differences of JCAS–Wi-Fi with respect to JCAS-Mobile (PMN).

Aspects	Wi-Fi networks (with respect to Mobile networks)	Different implications on sensing in Wi-Fi–JCAS (with respect to PMN)
Signal format and transmission	Simpler and flexible packet structure, while PMN has rigid timing and channel structure	Waveform optimization has more flexibility. Available sensing signals are more random in time
Multiuser access	Relatively simpler, while PMN has complicated resource allocation and mixed multiuser access methods	Sensing parameter estimation can be simpler with more optional algorithms
Deployment environment	Mostly indoor and low-speed movement	Richer multipath, but more stable clutter. Less challenging in sensing due to a simpler environment
Network infrastructure and scale	Smaller network. Less-powerful infrastructure such as smaller antenna array	Low potential for networked sensing. Lower-sensing resolution

1.2.3 Radar-Centric Design

Radar systems, particularly military radars, have the extraordinary capability of long-range operation, up to hundreds of kilometers. Therefore, a major advantage of implementing communications in radar systems is the possibility of achieving long-range communications, with much lower latency compared to satellite communications. However, the achievable data rates for such systems are typically limited, due to the inherent limitation in the radar waveform. In [42], the authors implemented a combined radar and communication system based on a software-defined radar platform, in which the radar pulses are used for communications. Research work in [43] and [44] shows that communication network can be potentially established for both static and moving radars used in the military and aviation domains. Adaptive transmit signals from airborne radar mounted on unmanned vehicles can also be used to simultaneously sense a scene and communicate sensed data to a receiver at the ground base station. The objective of such systems is to establish low latency, secure, and long-range communications on top of existing radar systems. Such JCAS systems have been mainly called as dual-function radar-communications (DFRC). For convenience, we will also use it interchangeably with radar-centric (RC) JCAS in this book.

Realization of communications in radar systems has traditionally been based on either pulsed or continuous-wave radar signals. Hence, information embedding is one of the major challenges. Various information embedding techniques have been investigated, and some reviews are available from [2, 4, 16]. The basic principle is that the embedded information should cause little impact on conventional radar operation, while high-speed information transmission is regarded as a design goal. Some of the information embedding techniques are summarized in Table 1.4.

Integrating communications into radar systems with new radar waveforms has also been investigated, such as the MIMO-OFDM radar [49] and frequency-agile (frequency hopping [FH]) radar [47, 48]. Their signal formats are closer to modern communication systems, and hence can be potentially better integrated for information transmission. Such systems typically apply *index modulation* to embed communication information into the radar waveform. Here, index modulation embeds information to various combinations and/or permutations of signal parameters over space, time, frequency, and code domains [4, 17, 50]. One example is to use the indexes of subcarriers and transmitting antennas to carry the information. The main advantage of applying it in RC JCAS is that index modulation does not change the basic radar waveform and signal structure and has negligible influence on radar operation. We will provide more discussions on this technology in Chapters 7–9.

Table 1.4 Summary of information embedding methods in radar-centric DFRC systems.

Modulations		Methods	Advantages	Disadvantages
Modified waveforms	Time-frequency embedding	Apply various combinations of amplitude, phase and/or frequency shift keying to radar chirp signals, or map data to multiple chirp subcarriers via the use of fractional Fourier Transform [45]	• The chirp signal form remains when the interpulse modulation is used, which is preferred in many radar applications • The waveform can be implemented in many existing radar systems with only modifications to the software	• The slow time coding is restricted by the pulse repetition frequency (PRF) of the radar, thereby limiting the maximum rate of communications
	Code-domain embedding	Modulate binary/poly-phased codes in radar signals using direct spread spectrum sequences [46]	• Naturally coexist with the code-division multiple access/direct sequence spread spectrum (CDMA/DSSS) communication signal form • Enables covert communications by spreading the signal over the bandwidth of radar	• Phase modulation will inevitably lead to spectrum alteration of the radar waveform, which may result in energy leakage outside the assigned bandwidth
	Spatial embedding	Modulate information bits to the sidelobes of the radar beampattern	• Has little impact on the radar-sensing performance in the mainlobe	• The performance is sensitive to the accuracy of array calibration and BF • The multipath of radar signal may incur interference to the communications
Index modulation (no waveform modification)		Represent information by the indexes of antennas, frequencies, and/or codes of the signals [47, 48]	• Naturally coexist with the radar functionality, with negligible impact on radar performance • Generally achieve higher-data rates compared to modulation with modified waveform	• Demodulation may be complicated • Demodulation performance is sensitive to channel if IM is applied to spatial domain • Codebook design could be a challenge

1.2.4 Joint Design without an Underlying System

Although there is no clear boundary between the third category of technologies and systems and the previous two categories, there is more freedom for the former in terms of signal and system design. That is, JCAS technologies can be developed without being limited to existing communication or radar systems. In this sense, they can be designed and optimized by considering the essential requirements for both communications and sensing, potentially providing a better trade-off between the two functions.

The mmWave and (sub-)Terahertz-wave JCAS systems are great examples of facilitating such joint design. On the one hand, with their large bandwidth and short wavelength, mmWave and Terahertz signals provide great potentials for high date-rate communications and high-accuracy sensing. On the other hand, mmWave and Terahertz systems are emerging. They are yet to be widely deployed, and the standards for Terahertz systems are yet to be developed. Millimeter and Terahertz JCAS can facilitate many new exciting applications, both indoor and outdoor. Existing research on mmWave JCAS has demonstrated its feasibility and potentials in indoor and vehicle networks [13, 23, 51]. The authors in [13] provide an in-depth discussion on the signal processing aspect of mmWave-based JCAS with an emphasis on waveform design for JCAS systems. Future mmWave JCAS for indoor sensing is envisioned in [52]. Hybrid beamforming design for mmWave JCAS systems is investigated in [53]. An adaptive mmWave waveform structure is designed in [54]. Design and selection of JCAS waveforms for automotive applications are investigated in [51], where comparisons between phase-modulated continuous-wave JCAS and OFDMA-based JCAS waveforms are provided by analyzing the system model and enumerating the impact of design parameters. In [23, 55], multibeam technologies are developed to allow communications and sensing at different directions, using a common transmitted signal. Beamforming vectors are designed and optimized to enable fast beam update and achieve balanced performance between communications and sensing. In [56], the beamforming design for Terahertz massive MIMO JCAS systems is investigated.

Another example is multichannel JCAS systems where one or more channels are used at a time, and multiple channels are occupied over a period of signal transmission. One specific example is the frequency hopping system, such as the existing Bluetooth system where the operating frequency channel is changed over different packets. Multichannel systems can offer an overall large signal bandwidth for sensing without increasing the instantaneous communication bandwidth. This can largely reduce the hardware cost and also match with the spectrum usage of communication systems well. Works on combining multichannel signals for sensing have been reported for passive radar in, e g. [57, 58]. The key challenge is how to remove or reduce the imperfections and distortions from the received signals for each channel and then concatenate them together for sensing. For JCAS, an additional important problem is how to design the signals to make such concatenation easier, while balancing the performance of communications and sensing.

1.2.5 Summary of Key Research Problems

We conclude this section by briefly summarizing some key research problems and the associated challenges for the three types of JCAS systems. The summary is presented in Tables 1.5 and 1.6.

1.3 Potential Sensing Applications of JCAS

With harmonized and integrated communications and sensing functions, JCAS systems are expected to have the following advantages:

- *Spectral efficiency*: Spectral efficiency can ideally be doubled by completely sharing the spectrum available for wireless communication and radar.
- *Beamforming efficiency*: Beamforming performance can be improved through exploiting channel structures obtained from sensing, for example, quick beam adaption to channel dynamics, and beam direction optimization.
- *Reduced cost/size*: Compared to two separated systems, the joint system can significantly reduce the cost and size of transceivers.
- *Mutual benefits to communications and sensing*: communications and sensing can benefit from each other with the integration. Communication links can provide better coordination between multiple nodes for sensing, and sensing provides environment-awareness to communications, with potentials for improved security and performance.

Motivated by these advantages, JCAS has been widely explored in various communications and radar systems and can facilitate a vast range of new applications. Many of such applications in CC JCAS systems have been demonstrated via Wi-Fi sensing [9]. Those applications include device-free localization and tracking, human activity recognition, breathing pattern detection, and gesture recognition. Recently, performing JCAS in the ubiquitous mobile networks, such as the fifth-generation (5G) and beyond, has attracted extensive attention. Such JCAS-capable mobile network is referred to the PMN in the literature [59]. Empowered by the ubiquitousness of mobile network infrastructures, PMN is promising for establishing a large-scale sensing network. We will provide more details on PMN in Chapter 4. Here, we use PMN as an example and explore the potential sensing applications.

Large-scale sensing is becoming increasingly important for the growth of our industry and society. It is a critical enabler for disruptive IoT applications and a diverse range of smart initiatives such as smart cities and smart transportation.

Table 1.5 Key research problems in three types of JCAS systems and the associated challenges.

Category	Key research problems	Challenges
Radar-centric	Embed information to radar signals, particularly those methods that can lead to higher data rates	– Without significantly affecting radar operation, such as the radar ambiguity function and PAPR – Modulation constellation and codebook design to maximize the Euclidean distance between constellation points
	Design frame structure and communication protocol based on the selected information embedding method	– Conventional preamble for communications has a special pattern, which may be a major problem when being realized with radar waveform – Communication protocols may introduce latency and low duty cycle to radar sensing
	Develop signal reception and processing technologies, including channel estimation, equalization, and demodulation schemes, particularly for JCAS based on new generations of radar systems	– Extraction of communication signals from the beat frequency signal of conventional FMCW radar is challenging, as the beat signal is range-dependent – Receiver processing needs to be tailored to the information embedding method and frame structure. They are better to be jointly designed
Communication-centric	Full-duplex technologies in the context of JCAS for colocated transmitter and receiver, an analogy to the monostatic radar	– Suppression of onboard leakage signals from the transmitter, particularly for MIMO systems
	Sensing with asynchronous transmitter and receiver, an analogy to the bistatic radar	– Resolution of sensing ambiguity in the estimation of propagation delay and Doppler frequency – Removal of random phase shift across discontinuous temporal measurements to enable coherent processing and sensing parameter estimation
	Sensing algorithms that can be adapted to communication signals and the propagation environment	– Estimation of continuous sensing parameters with potentially irregular measurements in time and frequency domains – How to effectively use data symbols, in addition to pilots, for sensing – How to remove unwanted multipath to improve sensing performance?

Table 1.6 Key research problems in three types of JCAS systems and the associated challenges.

Category	Key research problems	Challenges
Communication-centric	Joint signal design and optimization, such as beamforming optimization and sensing-assisted communications	– How to optimize signals jointly for communications and sensing across one or more domains of spatial, time, and frequency? – How to exploit environmental information to improve signal reception performance and the security of communications?
	Information theories and technologies for sensing in a networked environment	– Information theory for JCAS is very limited, particularly for a networked system, and almost needs to be started from a scratch – Prior research on multistatic radar is very limited – Cooperation and competition for sensing need to be jointly considered with the requirements for communications
Joint design	High-frequency systems such as mmWave and sub-Terahertz-wave JCAS systems that can potentially achieve both high-data-rate communications and high-accuracy sensing	– Design of effective signaling schemes that best exploit the large bandwidth – Beamforming design to meet the conflictive requirements for high directivity for communications and wide-range of scanning for sensing – Development of beamforming tracking techniques to support the communications and sensing of mobile nodes
	Multichannel JCAS systems which can offer an overall large signal bandwidth for sensing, while without increasing instantaneous communication bandwidth	– Multichannel stitching for sensing – Removal or reduction of the imperfections and distortions from the received signals across channels – Smart signal design to ease channel stitching

Unfortunately, its adoption is severely constrained by the high infrastructure cost due to the limited coverage areas of existing sensors. For example, seamless camera surveillance over expansive areas will be prohibitively expensive due to the sheer number of cameras and communication links required to connect them. In addition, there are significant privacy concerns.

PMN is able to provide simultaneous communications and radio-sensing services, and it can potentially become a ubiquitous solution for radio sensing because of its large broadband coverage and powerful infrastructure. Its joint and harmonized communications and sensing capabilities will increase the productivity of our society and facilitate the creation and adoption of a vast number of new applications that no existing sensors can efficiently enable. Some earlier works on passive sensing using mobile signals have demonstrated its potentials. For example, [60] and [61] used GSM-based radio signals for traffic monitoring, weather prediction, and remote sensing of rainfall, respectively. The perceptive network can be widely deployed for both communications and sensing applications in transport, communications, energy, precision agriculture, and security, where existing solutions are either infeasible or inefficient. It can also provide complementary sensing capabilities to existing sensor networks, with its unique features of day-and-night operation and see-through of fog, foliage, and even solid objects.

Compared to JCAS–Wi-Fi, the PMN has more advanced infrastructure, including larger antenna array, larger signal bandwidth, more powerful signal processing, and distributed and cooperative base stations. In particular, with massive MIMO, the PMN equivalently possesses a massive number of "pixels" for sensing. This enables radio devices to resolve numerous objects at a time and achieve sensing results with much better resolution.

Some of the sensing applications that can be enabled by PMN are illustrated in Figure 1.3. They may be classified into several major areas, such as smart transportation, smart city, smart home, industrial IoT, environmental sensing, and sensing-assisted communications. More specific examples of these applications are listed in Table 1.7.

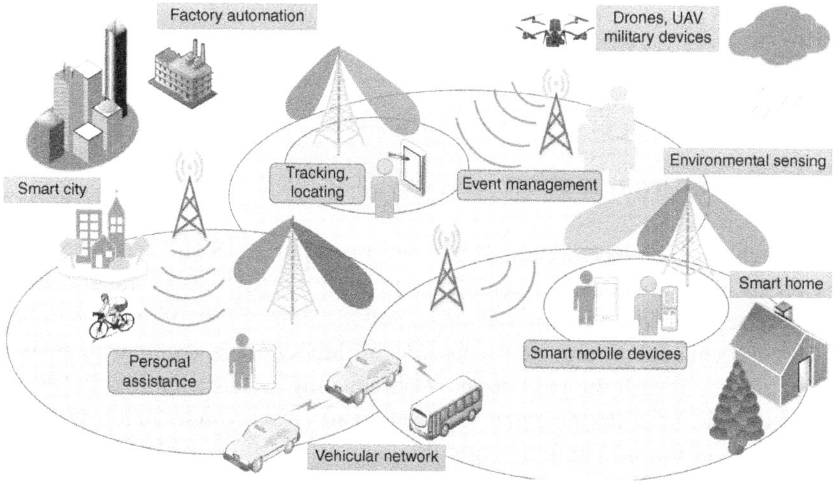

Figure 1.3 Applications and use cases of PMN, with integrated communications and sensing capabilities.

Table 1.7 Potential sensing applications of PMN.

Application areas	Cases and examples
Smart transportation	– Real-time citywide vehicle classification and tracking – Vehicle speed measurement – On-road parking space detection – Sensing assistant to autonomous driving – Drone monitoring and management
Smart city	– Extensive on-street and open space surveillance for security and safety – Low-cost automatic street lighting systems – Crowd management for major events and emergency evacuation – Integrated personal navigation and safety services provided by PMN and smart mobile devices
Smart home	– (Through-the-wall) localization and tracking – Human behavior recognition and fall detection – Monitoring of biomedical signals such as respiration patterns – Human presence detection and radio fence
Industrial IoT	– Localization and tracking of vehicles, equipment, and workers – Surveillance and proximity detection – Object recognition and authentication – Gesture recognition for equipment operation
Environmental sensing	– Factory emissions and pollution monitoring – Rainfall monitoring and flooding prediction – Animal migration monitoring – Monitoring of migratory birds and insects
Sensing-assisted communications	– Radio signal propagation mapping and site survey – Beam tracking and predictive beamforming – Sensing-seeded encrypted communications – Sensing-assisted resource optimization for communications

1.4 Book Organization

In this chapter, we have discussed the background of JCAS and the three categories of JCAS systems, i.e. communications-centric (CC) JCAS, radar-centric (RC) JCAS, and joint design and optimization. Since CC-JCAS and RC-JCAS incur less changes to an underlying system, less effort is required to realize them compared with the joint design. Consequently, we envision that CC-JCAS and RC-JCAS have the potential to speed up the market penetration of JCAS techniques in the near future, while the joint design offers a blank canvas for imaginative painting. As a stimulating book, we focus on CC-JCAS and RC-JCAS, introducing the basics and advanced recent progress to shed light on future JCAS technology development. The joint design and optimization of JCAS is still in its infancy. Interested readers may refer to the latest overview/survey articles, e.g. [2–5], to learn more on this topic.

This book is divided into three parts. Part I includes this and the next two chapters, introducing the background of JCAS and some common and basic signal processing issues. Part II, comprising of Chapters 4–6, introduces the latest advanced designs of CC-JCAS. Part III, including Chapters 7–9, introduces some latest RC-JCAS designs based on frequency hopping with high data rate, reliability, and communications secrecy. Their focuses are elaborated on below.

- Chapter 2 illustrates the signal processing basics of communications, radar, and JCAS systems. It starts with describing a unified signal model that suits both C&S systems. It then highlights the differences between the two systems and their signal features. The chapter also sequentially illustrates some most fundamental signal processing issues in CC-JCAS and RC-JCAS. Only signal processing basics essential to understand the JCAS techniques in this book are introduced. Readers may refer to relevant texts, e.g. [62–65], if they are not familiar with the fundamental signal processing theories, such as Fourier transform, filtering, uniform linear array response, the maximum likelihood detection, and estimation.
- Chapter 3 illustrates the parameter estimation problem in communications, radar, and their joint designs. Some efficient estimators that are based on the discrete Fourier transform (DFT) interpolations are introduced in this chapter. These estimators will be widely used in Chapters 5–8.
- Chapter 4 provides a detailed introduction to PMN, endeavoring to draw a clear picture on what the PMN will look like and how it may evolve from the current communications-only network from the viewpoints of both infrastructure and technology. We will also look at mobile-network-specific JCAS challenges and potential solutions, associated with heterogeneous network architecture and components, sophisticated mobile signal format, and complicated signal propagation environment.
- Chapter 6 introduces a low-complexity JCAS scheme using orthogonal time-frequency space (OTFS) modulation, which advocates the two promising techniques for future IoT systems. OTFS has been demonstrated to have higher reliability and energy efficiency than the currently popular IoT communication waveforms, e.g. OFDM. JCAS has also been highly recommended for IoT, since it saves cost, power, and spectrum compared to having two separate radio frequency systems. Performing JCAS based on OTFS, however, can be hindered by a lack of effective OTFS sensing. The designs/methods to be introduced in Chapter 6 can potentially fill this technology gap.

- Chapter 5 extends the designs in Chapter 6 and introduces a unified sensing framework integrating low-complexity and flexible sensing into existing multi- or single-carrier communication systems. Integrating sensing into standardized communication systems can potentially benefit many consumer applications that require both radio frequency functions. Existing sensing methods, which use communication payload signals, either have limited sensing performance or suffer from high complexity. In this chapter, we present a flexible sensing framework that has a complexity only dominated by Fourier transform and provides the flexibility in adapting to different sensing needs.
- Chapter 7 introduces the waveform design and channel estimation for frequency hopping (FH) MIMO radar-based communications. FH-MIMO enables communication symbol rate to exceed radar pulse repetition frequency, which requires accurate estimations of timing offset and channel parameters. The estimations, however, are challenging due to unknown, fast-changing hopping frequenciesand the multiplicative coupling between timing offset and channel parameters. In this chapter, we present some accurate methods for a single-antenna communication receiver to estimate timing offset and channel for FH-MIMO DFRC.
- Chapter 8 introduces a reliable FH-MIMO DFRC with a multiantenna array (c.f. single antenna in Chapter 7) equipped at the communications receiver. The single-antenna communication receiver can suffer from deep fading under multipath channels. This is unveiled in this chapter and addressed by introducing a mutiantenna array to the communications receiver. Moreover, we present in this chapter a multiantenna receiving scheme for FH-MIMO DFRC, addressing the channel estimation and synchronization issues in multipath channels.
- Chapter 9 introduces a secure FH-MIMO DFRC with improved data rate over the schemes in Chapters 7 and 8. Built on an orthogonal MIMO radar, FH-MIMO DFRC is prone to eavesdropping due to the spatially uniform illumination of the underlying radar. In this chapter, we reveal the potential of using permutations of hopping frequencies to achieve secure and high-speed FH-MIMO DFRC. Unlike combinations employed in Chapters 7 and 8, permutations-based communication symbols have substantially different demodulating performance. This will also be addressed by a series of waveform preprocessing to be illustrated in Chapter 9.

References

1 C. Sturm and W. Wiesbeck. Waveform design and signal processing aspects for fusion of wireless communications and radar sensing. *Proceedings of the IEEE*, 99(7):1236–1259, 2011. ISSN 0018-9219 1558-2256. doi: 10.1109/jproc.2011.2131110.

2 L. Zheng, M. Lops, Y. C. Eldar, and X. Wang. Radar and communication coexistence: An overview: A review of recent methods. *IEEE Signal Processing Magazine*, 36(5):85–99, September 2019. ISSN 1558-0792. doi: 10.1109/MSP.2019.2907329.

3 F. Liu, C. Masouros, A. P. Petropulu, H. Griffiths, and L. Hanzo. Joint radar and communication design: Applications, state-of-the-art, and the road ahead. *IEEE Transactions on Communications*, 68(6):3834–3862, 2020. doi: 10.1109/TCOMM.2020.2973976.

4 A. Hassanien, M. G. Amin, E. Aboutanios, and B. Himed. Dual-function radar communication systems: A solution to the spectrum congestion problem. *IEEE Signal Processing Magazine*, 36(5):115–126, September 2019. doi: 10.1109/msp.2019.2900571.

5 J. A. Zhang, Md. L. Rahman, K. Wu, X. Huang, Y. J. Guo, S. Chen, and J. Yuan. Enabling joint communication and radar sensing in mobile networks–a survey. *IEEE Communication Surveys and Tutorials*, 24(1):306–345, 2022. doi: 10.1109/COMST.2021.3122519.

6 B. Li, A. P. Petropulu, and W. Trappe. Optimum co-design for spectrum sharing between matrix completion based MIMO radars and a MIMO communication system. *IEEE Transactions on Signal Processing*, 64(17):4562–4575, September 2016.

7 D. E. Hack, L. K. Patton, B. Himed, and M. A. Saville. Detection in passive MIMO radar networks. *IEEE Transactions on Signal Processing*, 62(11): 2999–3012, June 2014. ISSN 1053-587X. doi: 10.1109/TSP.2014.2319776.

8 H. Kuschel, D. Cristallini, and K. E. Olsen. Tutorial: Passive radar tutorial. *IEEE Aerospace and Electronic Systems Magazine*, 34(2):2–19, 2019. doi: 10.1109/MAES.2018.160146.

9 J. Liu, H. Liu, Y. Chen, Y. Wang, and C. Wang. Wireless sensing for human activity: A survey. *IEEE Communication Surveys and Tutorials*, 1, 2019. ISSN 2373-745X. doi: 10.1109/COMST.2019.2934489.

10 S. Wang and G. Zhou. A review on radio based activity recognition. *Digital Communications and Networks*, 9, March 2015. doi: 10.1016/j.dcan.2015.02.006.

11 J. A. Zhang, A. Cantoni, X. Huang, Y. J. Guo, and R. W. Heath Jr. Framework for an innovative perceptive mobile network using joint communication and sensing. In *2017 IEEE 85th Vehicular Technology Conference (VTC Spring)*, pages 1–5. IEEE, June 2017. doi: 10.1109/VTCSpring.2017.8108564.

12 J. A. Zhang, X. Huang, Y. J. Guo, and R. W. Heath Jr. Joint communications and sensing using two steerable analog antenna arrays. In *2017 IEEE 85th Vehicular Technology Conference (VTC Spring)*, pages 1–5, June 2017.

13 K. V. Mishra, M. R. Bhavani Shankar, V. Koivunen, B. Ottersten, and S. A. Vorobyov. Toward millimeter-wave joint radar communications: A signal processing perspective. *IEEE Signal Processing Magazine*, 36(5):100–114, September 2019. ISSN 1558-0792. doi: 10.1109/MSP.2019.2913173.

14 Z. Feng, Z. Fang, Z. Wei, X. Chen, Z. Quan, and D. Ji. Joint radar and communication: A survey. *China Communications*, 17(1):1–27, 2020. doi: 10.23919/JCC.2020.01.001.

15 P. Kumari, S. A. Vorobyov, and R. W. Heath. Adaptive virtual waveform design for millimeter-wave joint communication-radar. *IEEE Transactions on Signal Processing*, 68:715–730, 2020. doi: 10.1109/TSP.2019.2956689.

16 A. Hassanien, M. G. Amin, Y. D. Zhang, and F. Ahmad. Signaling strategies for dual-function radar communications: An overview. *IEEE Aerospace and Electronic Systems Magazine*, 31(10):36–45, 2016. doi: 10.1109/MAES.2016.150225.

17 D. Ma, N. Shlezinger, T. Huang, Y. Liu, and Y. C. Eldar. Joint radar-communication strategies for autonomous vehicles: Combining two key automotive technologies. *IEEE Signal Processing Magazine*, 37(4):85–97, 2020. doi: 10.1109/MSP.2020.2983832.

18 L. Han and K. Wu. Joint wireless communication and radar sensing systems - state of the art and future prospects. *IET Microwaves, Antennas and Propagation*, 7(11):876–885, 2013. ISSN 1751-8725 1751-8733. doi: 10.1049/iet-map.2012.0450.

19 F. Liu, C. Masouros, A. Li, and T. Ratnarajah. Robust MIMO beamforming for cellular and radar coexistence. *IEEE Wireless Communications Letters*, 6(3):374–377, June 2017. ISSN 2162-2345. doi: 10.1109/LWC.2017.2693985.

20 S. Gogineni, M. Rangaswamy, B. D. Rigling, and A. Nehorai. Cramér-Rao bounds for UMTS-based passive multistatic radar. *IEEE Transactions on Signal Processing*, 62(1):95–106, January 2014. ISSN 1053-587X. doi: 10.1109/TSP.2013.2284758.

21 R. Syamsul A. R. Abdullah, A. A. Salah, A. Ismail, F. Hashim, N. E. Abdul Rashid, and N. H. Abdul Aziz. LTE-based passive bistatic radar system for detection of ground-moving targets. *ETRI Journal*, 38(2):302–313, 2016. doi: 10.4218/etrij.16.0115.0228.

22 R. Saruthirathanaworakun, J. M. Peha, and L. M. Correia. Opportunistic sharing between rotating radar and cellular. *IEEE Journal on Selected Areas in Communications*, 30(10):1900–1910, November 2012. ISSN 1558-0008. doi: 10.1109/JSAC.2012.121106.

23 J. A. Zhang, X. Huang, Y. J. Guo, J. Yuan, and R. W. Heath. Multibeam for joint communication and radar sensing using steerable analog antenna arrays. *IEEE Transactions on Vehicular Technology*, 68(1):671–685, January 2019. ISSN 1939-9359. doi: 10.1109/TVT.2018.2883796.

24 N. C. Luong, X. Lu, D. T. Hoang, D. Niyato, and D. I. Kim. Radio resource management in joint radar and communication: A comprehensive survey. *IEEE Communication Surveys and Tutorials*, 23(2):780–814, 2021. doi: 10.1109/COMST.2021.3070399.

25 J. Li and P. Stoica. *MIMO Radar Signal Processing*. John Wiley & Sons, Ltd, 2008. ISBN 9780470391488. doi: 10.1002/9780470391488.fmatter.

26 M. Weib. Synchronisation of bistatic radar systems. In *IGARSS 2004. 2004 IEEE International Geoscience and Remote Sensing Symposium*, volume 3, pages 1750–1753, 2004. doi: 10.1109/IGARSS.2004.1370671.

27 Y. L. Sit, C. Sturm, and T. Zwick. Doppler estimation in an OFDM joint radar and communication system. In *2011 German Microwave Conference*, pages 1–4, March 2011.

28 P. Kumari, J. Choi, N. G. Prelcic, and R. W. Heath. IEEE 802.11ad-based radar: An approach to joint vehicular communication-radar system. *IEEE Transactions on Vehicular Technology*, PP(99):1, 2017. ISSN 0018-9545. doi: 10.1109/TVT.2017.2774762.

29 R. C. Daniels, E. R. Yeh, and R. W. Heath. Forward collision vehicular radar with IEEE 802.11: Feasibility demonstration through measurements. *IEEE Transactions on Vehicular Technology*, 67(2):1404–1416, February 2018. ISSN 0018-9545. doi: 10.1109/TVT.2017.2758581.

30 D. H. N. Nguyen and R. W. Heath. Delay and Doppler processing for multi-target detection with IEEE 802.11 OFDM signaling. In *2017 IEEE International Conference on Acoustics, Speech and Signal Processing (ICASSP)*, pages 3414–3418, March 2017. doi: 10.1109/ICASSP.2017.7952790.

31 M. Braun. *OFDM Radar Algorithms in Mobile Communication Networks*. PhD thesis, Institut fur Nachrichtentechnik des Karlsruher Instituts fur Technologie, Karlsruhe, 2014.

32 X. Wang, A. Hassanien, and M. G. Amin. Dual-function MIMO radar communications system design via sparse array optimization. *IEEE Transactions on Aerospace and Electronic Systems*, 55(3):1213–1226, 2019. doi: 10.1109/TAES.2018.2866038.

33 X. Wang, A. Hassanien, and M. G. Amin. Sparse transmit array design for dual-function radar communications by antenna selection. *Digital Signal Processing*, 83:223–234, 2018.

34 Y. Liu, G. Liao, J. Xu, Z. Yang, and Y. Zhang. Adaptive OFDM integrated radar and communications waveform design based on information theory. *IEEE Communications Letters*, 21(10):2174–2177, 2017.

35 F. Liu, C. Masouros, A. Li, H. Sun, and L. Hanzo. MU-MIMO communications with MIMO Radar: From co-existence to joint transmission. *IEEE Transactions on Wireless Communications*, 17(4):2755–2770, April 2018. ISSN 1536-1276. doi: 10.1109/TWC.2018.2803045.

36 Y. Rong, A. R. Chiriyath, and D. W. Bliss. Multiple-antenna multiple-access joint radar and communications systems performance bounds. In *2017 51st Asilomar Conference on Signals, Systems, and Computers*, pages 1296–1300, October 2017. doi: 10.1109/ACSSC.2017.8335562.

37 M. Bică and V. Koivunen. Multicarrier radar-communications waveform design for RF convergence and coexistence. In *ICASSP 2019 - 2019 IEEE International Conference on Acoustics, Speech and Signal Processing (ICASSP)*, pages 7780–7784, May 2019. doi: 10.1109/ICASSP.2019.8683655.

38 J. A. Zhang, X. Huang, Y. J. Guo, and Md. L. Rahman. Signal stripping based sensing parameter estimation in perceptive mobile networks. In *2017 IEEE-APS Topical Conference on Antennas and Propagation in Wireless Communications (APWC)*, pages 67–70. IEEE, 2017.

39 M. L. Rahman, J. A. Zhang, X. Huang, and Y. J. Guo. Analog antenna array based sensing in perceptive mobile networks. In *2017 IEEE-APS Topical Conference on Antennas and Propagation in Wireless Communications (APWC)*, pages 199–202, September 2017. doi: 10.1109/APWC.2017.8062279.

40 M. L. Rahman, P. Cui, J. A. Zhang, X. Huang, Y. J. Guo, and Z. Lu. Joint communication and radar sensing in 5G mobile network by compressive sensing. In *2019 19th International Symposium on Communications and Information Technologies (ISCIT)*, pages 599–604, September 2019. doi: 10.1109/ISCIT.2019.8905229.

41 M. L. Rahman, J. A. Zhang, X. Huang, Y. J. Guo, and R. W. Heath Jr. Framework for a perceptive mobile network using joint communication and radar sensing. *IEEE Transactions on Aerospace and Electronic Systems*, 56(3):1926–1941, 2020.

42 C. W. Rossler, E. Ertin, and R. L. Moses. A software defined radar system for joint communication and sensing. In *2011 IEEE Radar Conference (RADAR)*, pages 1050–1055. IEEE, 2011.

43 A. R. Chiriyath, B. Paul, and D. W. Bliss. Radar-communications convergence: Coexistence, cooperation, and co-design. *IEEE Transactions on Cognitive Communications and Networking*, 3(1):1–12, 2017. ISSN 2332-7731. doi: 10.1109/tccn.2017.2666266.

44 B. Paul. *RF Convergence of Radar and Communications: Metrics, Bounds, and Systems*. PhD thesis, Electrical Engineering, 2017.

45 D. Gaglione, C. Clemente, C. V. Ilioudis, A. R. Persico, I. K. Proudler, and J. J. Soraghan. Fractional Fourier based waveform for a joint radar-communication system. In *2016 IEEE Radar Conference (RadarConf)*, pages 1–6, 2016. doi: 10.1109/RADAR.2016.7485314.

46 M. Jamil, H. Zepernick, and M. I. Pettersson. On integrated radar and communication systems using oppermann sequences. In *MILCOM 2008 - 2008 IEEE Military Communications Conference*, pages 1–6, 2008. doi: 10.1109/MILCOM.2008.4753277.

47 T. Huang, N. Shlezinger, X. Xu, Y. Liu, and Y. C. Eldar. MAJoRCom: A dual-function radar communication system using index modulation. *IEEE*

Transactions on Signal Processing, 68:3423–3438, 2020. doi: 10.1109/TSP.2020.2994394.

48 K. Wu, J. A. Zhang, X. Huang, Y. Jay Guo, and R. W. Heath. Waveform design and accurate channel estimation for frequency-hopping MIMO radar-based communications. *IEEE Transactions on Communications*, 1, 2020. doi: 10.1109/TCOMM.2020.3034357.

49 Z. Lin and Z. Wang. Interleaved OFDM signals for MIMO radar. *IEEE Sensors Journal*, 15(11):6294–6305, 2015. doi: 10.1109/JSEN.2015.2458178.

50 S. D. Tusha, A. Tusha, E. Basar, and H. Arslan. Multidimensional index modulation for 5G and beyond wireless networks. *Proceedings of the IEEE*, 109(2):170–199, 2021. doi: 10.1109/JPROC.2020.3040589.

51 S. H. Dokhanchi, B. S. Mysore, K. V. Mishra, and B. Ottersten. A mmWave automotive joint radar-communications system. *IEEE Transactions on Aerospace and Electronic Systems*, 55(3):1241–1260, June 2019. ISSN 2371-9877. doi: 10.1109/TAES.2019.2899797.

52 M. Alloulah and H. Huang. Future millimeter-wave indoor systems: A blueprint for joint communication and sensing. *Computer*, 52(7):16–24, July 2019. ISSN 1558-0814. doi: 10.1109/MC.2019.2914018.

53 F. Liu and C. Masouros. Hybrid beamforming with sub-arrayed MIMO radar: Enabling joint sensing and communication at mmWave band. In *ICASSP 2019 - 2019 IEEE International Conference on Acoustics, Speech and Signal Processing (ICASSP)*, pages 7770–7774, May 2019. doi: 10.1109/ICASSP.2019.8683591.

54 P. Kumari, S. A. Vorobyov, and R. W. Heath Jr. Adaptive virtual waveform design for millimeter-wave joint communication-radar, 2019.

55 Y. Luo, J. A. Zhang, X. Huang, W. Ni, and J. Pan. Optimization and quantization of multibeam beamforming vector for joint communication and radio sensing. *IEEE Transactions on Communications*, 67(9):6468–6482, September 2019. ISSN 1558-0857. doi: 10.1109/TCOMM.2019.2923627.

56 A. M. Elbir, K. V. Mishra, and S. Chatzinotas. Terahertz-band joint ultra-massive MIMO radar-communications: Model-based and model-free hybrid beamforming, 2021. arXiv: 2103.00328.

57 J. Wei, J. Li, Z. Cao, Q. Chen, C. Song, and Z. Xu. A passive radar prototype based on multi-channel joint detection and its test results. In *2020 IEEE 11th Sensor Array and Multichannel Signal Processing Workshop (SAM)*, pages 1–5, 2020. doi: 10.1109/SAM48682.2020.9104263.

58 Y. Xie, Z. Li, and M. Li. Precise power delay profiling with commodity Wi-Fi. *IEEE Transactions on Mobile Computing*, 18(6):1342–1355, 2019. doi: 10.1109/TMC.2018.2860991.

59 J. A. Zhang, Md. L. Rahman, K. Wu, X. Huang, Y. J. Guo, S. Chen, and J. Yuan. Enabling joint communication and radar sensing in mobile networks

-a survey. *IEEE Communication Surveys and Tutorials*, 1, 2021. doi: 10.1109/COMST.2021.3122519.

60 R. Harris, J.-K. Lam, and E. F. Burroughs. The potential of cell phone radar as a tool for transport applications. Report, Association for European Transport and contributors, 2005.

61 N. David, O. Sendik, H. Messer, and P. Alpert. Cellular network infrastructure: The future of fog monitoring? *Bulletin of the American Meteorological Society*, 96(10):1687–1698, 2015. ISSN 0003-0007 1520-0477. doi: 10.1175/bams-d-13-00292.1.

62 A. V. Oppenheim. *Discrete-time Signal Processing*. Pearson Education India, 1999.

63 D. Tse and P. Viswanath. *Fundamentals of Wireless Communication*. Cambridge University Press, 2005.

64 M. A. Richards, J. Scheer, W. A. Holm, and W. L. Melvin. *Principles of Modern Radar*. CiteSeer, 2010.

65 H. L. Van Trees. *Optimum Array Processing: Part IV of Detection, Estimation, and Modulation Theory*. John Wiley & Sons, 2004.

2

Signal Processing Fundamentals for JCAS

In Chapter 1, we outlined the basic JCAS system architectures and applications. In this chapter, we present the basic signal and system models for JCAS. The channel models for communications and radar are described first in Section 2.1. Then signals and system models for MIMO communications and MIMO radar systems are explained in Sections 2.2 and 2.3, respectively. Owing to its numerous advantages and popularity, the discussions are mainly focused on systems based on orthogonal frequency division multiplexing modulation (OFDM). Section 2.4 provides a brief summary of signal-processing techniques for radar sensing. Section 2.5 introduces signal processing basics for communications-centric JCAS systems. Section 2.6 is devoted to basic signal processing techniques for radar-centric JCAS systems, or dual functional radar and communications (DFRC). Section 2.7 concludes the chapter. The objective of the chapter is to provide basic but sufficient theoretical background for understanding the operations of JCAS systems, leaving advanced topics to Chapters 5–9.

2.1 Channel Model for Communications and Radar

Both communications and radio sensing rely on the transmission of wireless signals through the surrounding environment. Regardless of the substantial differences in their system structures and functions, radar and communication signals interact with the propagation media in similar manners. This enables us to employ a united input–output relationship, the wireless channel model, to describe how a signal is changed from a transmitter to a receiver, for both communications and sensing.

Consider a general system with $Q \geq 2$ nodes, and each node has a uniform linear antenna array (ULA). Let B denote the total signal bandwidth. For MIMO–OFDM

Joint Communications and Sensing: From Fundamentals to Advanced Techniques, First Edition.
Kai Wu, J. Andrew Zhang, and Y. Jay Guo.

systems (for either communications or radar), B is divided into N subcarriers, and the subcarrier interval is $f_0 = B/N$. The OFDM symbol period is $T_s = T_0 + T_{cp}$, where $T_0 = N/B$ and T_{cp} is the period of cyclic prefix. For the clarity of illustration, we describe the signal models with reference to a general transmitter and receiver without indexing them. When these models need to be differentiated for different nodes, we will add a subscript of the node index to the variables in the models.

Let the angle-of-departure (AoD) and angle-of-arrival (AoA) of a multipath be θ_ℓ and $\phi_\ell, \ell \in [1, L]$, respectively. Assume a planar wave-front in signal propagation. The array steering/response vector of a ULA is given by

$$\mathbf{a}(M, \alpha) = [1, e^{j2\pi d/\lambda \sin(\alpha)}, \dots, e^{j(M-1)2\pi d/\lambda \sin(\alpha)}]^T,$$

where M is the number of antennas, λ is the wavelength, d is the interval of antennas, and α is either AoD or AoA.

For M_T transmitting and M_R receiving antennas, the $M_R \times M_T$ time-domain baseband channel matrix at time t can be represented as follows:

$$\mathbf{H}(t) = \sum_{\ell=1}^{L} b_\ell \delta(t - \tau_\ell - \tau_o(t)) e^{j2\pi(f_{D,\ell} + f_o(t))t} \cdot \mathbf{a}(M_R, \phi_\ell) \mathbf{a}^T(M_T, \theta_\ell), \tag{2.1}$$

where for the ℓ-th multipath, b_ℓ is a complex path gain, accounting for both signal attenuation and initial phase difference; τ_ℓ is the propagation delay; $f_{D,\ell}$ is the associated Doppler frequency; and $\tau_o(t)$ and $f_o(t)$ denote the potential time-varying timing offset and carrier frequency offset (CFO) due to possibly unlocked clocks between transmitter and receiver, respectively. Here, we ignore the "beam squint" effect [1] in beamforming and assume b_ℓ is frequency independent.

Equation (2.1) represents a general channel model that can be used for both communications and radar, although the physical meaning and names of these parameters are slightly different. The description above is mainly based on the terminologies in communications. For radar sensing, $\{\tau_\ell, f_{D,\ell}, \phi_\ell, \theta_\ell, b_\ell\}$ are the *sensing parameters* to be estimated. These parameters can be used to determine a target/reflector's spatial and moving information. In particular, ϕ_ℓ and θ_ℓ are the AoA and AoD of the target in relation to the receiving and transmitting ULA, respectively; $\tau_\ell = R_\ell/c$ and $f_{D,\ell} = v_\ell f_c/c$, where R_ℓ is the signal propagation distance, c is the speed of light, v_ℓ is the radial velocity of the reflector, and f_c is the carrier frequency; and b_ℓ is related to the radar cross-section (RCS) and material property of the target. To improve radar performance, radar echo signals are generally jointly performed in a so-called *coherent processing interval (CPI)* within which all these parameters remain almost unchanged. The length of CPI depends on the mobility of objects in the channels and is typically tens of milliseconds when objects move at speeds of tens of meters per second.

For a broadband OFDM system, the frequency-domain channel matrix at the n-th subcarrier corresponding to (2.1) is

$$\tilde{\mathbf{H}}_n(t) = \sum_{\ell=1}^{L} b_\ell e^{-j2\pi n(\tau_\ell+\tau_o(t))f_0} e^{j2\pi(f_{D,\ell}+f_o(t))t} \cdot \mathbf{a}(M_R, \phi_\ell)\mathbf{a}^T(M_T, \theta_\ell), \tag{2.2}$$

where we have approximated the slightly varying phases due to Doppler frequency and CFO over one OFDM block as a constant value. For the k-th OFDM symbol with $t = kT_s$, we denote $\tilde{\mathbf{H}}_{n,k} = \tilde{\mathbf{H}}_n(t)_{t=kT_s}$.

Note that for communications, we generally only need to know the composite values of the matrix $\mathbf{H}(t)$ or $\tilde{\mathbf{H}}_n(t)$. They can typically be obtained by directly estimating channel coefficients, or for OFDM, directly estimating at some subcarriers and obtaining the rest via interpolation. For radar sensing, however, the system needs to resolve the detailed channel structure and estimate the sensing parameters. Note that this beam-space channel model is also widely used in millimeter wave (mmWave) communication systems.

When the oscillator clocks of the transmitter and the receiver are not locked/synchronized, both the timing offset $\tau_0(t)$ and CFO $f_0(t)$ are nonzero. Their values can also be fast time-varying due to crystal oscillator's instability. For example for a typical clock stability of 20 parts-per-million (PPM), the accumulated maximal variation of $\tau_0(t)$ over 1 millisecond can be 20 nanoseconds, which translates to a ranging error of 6 meters. For communications, these offsets are generally not a big problem, as $\tau_0(t)$ can be absorbed into channel estimates after synchronization, and $f_0(t)$ can be estimated and compensated. Their residual values become relatively small compared to the baseband signal parameters. However, for radar sensing, when these offsets are unknown to the receiver, they can cause ambiguity in range and speed estimation and become obstacles for processing signals across packets coherently.

2.2 Basic Communication Signals and Systems

We consider a node transmitting M_S spatial streams with $M_T \geq M_S$ antennas. The description here is with reference to a single-user MIMO system, and it will be extended to multiuser MIMO later.

2.2.1 Single-Carrier MIMO

For a general single-carrier (SC) MIMO system, we can represent the baseband signal vector at time t from the transmitter as follows:

$$\mathbf{x}(t) = \mathbf{P}\mathbf{s}(t), \tag{2.3}$$

where $\mathbf{P}$ is the spatial precoding matrix of size $M_T \times M_S$ and is typically fixed over a packet, and $\mathbf{s}(t)$ is the data vector of $M_S \times 1$. The symbols in $\mathbf{s}(t)$ can be either the directly modulated constellation points that are unknown to the receiver or known pilots. In the case of spread spectrum signals, each element in $\mathbf{s}(t)$ can be the product of a pseudorandom code and the constellation point (or pilot).

For a narrowband system, after propagating over the channel $\mathbf{H}(t)$, the received signals are given by

$$\mathbf{y}(t) = \mathbf{H}(t) \circledast \mathbf{x}(t) + \mathbf{z}(t), \tag{2.4}$$

where $\circledast$ denotes convolution, and $\mathbf{z}(t)$ is the AWGN.

2.2.2 MIMO-OFDM

For a MIMO-OFDM system, the baseband transmitting signals at subcarrier n in the k-th OFDM symbol over all antennas can be represented as follows:

$$\tilde{\mathbf{x}}_{n,k} = \mathbf{P}_{n,k}\mathbf{s}_{n,k}, \tag{2.5}$$

where these variables are similarly defined as those in (2.3), but could have different values for different subcarriers. In the case of MIMO-OFDMA, each node may be allocated to a resource block occupying groups of the antennas, subcarriers, and OFDM symbols, that are typically nonoverlapping. These blocks can be discontinuous and irregular in these domains, which can cause significant challenges in sensing parameter estimation, as will be discussed in Section 2.5.3.

After propagating over the channel, the received frequency-domain baseband signals at subcarrier n over all antennas are given by

$$\tilde{\mathbf{y}}_{n,k} = \tilde{\mathbf{H}}_{n,k}\tilde{\mathbf{x}}_{n,k} + \tilde{\mathbf{z}}_{n,k}, \tag{2.6}$$

where $\tilde{\mathbf{z}}_{n,k}$ is the AWGN.

2.2.3 Transmitter and Receiver Signal Processing in Communications

Figure 2.1 shows the block diagram for the transceiver of a typical MIMO-OFDM system with packet communications. At the transmitter, information bits are padded and input to the modulator, and the modulated symbols are then mapped to subcarriers and spatial streams. The optional phase rotation module multiplies some phase shifting terms to the symbols to avoid undesired beamforming effect. Spatial precoding is then applied to spatial streams at each subcarrier. An inverse fast Fourier transform (IFFT) is applied to each stream to convert the signals to the time domain. The framing module adds additional signals such as training sequences and forms a packet for transmission. The baseband signal is then up-converted to a carrier frequency for transmission.

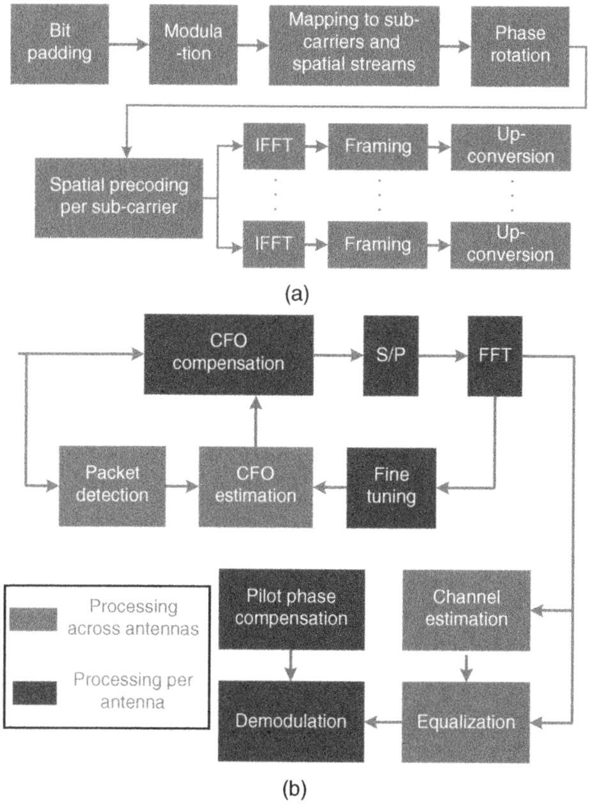

Figure 2.1 Block diagram of a MIMO-OFDM communication (a) transmitter and (b) receiver.

At the receiver, the packet detection module monitors the channel and can detect the packet when it is received, typically based on an autocorrelation operation. CFO is being estimated and compensated. The fine timing module synchronizes the receiver to the transmitter at the symbol level, typically based on cross-correlating the received signals with a local template signal. The serial synchronized signals are then converted to parallel in the S/P module, and the FFT operation is applied to each stream. Channel estimation is performed to get the channel coefficients in the MIMO channel matrix, and signals across all streams are then jointly equalized to remove the channel impact. Pilot phase compensation may be applied to the equalized symbols, before they are demodulated.

Note that the above processes may be slightly different across different systems. For example, in mobile networks, there are many logical channels, and the frame structure is more flexible and the signal format could also be more complicated.

2.3 MIMO Radar Signals and Systems

2.3.1 Single-Carrier MIMO Radar

In a MIMO radar, the waveforms transmitted from different antennas are typically orthogonal, and they can be either pulsed or continuous. The MIMO radar baseband waveform at a transmitter with M_T antennas can be expressed as follows:

$$\mathbf{x}_R(t) = \sum_{m=1}^{M_T} \mathbf{w}_m \psi_m(t) = \mathbf{W}\psi(t), \tag{2.7}$$

where $\mathbf{x}_R(t) = \{x_{R,m}(t)\}, m = 1, \dots, M_T$ with $x_{R,m}(t)$ being the signal at the m-th antenna, $\psi_m(t)$ is the basic radar waveform, $\mathbf{w}_m$ is the $M_T \times 1$ precoding/beamforming vector, and $\mathbf{W} = [\mathbf{w}_1, \dots, \mathbf{w}_{M_T}]$ and $\psi(t) = \{\psi_m(t)\}, m = 1, \dots, M_T$. Note that in radar, the peak-to-average power ratio (PAPR) of the transmitted signal is typically required to be very low, so that high-power efficiency can be achieved. Thus, $\mathbf{W}$ is typically an identity matrix.

The basic waveforms $\psi_m(t)$ can be any set of signals that are orthogonal in either one or multiple domains of time, frequency, space, and code. It can also take any waveform of pulsed and continuous-wave radars [2, 3]. In pulsed radar systems, short pulses of large bandwidth are transmitted either individually or in a group, followed by a silent period for receiving the echoes of the pulses. Continuous-wave radars transmit waveforms continuously, typically scanning over a large range of frequencies. In both systems, the waveforms are typically nonmodulated.

Referring to the channel model in (2.1), $\tau_o(t)$, and $f_o(t)$ are zeros for radar since the clocks for transmitter and receiver are typically locked. With the transmitting signal in (2.7), the received noise-free radar signal is given by

$$\mathbf{y}_R(t) = \sum_{m=1}^{M_T}\sum_{\ell=1}^{L} b_\ell \psi_m(t-\tau_\ell)e^{j2\pi f_{D,\ell}t} \cdot \mathbf{a}(M_R, \phi_\ell)\mathbf{a}^T(M_T, \theta_\ell)\mathbf{w}_m, \tag{2.8}$$

Applying matched filtering with $\psi_{m'}(t)$ to $\mathbf{y}_R(t)$, we obtain

$$\mathbf{r}_m(t) = \sum_{\ell=1}^{L} b_\ell \rho_m(t-\tau_\ell)e^{j2\pi f_{D,\ell}t}\mathbf{a}(M_R, \phi_\ell)\mathbf{a}^T(M_T, \theta_\ell)\mathbf{w}_m,$$

where $\rho_m(t)$ is the nonzero output of the matched filtering of $\psi_m(t)$, when $m = m'$, as all other outputs for $m \neq m'$ are zeros.

Assume that $\rho_m(t) = \rho(t)$ is the same for all $m \in [1, M_T]$. Staking all $\mathbf{r}_m(t)$, $m = 1, \dots, M_T$ to a matrix, we get

$$\mathbf{R}(t) = \sum_{\ell=1}^{L} b_\ell \rho(t-\tau_\ell)e^{j2\pi f_{D,\ell}t}\mathbf{a}(M_R, \phi_\ell)\mathbf{a}^T(M_T, \theta_\ell)\mathbf{W}.$$

When $\mathbf{W}$ is an identity matrix (if not, multiplying both sides with $\mathbf{W}^{-1}$), we can vectorize $\mathbf{R}(t)$ and get

$$\mathrm{vec}(\mathbf{R}(t)) = \sum_{\ell=1}^{L} b_\ell \rho(t-\tau_\ell)e^{j2\pi f_{D,\ell}t}\mathbf{a}(M_R,\phi_\ell)\otimes \mathbf{a}^T(M_T,\theta_\ell),$$

where $\otimes$ denotes the Kronecker product.

In MIMO radars, particularly monostatic radars, the antenna intervals of transmitter and receiver, d_T and d_R, are typically set as $d_T = M_T d_R$ or $d_R = M_T d_T$. Then when $\phi_\ell = \theta_\ell$, which is generally true for a monostatic radar, we have $\mathbf{a}(M_R,\phi_\ell)\otimes \mathbf{a}^T(M_T,\theta_\ell) = \mathbf{a}(M_R M_T,\phi_\ell)$. This enables a MIMO radar to achieve the spatial resolution corresponding to a virtual ULA with $M_T M_R$ antennas [3]. Note that the increased aperture of the virtual array is only meaningful when ϕ_ℓ is related to θ_ℓ in some way, although $\phi_\ell = \theta_\ell$ is not a necessary condition.

Here, we mainly consider the following emerging MIMO radars, which have not been widely implemented in practice, but have great potentials for realizing JCAS systems with balanced communications and radar performance.

2.3.2 MIMO-OFDM Radar

In a MIMO-OFDM radar, the waveform $\psi_m(t)$ is in the form of time-domain OFDM signals. Without considering the cyclic prefix, the baseband signal of $\psi_m(t)$ can be represented as follows:

$$\psi_m(t) = \sum_{n\in S_m} \tilde{w}_{m,n} e^{j2\pi n f_0 t} g(t-kT_0), \tag{2.9}$$

where S_m is the set of used subcarriers, $g(t)$ is a windowing function, and $\tilde{w}_{m,n}$ can be a complex variable combining orthogonal coding, subcarrier-dependent spatial precoding/beamforming, and/or other processing such as PAPR reduction.

To achieve orthogonality, S_m's are typically selected to be orthogonal for different m. The set of subcarriers can be allocated in various forms, such as the interleaved pattern [4], and nonequidistant subcarrier interleaving [5]. Different allocations may lead to different signal properties and ambiguity functions. Alternatively, the orthogonality may also be achieved over time domain via the use of orthogonal codes, while different antennas share the subcarriers.

MIMO-OFDM signals for radar are very similar to those for communications, except that they are typically nonmodulated and/or orthogonal across antennas. Hence, MIMO-OFDM signals are excellent options for JCAS systems. In particular, the training sequence in the preamble of MIMO-OFDM communication systems holds the desired characteristic of orthogonality and can be directly used for radar sensing.

2.3.3 FH-MIMO Radar

The frequency-hopping (FH) MIMO radar [6] and the frequency agile radar (FAR) [7], as well as its extensions to multicarrier [8], all use the FH technologies – the total bandwidth is divided into many subbands, only a subset of subbands are used at a time, and the subset at each antenna randomly varies over time. FH can be implemented in either fast time or slow time and in the form of either pulse or continuous-wave. We consider pulsed fast FH here, i.e. the signals are continuously transmitted with frequencies being changed rapidly and in multiple times over a pulse repetition interval (PRI), followed by a silent period. Using FH leads to major advantages, such as better security, and lower implementation cost by avoiding the use of costly instantaneous wideband components, while with negligible degradation in sensing performance, compared to using full bandwidth signals. FH-MIMO radar and FAR and its variants differ in how the frequencies are used at each antenna. Next, we briefly present their signal models, using notations similar to those for OFDM: B for the radar bandwidth, N for the number of sub-bands, and T for the hop duration.

In a FAR, all antennas use one common frequency at a hop, and a beamforming weight is applied to each antenna so that the array forms steerable beam [7]. The basic concept of FAR is extended to multisubband signaling in [8], that is a subset of more than one frequencies are used in each hop. In particular, in the multicarrier AgilE-phaSed ArrayRadar (CAESAR) scheme proposed in [8], the whole array is divided into multiple nonoverlapped subarrays, and each antenna in one subarray only uses one common frequency from the frequency subset. CAESAR randomizes both the frequencies and their allocation among the antenna elements and induces both frequency and spatial agility. It also maintains narrowband transmission from each antenna and introduces the beamforming capability. These capabilities make CAESAR more attractive than the original FAR.

Although FH-MIMO was developed before CAESAR, it can be treated as a special case of CAESAR, where each subarray has only one antenna and each frequency is only used by one antenna, that is $S = M_T$ and $|\mathcal{A}_s| = 1$. CAESAR can also be regarded as a generalization of FH-MIMO radar by introducing the beamforming capability.

In an FH-MIMO radar, each pulse is divided into H subpulses, i.e. *hops* [9]. The baseband frequency of the transmitted signal changes over hops and antennas, as illustrated in Figure 2.2. Denote the radar bandwidth as B. By dividing the frequency band evenly into K subbands, the baseband frequency of the k-th subband can be given by $f_k = \frac{kB}{K}$ $(k = 0, 1, \dots, K-1)$. Let M denote the number of antennas in the radar transmitter array, where $M < K$ is satisfied in general. Out of the K subbands, M are selected to be the hopping frequencies at a hop, one per antenna. Denote the hopping frequency at hop h and antenna m as f_{hm} that

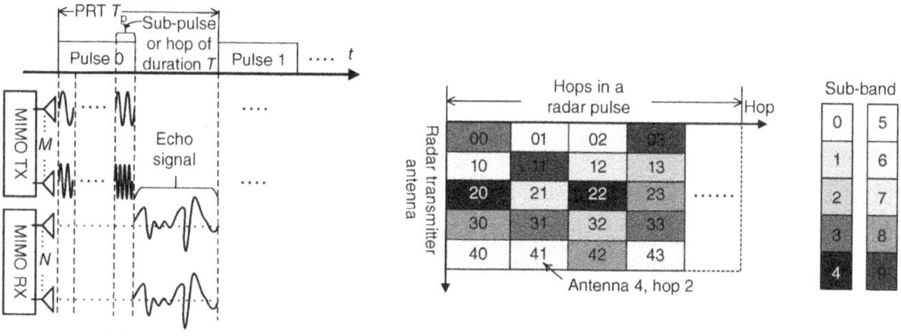

(a) (b)

Figure 2.2 (a) Illustration on the system diagram of an FH-MIMO DFRC; (b) The signal frame structure for the downlink communication in Figure 7.1a.

satisfies $f_{hm} \in \{f_k\ \forall k\}$. To facilitate FH-MIMO DFRC, the following waveform constraints are generally enforced [10, 11]

$$f_{hm} \neq f_{hm'}\ (\forall m \neq m',\ \forall h);\quad BT/K \in \mathbb{I}_+, \tag{2.10}$$

where $\mathbb{I}_+$ denotes the set of positive integers and T is hop duration. At hop h, the m-th antenna of the FH-MIMO radar transmitter transmits

$$s_{hm}(t) = e^{-j2\pi f_{hm} t},\ 0 \leq t - hT \leq T. \tag{2.11}$$

Based on (2.11), the m-th radar antenna transmits

$$s_m(t) = \sum_{h=0}^{H-1} s_{hm}(t) g_T(t - hT), \tag{2.12}$$

where $g_T(t)$ is the step function taking unit one for $0 \leq t \leq T$ and 0 elsewhere. Assume that there are N_{RT} radar targets. In a colocated MIMO radar, the transmitter and receiver arrays see the r-th target from the same direction, denoted by θ_r. Let $\mathbf{a}(\theta_r)$ and $\mathbf{b}(\theta_r)$ denote the array steering vectors of the radar transmitter and receiver arrays, respectively. Assuming uniform linear array, the antenna spacing is $\frac{\lambda}{2}$ and $\frac{M\lambda}{2}$ for transmitter and receiver, respectively, where λ is radar wavelength. The m-th element of $\mathbf{a}(\theta_r)$ is $e^{-jm\pi \sin\theta_r}$, and the n-th element of $\mathbf{b}(\theta_r)$ is $e^{-jnM\pi \sin\theta_r}$. Given the round-trip propagation delay of the r-th radar target, τ_r, and the target reflection coefficient, α_r, the signals received by antenna n at PRI p is

$$x_{pn}(t) = \sum_{r=0}^{N_{RT}-1} \sum_{m=0}^{M-1} \alpha_r s_m(t - \tau_r) e^{j2\pi f_{D,r} p T_P} e^{-j(m+nM)u_r}, \tag{2.13}$$

where $u_r = \pi \sin \theta_r$, $f_{D,r}$ is the Doppler frequency of target r and T_p denotes the PRI. For notational simplicity, the noise term is dropped here.

Remark 2.1 In (2.13), the "stop-and-hop" model is employed to depict Doppler impact, i.e. omitting the intrapulse phase change caused by Doppler impact. While the "stop-and-hop" is widely used, it can be subject to the condition that $\frac{4\pi f_{D,r}^2 T_{\mathrm{CPI}}}{\lambda \mathrm{C}}$ is less than a fraction of π radians, e.g. $\pi/4$ [12]. Otherwise, range migration can happen, possibly making the range bin of the same target different over PRTs. Here, T_{CPI} is a coherent processing time interval and C is the light speed.

2.4 Basic Signal Processing for Radar Sensing

For illustration convenience, we base the descriptions on the FH-MIMO radar mentioned earlier. A major reason is that DFRC systems based on such a radar will be the main focus in this book (Chapters 7–9). Nevertheless, it is noteworthy that the sensing steps to be introduced can also be applied to other sensing systems, such as the one to be illustrated in Chapter 5.

As a MIMO radar, we can perform the following typical processing steps [11, 13] at an FH-MIMO radar receiver. Note that these steps are common in classical radar but not unique. One may adjust the order of different steps to achieve different sensing performance. There are also more advanced and new sensing technologies that are more suitable for communications-centric JCAS, as will be detailed in Chapter 5.

2.4.1 Matched Filtering

This is a typical first step in radar signal processing. The transmitted signals are used as matched filtering coefficients, and these filters are performed on the received echo signal of each antenna. Specifically, for $x_{pn}(t)$ given in (2.13), the m-th matched filtering output can be given by

$$\tilde{x}_{pnm}(t) = s_m^*(-t) \circledast x_{pn}(t) = \int_{-\infty}^{\infty} s_m^*(-(t-\tau))x_{pn}(\tau)\mathrm{d}\tau, \tag{2.14}$$

where $s_m(t)$ is given in (2.12), $()^*$ takes conjugate, and $\circledast$ calculates the linear convolution.

Remark 2.2 Here, we provide a rationale for constructing the matched filtering coefficient as $s_m^*(-t)$ in (2.14). The linear convolution of two signals in the time domain is equivalent to the pointwise product of their respective Fourier transforms. So let us check the Fourier transforms of the two terms on the right-hand side of (2.14). The Fourier transform of $s_m^*(-t)$ becomes $S_m^*(f)$, where $S_m(f)$ denotes the Fourier transform of $s_m(t)$. The Fourier transform of $x_{pn}(t)$, as given in (2.13), can be written as follows:

$$X_{pn}(f) = \sum_{r=0}^{N_{\mathrm{RT}}-1} \alpha_r S_m(f) e^{-\mathrm{j}2\pi f \tau_r} e^{\mathrm{j}2\pi f_{\mathrm{D},r} p T_{\mathrm{P}}} e^{-\mathrm{j}(m+nM)u_r} + \sum_{r=0}^{N_{\mathrm{RT}}-1} \sum_{\substack{m'=0 \\ m' \neq m}}^{M-1} \alpha_r S_{m'}(f) e^{-\mathrm{j}2\pi f \tau_r} e^{\mathrm{j}2\pi f_{\mathrm{D},r} p T_{\mathrm{P}}} e^{-\mathrm{j}(m'+nM)u_r}. \tag{2.15}$$

As said above, the linear convolution in (2.14) is equivalent to $S_m^*(f) X_{pn}(f)$, which is now

$$S_m^*(f) X_{pn}(f) = \sum_{r=0}^{N_{\mathrm{RT}}-1} \alpha_r |S_m(f)|^2 e^{-\mathrm{j}2\pi f \tau_r} e^{\mathrm{j}2\pi f_{\mathrm{D},r} p T_{\mathrm{P}}} e^{-\mathrm{j}(m+nM)u_r} + \sum_{r=0}^{N_{\mathrm{RT}}-1} \sum_{\substack{m'=0 \\ m' \neq m}}^{M-1} \alpha_r S_m^*(f) S_{m'}(f) e^{-\mathrm{j}2\pi f \tau_r} e^{\mathrm{j}2\pi f_{\mathrm{D},r} p T_{\mathrm{P}}} e^{-\mathrm{j}(m'+nM)u_r}. \tag{2.16}$$

The inverse Fourier transform of $S_m^*(f) X_{pn}(f)$ results in $\tilde{x}_{pnm}(\tau)$. From the right-hand side of (2.16), we see that the inverse Fourier transform of $S_m^*(f) X_{pn}(f)$ is locally maximized at $t = \tau_r$ $(\forall r)$, provided that the inverse Fourier transform of$\sum_{\substack{m'=0 \\ m' \neq m}}^{M-1} \alpha_r S_m^*(f) S_{m'}(f) e^{-\mathrm{j}2\pi f \tau_r}$ is negligible. This negligible assumption can be well approximated if the waveforms of different transmitted antennas are nearly orthogonal.

2.4.2 Moving Target Detection (MTD)

This is often done in radar signal processing after the matched filtering, i.e. based on the signal $\tilde{x}_{pnm}(t)$ obtained in (2.14). For convenience, let us replace t with its discrete version iT_{S}. Then we write $\tilde{x}_{pnm}(t)$ as $\tilde{x}_{pnm}[i]$. Note that p, n, m, i is the index of PRT, receiver antenna, transmitter antenna, and sample, respectively. At any n, m and i, taking the DFT of $\tilde{x}_{pnm}[i]$ over p leads to

$$\tilde{X}_{qnm}[i] = \sum_{p=0}^{P-1} \tilde{x}_{pnm}[i] e^{-\mathrm{j}\frac{2\pi pq}{P}}, \quad q = 0, 1, \ldots, P-1, \tag{2.17}$$

where P is the number of PRTs in a radar coherent processing interval. Although we have not given the analytical expression of $\tilde{x}_{pnm}[i]$, we see from (2.13) that the only Doppler-related term is $e^{\mathrm{j}2\pi f_{\mathrm{D},r} p T_{\mathrm{P}}}$. This term is approximately constant within each PRT and hence the matched filtering performed above does not change the term. Thus, the DFT performed in (2.17) will result in local peaks at

$$q = \lfloor P f_{D,r} T_{\mathrm{P}} \rceil, \quad r = 0, 1, \ldots, N_{\mathrm{RT}} - 1, \tag{2.18}$$

where $\lfloor \cdot \rceil$ denotes rounding.

2.4.3 Spatial-Domain Processing

The result $\tilde{X}_{qnm}[i]$ spans over range, Doppler, and spatial domains, as represented by i, q, and (n, m), respectively. We can establish a maximum likelihood detector over the three domains, which, however, will incur too large a searching space and hence is difficult to perform in real-time applications. There are two common ways of reducing complexity, as illustrated below.

Option 1. Discretizing the spatial domain: Given any q and i, stacking $\tilde{X}_{qnm}[i]$ into a vector in the order of first m and then n, we can obtain the following vector:

$$\mathbf{y}_{qi} = \left[\tilde{X}_{q0}[i], \tilde{X}_{q1}[i], \dots, \tilde{X}_{q\tilde{n}}[i], \dots, \tilde{X}_{q(NM-1)}[i]\right]^{\mathrm{T}}, \tag{2.19}$$

where the subscripts n and m are replaced with $\tilde{n}$ via the relation $\tilde{n} = nM + m$. Corresponding to $\mathbf{y}_{qi}$, the spatial steering vector of the r-th target can be obtained based on (2.16); specifically, we have

$$\mathbf{a}_r = \left[1, e^{-\mathrm{j}u_r}, \dots, e^{-\mathrm{j}(NM-1)u_r}\right]^{\mathrm{T}}. \tag{2.20}$$

Since $\mathbf{y}_{qi}$ is essentially a linear combination of $\mathbf{a}_r$ $(r = 0, 1, \dots, N_{\mathrm{RT}} - 1)$, we can focus $\mathbf{y}_{qi}$ onto different beams through digital beamforming. A special case is to take the number of beams as NM with the $\tilde{n}$-th beam pointing at $\frac{2\pi\tilde{n}}{NM}$. Then, the digital beamforming can be performed through a NM-dimensional DFT on $\mathbf{y}_{qi}$ $(\forall q, i)$. Let $\mathbf{W}$ denote the DFT matrix and the beamforming output can be given by $\mathbf{z}_{qi} = \mathbf{W}\mathbf{y}_{qi}$. Let the $\tilde{n}$-th entry of $\mathbf{z}_{qi}$ be denoted by $z_{\tilde{n}qi}$. Stacking $z_{\tilde{n}qi}$ into a matrix with the row indexed by q and column indexed by i, we have

$$\mathbf{Z}_{\tilde{n}} = \begin{bmatrix} z_{\tilde{n}00} & z_{\tilde{n}10} & \cdots & z_{\tilde{n}(Q-1)0} \\ z_{\tilde{n}01} & z_{\tilde{n}11} & \cdots & z_{\tilde{n}(Q-1)1} \\ \vdots & \vdots & \ddots & \vdots \\ z_{\tilde{n}0(I-1)} & z_{\tilde{n}1(I-1)} & \cdots & z_{\tilde{n}(Q-1)(I-1)} \end{bmatrix}, \tag{2.21}$$

which is generally referred as the range-Doppler matrix (RDM).

Option 2. Suppressing the spatial domain: This is a relatively easier way of reducing complexities. In particular, we can incoherently accumulate $\tilde{X}_{qnm}[i]$ over n and m, obtaining $\tilde{z}_{qi} = \sum_{n,m} \left|\tilde{X}_{qnm}[i]\right|$. Similar to (2.21), we can also stack $\tilde{z}_{qi}$ into an $I \times Q$ RDM, as denoted by $\tilde{\mathbf{Z}}$.

Some comparisons between the two options are made below:

1. Option 1 incurs more complexities than Option 2, not only in the way the RDM is calculated but also in the sequential processing. Take target detection for example. All RDMs in Option 1 need to be processed by some detecting algorithm for target detection, while there is only a single RDM generated from Option 2.

2. As more RDMs are generated, more storage would be required by Option 1 as well, compared with Option 2.
3. A disadvantage of Option 2 is that interference among targets can be stronger than that in Option 1. This is because targets in the whole angular region are merged together for target detection in Option 2.

2.4.4 Target Detection

The RDM obtained in the previous step can be further used for target detection. The magnitude of each entry of the RDM is the echo strength (or power) at a range-Doppler bin. The task of target detection is to tell whether there exists a target at a range-Doppler bin or not. Numerous algorithms are available for this task. Here, we briefly describe the cell-average constant false-alarm rate (CA-CFAR) detection given its wide applicability in radar systems. The basic principle of CA-CFAR is to compare the power of the range-Doppler bin under test (BUT) with a threshold that is the scaled averaged power of the surrounding range-Doppler bins of the BUT.

As illustrated in Figure 2.3, at a BUT, the direct neighboring range-Doppler bins form the so-called "guard interval." These bins are excluded from calculating the detecting threshold. Let the numbers of bins in the guard interval be E and F in the Doppler and range dimensions, respectively. Then, let A and B be the numbers of reference bins over Doppler and range dimensions, respectively. These bins will be used for estimating the interference-plus-noise (IN) power. Averaging the power of the signals in reference bins leads to P_{ref}. Then the CA-CFAR threshold can be constructed as follows: CA-CFAR threshold as [14, (16.23)],

$$\mathsf{T} = \beta P_{\text{ref}},\ \beta = (AB - EF)\left(\mathsf{P}_{\mathrm{F}}^{-1/(AB-EF)} - 1\right), \tag{2.22}$$

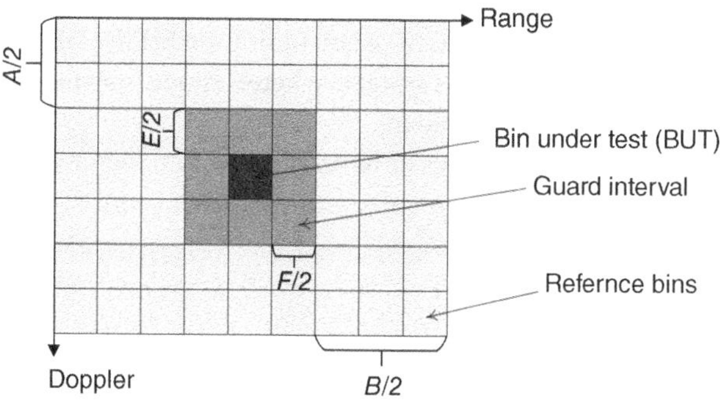

Figure 2.3 (a) Illustration on the system diagram of an FH-MIMO DFRC; (b) The signal frame structure for the downlink communication in Figure 7.1(a).

where $\mathrm{P_F}$ denotes the expected false-alarm rate, e.g. 10^{-4}. Let P_{BUT} denote the power of BUT. If $P_{\mathrm{BUT}} \geq \mathrm{T}$, we report the presence of a target at the BUT; otherwise, we report no target.

2.4.5 Spatial Refinement

Performing CA-CFAR based on the RDM $\tilde{\mathbf{Z}}$ obtained by Option 2 in Section 2.4.3. We can only detect targets in range and Doppler dimensions, as the spatial information is suppressed over the incoherent accumulation. Now, assume that there is a target detected at the (q^*, i^*)-th bin. Substituting $q = q^*$ and $i = i^*$ into (2.19), we obtain $\mathbf{y}_{q^*i^*}$. This $NM \times 1$ vector is underlain by the steering vector given in (2.20). This enables us to estimate the angle information of the target. Numerous angle estimation methods, e.g. the classical multiple signal classification (MUSIC), can be used for the task.

2.5 Signal Processing Basics for Communication-Centric JCAS

In communication-centric JCAS systems, radar sensing is integrated into existing communication systems as a secondary function. Revision and enhancement to communication infrastructure and systems may be required, but the primary communication signals and protocols largely remain unchanged.

Considering the topology of communication networks, communication-centric JCAS systems can be classified into two types, namely, those realizing sensing in point-to-point communication systems and in large multiple-access networks such as mobile networks. Two good examples are the IEEE 802.11ad JCAS systems for vehicular networks [15–17] and the perceptive mobile networks (PMNs) [18, 19], respectively. The former can use both of the single-carrier and OFDM signals as described in Section 2.2, and the latter is based on multiuser-MIMO OFDM currently and in the near future. Both are JCAS systems where a single transmitted signal is used for both communications and radar.

In this section, we first describe the two types of JCAS systems, and then review the core signal processing techniques for general communication-centric JCAS systems.

2.5.1 802.11ad JCAS Systems

The 802.11ad standard defines a millimeter-wave packet communication system operating in the 60 GHz unlicensed band. There are three types of physical-layer (PHY) packets: single-carrier, OFDM, and control, with OFDM being optional.

The preamble in each packet is the main signal that has been exploited for radar sensing [15–17, 20, 21], particularly for vehicular networks. In a typical setup, the sensing receiver is colocated with the JCAS transmitter, using two separated analog arrays. The JCAS device can be located either on the road side unit (RSU) or on a vehicle.

Referring to (2.1), the noise-free, time-domain echo signal at the sensing receiver can be represented as follows:

$$y(t) = \sum_{\ell=1}^{L} h_\ell(t)s(t-\tau_\ell)e^{j2\pi f_{D,\ell} t}, \tag{2.23}$$

where $h_\ell(t) \triangleq b_\ell \mathbf{w}_R(t)^T\mathbf{a}(M_R, \theta_\ell)\mathbf{a}^T(M_T, \theta_\ell)\mathbf{w}_T(t)$, $\mathbf{w}_T(t)$, and $\mathbf{w}_R(t)$ are the beamforming vector in the transmitter and the receiver, respectively, and the AoA and the AoD are assumed to be the same. Note that the clock between the transmitter and the sensing receiver is locked; therefore, $\tau_o(t) = 0$ and $f_o(t) = 0$ in (2.1). The primary goal of sensing here is estimating the location and velocity of objects via estimating τ_ℓ, θ_ℓ, and $\mathbf{f}_{D,\ell}$.

The three PHYs in 802.11ad have a similar preamble structure, consisting of short training field (STF) and channel estimation field (CEF). The STF consists of tens of repeated 128-sample Golay sequences, followed by its binary complement. The CEF consists of two 512-sample Golay complementary pair, which has the property of perfect aperiodic autocorrelation, i.e. $s(t-\tau) \circledast s(t) \neq 0$ if and only if $\tau = 0$. Both the STF and the CEF can be used for sensing, in either a hierarchical or a joint manner [20]. The hierarchical strategy processes the STF and the CEF separately, exploiting their respective properties. For example, the repetition pattern of STF is typically used for packet detection in communications, and hence, it is ideal for target detection in sensing; while the perfect aperiodic autocorrelation of CEF can lead to excellent channel estimation and sensing performance, based on, e.g. the generalized likelihood ratio test [21]. The joint strategy uses both STF and CEF for common tasks of sensing, based on, e.g. matched filtering [20]. The sensing performance bounds are also derived in [20, 21]. More advanced sensing algorithms will be discussed in Section 2.5.3.

Both the single-carrier PHY, which has an identical preamble with the OFDM PHY, and the control PHY have been explored for sensing [20, 21]. There are some differences between their sensing efficiency. On the one hand, in the standard, a beamforming training protocol is defined to align the transmit and receive beams, using beam scanning and the control-PHY signals. Single-carrier or OFDM PHY signals are typically used after beamforming training. On the other hand, the control PHY has a longer STF. So in terms of sensing, the control PHY enables a wider field-of-view (FoV) and potentially better accuracy, while the other two are generally limited to the fixed direction of communications.

2.5.2 Mobile Network with JCAS Capabilities

In [18], the framework of PMNs is introduced by applying the JCAS techniques to cellular networks. Downlink sensing and uplink sensing are defined, corresponding to sensing using the received downlink and uplink communication signals, respectively. In the scenario of cloud radio access networks (CRANs), where distributed remote radio units (RRUs) cooperatively communicate with user equipment (UE), the received downlink communication signals from one RRU itself and other cooperative RRUs can be used for downlink active sensing and downlink passive sensing, respectively.

Extend the single-user MIMO-OFDM model in Section 2.2 to multiuser MIMO-OFDMA. Suppose that one node receives signals transmitted from a set of nodes $q, q \in \mathcal{Q}_T$, and uses the signals for sensing. Let Q_T be the cardinality of $\mathcal{Q}_T$. Referring to the transmitting signal model in (2.5) and the channel model in (2.2), we can represent the received noise-free k-th OFDM symbol at the n-th subcarrier as follows:

$$\tilde{\mathbf{y}}_{n,k} = \sum_{q \in \mathcal{Q}_T} \sum_{\ell=1}^{L_q} b_{q,\ell} e^{-j2\pi n(\tau_{q,\ell}+\tau_{o,q,k})f_0} e^{j2\pi k(f_{D,q,\ell}+f_{o,q,k})T_s} \cdot \mathbf{a}(M_R, \phi_{q,\ell}) \mathbf{a}^T(M_q, \theta_{q,\ell}) \tilde{\mathbf{x}}_{q,n,k}. \tag{2.24}$$

Scenarios represented by this model are exemplified below:

1. *Downlink sensing in a standalone base station (BS)*: This is the case where the BS uses its own reflected transmitted signals for sensing, similar to a monostatic radar. In this case, $Q_T = 1$ and $\tau_{o,q,k} = f_{o,q,k} = 0$;
2. *Uplink sensing in a standalone BS*: $\mathcal{Q}_T$ denotes the set of Q_T UEs sharing the same subcarriers via SDMA, and $\tau_{o,q,k} \neq 0$ and $f_{o,q,k} \neq 0$. Each UE only occupies partial of the total subcarriers;
3. *Downlink sensing in an RRU*: $\mathcal{Q}_T$ denotes the set of RRUs whose downlink communication signals are seen by the sensing RRU, including its own echo signals.

The sensing can be based on (2.24) with the signals $\tilde{\mathbf{x}}_{q,n,k}$ corresponding to both the pilots and data symbols, as will be detailed in Chapter 4.

2.5.3 Sensing Parameter Estimation

Communications and sensing can share a number of processing modules. This is illustrated in Figure 2.4, referring to the widely employed MIMO-OFDM transceiver. The whole transmitter and many modules in the receiver that are shown in purple can be shared by communication and sensing functionalities.

Here, we briefly discuss the core techniques of sensing parameter estimation for communication-centric JCAS systems, while providing a comprehensive review on signal processing with reference to PMN in Chapter 4.

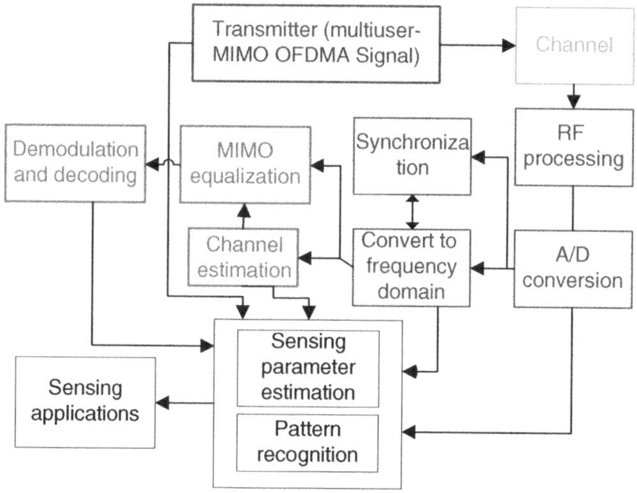

Figure 2.4 A block diagram of a transceiver showing the components that can be shared by communications and sensing in an MIMO-OFDM JCAS system. The blocks, with the key words "Transmitter", "Synchronization", "Convert", "RF" and "A/D", are shared by communications and sensing; blocks with the key words "Demodulation", "MIMO equalization" and "Channel estimation" are for communications only; whereas, blocks with the key words "Sensing" and "Pattern" are for sensing only. The sensing unit, including sensing parameter estimation and pattern recognition, can use signals outputs from multiple modules. Some additional modules, such as clutter suppression, sensing, and communication cooperation, are not shown in the figure.

Sensing parameter estimation in communications-centric JCAS is generally different to that in traditional radar systems, due to the significant differences between the two types of signals as described in Chapter 1. Next, referring to the MIMO-OFDM signal models in Section 2.5.2, we review key techniques in sensing parameter estimation. Most of them are also applicable to single-carrier systems such as the IEEE802.11ad JCAS. We first discuss two important problems to be resolved, before we review optional sensing algorithms. Note that for OFDM, sensing parameter estimation can be done in both the time domain and the frequency domain. The sensing applications may demand either sensing parameter estimation or pattern recognition results, or both.

2.5.3.1 Direct and Indirect Sensing

The first problem is how to deal with $\tilde{\mathbf{x}}_{q,n,k}$ in the received signals, which can represent either known pilots or (unknown) data payload in a packet. Both can be used for sensing. For unknown data payload, it can be demodulated after channel estimation, as in conventional communications. Using data payload can significantly extend the sensing capability such as the range, as it is much longer than the pilot. Here, the receiver is assumed to know $\tilde{\mathbf{x}}_{q,n,k}$ or its estimate via demodulating $\tilde{\mathbf{s}}_{q,n,k}$. For multiuser-MIMO signals, for example, signals received at an RRU from multiple RRUs in downlink sensing, or signals received at a standalone BS from multiple UEs, we can use two methods to formulate the estimation problem.

One method, which may be called as *direct sensing*, directly feeds the received signals to sensing algorithms. In some cases, e.g. sensing using the data payload in a MIMO system, this is the only option as $\tilde{\mathbf{x}}_{q,n,k}$ cannot be readily removed even when they are known. The presence of $\tilde{\mathbf{x}}_{q,n,k}$ often limits the optional algorithms for sensing parameter estimation.

Let us have a look at one example in [18], where direct sensing is conducted via the block compressive sensing (CS) techniques [22], and the symbols $\tilde{\mathbf{x}}_{q,n,k}$ are used as part of the sensing matrix. For the clarify of presentation, we consider the case of $Q_T = 1$ in (2.24) and ignore the timing offset $\tau_{o,q,k}$. Ignore the noise and rewrite (2.24) in a more compact matrix form as follows:

$$\tilde{\mathbf{y}}_{n,k} = \mathbf{A}(M_R, \boldsymbol{\phi}_q)\mathbf{C}_n\Delta_k\mathbf{A}^T(M_q, \boldsymbol{\theta}_q)\tilde{\mathbf{x}}_{q,n,k}, \tag{2.25}$$

where $\mathbf{A}(M_R, \boldsymbol{\phi}_q)$ and $\mathbf{A}(M_q, \boldsymbol{\theta}_q)$ are matrices with the ℓ-th column being $\mathbf{a}(M_R, \phi_{q,\ell})$ and $\mathbf{a}(M_q, \theta_{q,\ell})$), respectively, and Δ_k and $\mathbf{C}_n$ are diagonal matrices with the diagonal element being $b_{q,\ell}e^{j2\pi k f_{D,q,\ell}T_s}$ and $e^{-j2\pi n\tau_{q,\ell}f_0}$, respectively.

In order to apply sensing algorithms, we need to reorganize signals so that we can stack more measurements over the same domain. Consider the case of collecting samples from all the subcarriers for the estimation of delay and AoA. Take the transpose of $\tilde{\mathbf{y}}_{n,k}$ in (2.25), and rewrite it as follows:

$$\tilde{\mathbf{y}}_{n,k}^T = \tilde{\mathbf{x}}_{q,n,k}^T(\mathbf{c}_n^T \otimes \mathbf{I}_{M_q})\mathbf{V}_q\mathbf{A}^T(M_R, \boldsymbol{\phi}_q). \tag{2.26}$$

where $\mathbf{c}_n$ is a column vector containing the diagonal elements of $\mathbf{C}_n$, $\mathbf{I}_{M_q}$ is an $M_q \times M_q$ identity matrix, and $\mathbf{V}_q$ is a block diagonal matrix

$$\mathbf{V}_q = \mathrm{diag}\{b_\ell e^{-j2\pi k f_{D,q,\ell}T_s}\mathbf{a}(M_q, \theta_{q,\ell})\}, \ell = 1, \dots, L_q.$$

We have now separated signals $\tilde{\mathbf{x}}_{q,n,k}^T(\mathbf{c}_n^T \otimes \mathbf{I}_{M_q})$ that depend on n from those on other variables. We can then stack the row vectors $\tilde{\mathbf{y}}_{n,k}^T$ from all available subcarriers to a matrix and obtain

$$\tilde{\mathbf{Y}}_k \triangleq [\tilde{\mathbf{y}}_{1,k}, \dots, \tilde{\mathbf{y}}_{n,k}, \dots]^T = \mathbf{W}\mathbf{V}_q\mathbf{A}^T(M_R, \boldsymbol{\phi}_q), \tag{2.27}$$

where the n-th row of $\mathbf{W}$ is $\tilde{\mathbf{x}}_{q,n,k}^T(\mathbf{c}_n^T \otimes \mathbf{I}_{M_q})$.

The signal model in (2.27) enables the applications of both 1-D multimeasurement vector (MMV) CS and 2-D CS techniques. For 1-D MMV CS, $\mathbf{c}_n$ is expanded to a quantized on-grid vector, and then $\mathbf{W}$ is used as the sensing matrix. The delay and $\mathbf{V}_q\mathbf{A}^T(M_R, \boldsymbol{\phi}_q)$ will then be the outputs of the algorithm, and the Doppler frequency and AoA can be further estimated from the estimates of $\mathbf{V}_q\mathbf{A}^T(M_R, \boldsymbol{\phi}_q)$. For 2D CS, both delay and AoD can be estimated together by expanding both $\mathbf{c}_n$ and $\mathbf{A}(M_R, \boldsymbol{\phi}_q)$

to on-grid models. It is easy to see that if we swap the terms of Doppler frequency and delay in (2.26), samples across OFDM symbols can be stacked and the Doppler frequencies can be estimated first.

The direct sensing method has a high-computational complexity. Due to the presence of $\tilde{\mathbf{x}}_{q,n,k}$, the applicable sensing solutions are also limited. However, it is the only option, when $\tilde{\mathbf{x}}_{q,n,k}$ cannot be removed due to, e.g. insufficient measurements.

The other method, *indirect sensing*, first estimate the elements of channel matrix for each node. It decorrelates signals from multiple nodes, removes $\tilde{\mathbf{x}}_{q,n,k}$ or $\tilde{\mathbf{s}}_{q,n,k}$ from the received signals, and applies sensing parameter estimation to the estimated channel matrix. In multiuser MIMO systems, referring to (2.24), this can be achieved by decorrelating signals collected from K $\tilde{\mathbf{y}}_{n,k}$s, $k = k', k'+1, \dots, k'+K-1, K \geq M_T Q_T$ at subcarrier n. Mathematically, this can be represented as follows:

$$\left[\tilde{\mathbf{H}}_{1,n,k}, \dots, \tilde{\mathbf{H}}_{Q_T,n,k}\right] = \left[\check{\mathbf{x}}_{n,1}, \dots, \check{\mathbf{x}}_{n,K}\right]^{-1} \left[\tilde{\mathbf{y}}_{n,1}, \dots, \tilde{\mathbf{y}}_{n,K}\right], \tag{2.28}$$

where $\check{\mathbf{x}}_{n,k} = [\tilde{\mathbf{x}}_{1,n,k}^T, \dots, \tilde{\mathbf{x}}_{Q_T,n,k}^T]^T$ is a $M_T Q_T \times 1$ vector. Note that the channel matrix is assumed to be constant over this interval. Equation (2.28) indicates that the decorrelation is only possible when $K \geq M_T Q_T$ $\check{\mathbf{x}}_{n,k}$s are available during a CPI and the inversion of $\left[\check{\mathbf{x}}_{n,1}, \dots, \check{\mathbf{x}}_{n,K}\right]$ exists.

The decorrelation involves high-computational complexity due to matrix inversion and may cause significant noise enhancement, unless $\tilde{\mathbf{x}}_{q,n,k}$ or $\tilde{\mathbf{s}}_{q,n,k}$ are orthogonal. Hence, indirect sensing is particularly suitable for training and pilot symbols which are typically orthogonal. After $\tilde{\mathbf{x}}_{q,n,k}$ is removed, we can equivalently work on the single-user channel matrix $\tilde{\mathbf{H}}_n(t)$ in (2.2). This can largely simplify sensing parameter estimation and offer great flexibility in problem formulation. Note that if the precoding matrix $\mathbf{P}_{n,k}$ is unknown to the receiver, $\mathbf{a}^T(M_T, \theta_\ell)$ in (2.2) will be replaced by $\mathbf{a}^T(M_T, \theta_\ell)\mathbf{P}_{n,k}$. This will make the estimation of AoD challenging.

2.5.3.2 Sensing Algorithms

We now discuss options for sensing algorithms based on the indirect method. Traditional radar typically applies matched filtering for sensing parameter estimation, which has also been adopted in some JCAS systems, e.g., 802.11ad JCAS [15, 17, 20]. However, the accuracy and resolution capability of these methods largely depend on the signal correlation properties (i.e. ambiguity functions). More options that are less affected by the correlation property can be explored for communication-centric JCAS signals.

From the decorrelated estimates of $\tilde{\mathbf{H}}_{n,k}$s in (2.2), we can represent the (m_R, m_T)-th element in $\tilde{\mathbf{H}}_{n,k}$ as follows:

$$\tilde{h}_{n,k,m_R,m_T} = \sum_{\ell=1}^{L} b_\ell e^{-j2\pi n \tau_\ell f_0} e^{j2\pi k f_{D,\ell} T_s} e^{j2\pi d_R m_R \sin(\phi_\ell)/\lambda} e^{j2\pi d_T m_T \sin(\theta_\ell)/\lambda}, \tag{2.29}$$

where m_R and m_T represent the indexes of the receiving and transmitting antennas, respectively. This is also known as a 4-D *Harmonic retrieval* problem [23], where the observation signals in each domain can be represented as a Vandermonde matrix when the samples are equally spaced. The 4-D harmonic retrieval problem can be reduced to multiple-snapshot, lower-dimensional problems by combining one or more of the exponential functions with the unknown variable b_ℓ. Thus, we can rewrite them to different matrix and Tensor forms so that sensing parameters can be estimated in different ways and orders. Classical techniques, such as periodogram and MUSIC, and modern techniques, such as CS and Tensor, can then be applied to solve this Harmonic retrieval problem. More details will be provided in Chapter 4.

2.6 Signal Processing Basics for DFRC

Radar systems, particularly military radar, have the extraordinary capability of long-range operation, up to hundreds of kilometers. Therefore, a major advantage of implementing communication in radar systems is the possibility of achieving very long-range communications, with much lower latency compared to satellite communications. However, the achievable data rates for such systems are typically limited due to the inherent limitation in the radar waveform [24–27].

Research on radar-centric JCAS, DFRC, has been mainly focused on the information-embedding technologies, and there are only limited works on other aspects such as communication protocol and receiver design based on the radar-centric JCAS signals. In this section, we concentrate on more recent JCAS systems based on MIMO-OFDM, CAESAR, and FH-MIMO radar, because of the remarkable benefits they can offer as described in Section 2.3.

2.6.1 Embedding Information in Radar Waveform

Realization of communication in radar systems needs to be based on either pulsed or continuous-wave radar signals. Hence, information embedding with little interference on radar operation is one of the major challenges. This topic has been widely investigated, as reviewed in, e.g. [26–28], as summarized in Table 1.4 in Chapter 1.

One of the particular techniques of interest is index modulation (IM). IM embeds information to different combinations of radar signals' parameters, over one or more domains of space, time, frequency, and code [27, 29, 30]. Thus IM does not change the basic radar waveform and signal structure and has negligible influence on radar operation. For MIMO-OFDM, CAESAR, and FH-MIMO radar, IM can be realized via frequency combination and/or antenna permutation [31–34]. Frequency combination selects different sets of frequencies, and antenna permutation allocates the selected frequencies to different antennas. Information is represented by the combinations and permutations. Let the number of combinations and permutations be N_c and N_p, respectively. Then the number of bits can be represented is $\log_2 N_c$ and $\log_2 N_p$, respectively. Mathematically, frequency combination and antenna permutation can generally be combined. However, decoupling them is consistent with the way that the information is demodulated, as will be discussed later.

For MIMO-OFDM radar with orthogonal frequency allocation, frequency combination allocates the total N subcarriers to M_T groups without repetition, with each group having at least one subcarrier [31, 35]. If without additional constraint on subcarrier allocation, there are a total of $N_c = C_{M_T}^N M_T^{N-M_T}$ combinations (i.e. selecting M_T out of N subcarriers first to ensure each group to have at least one, and then the remained $N - M_T$ subcarriers can go to any of the M_T groups); if each antenna needs to have the same number of $L_s = N/M_T$ subcarriers, the total number of combinations is $N_c = C_{L_s}^N C_{L_s}^{N-L_s} \dots C_{L_s}^{L_s} = N!/(L_s!)^{M_T}$. The number of permutations of allocating M_T groups of subcarriers to M_T antennas is $N_p = M_T!$.

The DFRC system extended from CAESAR is proposed in [32], where each virtual subarray is assumed to have the same number of antennas and use one frequency. Hence, frequency combination selects S out of N frequencies, and $N_c = C_S^N$. The number of total antenna permutations is shown to be $N_p = M_T!/((M_T/S)!)^S$.

For FH-MIMO DFRC systems [33, 36], the total number of frequency combinations and antenna permutations are $N_c = C_{M_T}^N$ and $N_p = M_T!$, respectively. This corresponds to $S = M_T$ in [32]. Let B_k denote the $M_T \times M_T$ antenna permutation matrix at hop k, which has only a single nonzero element, 1, in each row and each column. The transmitted signal of the FH-MIMO DFRC can be represented as follows:

$$\mathbf{x}_R(t) = B_k \boldsymbol{\psi}(t). \tag{2.30}$$

Note that both B_k and the frequency set $\mathcal{F}_k$ vary with k and are determined by the information bits. One simple example of information embedded FH-MIMO DFRC for packet communication is shown in Figure 2.5.

It is noted that the values of N_c and N_p above define the maximum achievable bit rates only, without considering the communication reception performance. In practice, the number of actually used combinations and permutations may be reduced due to the overall system design and the consideration of the demodulation complexity and performance. In particular, demodulating the bits embedded in antenna permutation is much more difficult and subject to higher-demodulation error, compared to demodulating those embedded in frequency combination. These issues will be discussed in detail in Section 2.6.2 and 2.6.3.

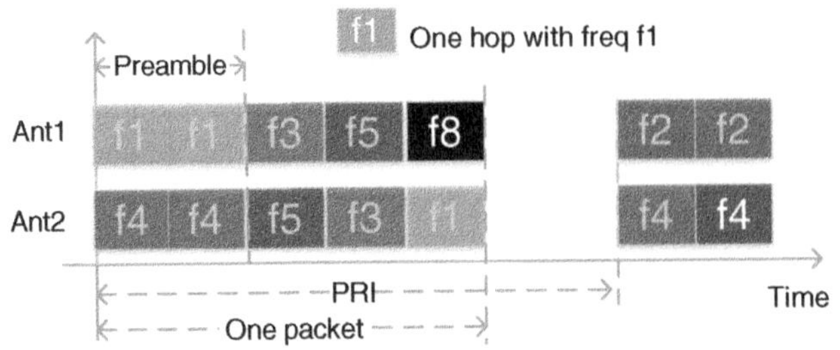

Figure 2.5 A simple example showing the packet structure of FH-MIMO DFRC with $N = 8$ and $M_T = 2$, consisting of a preamble with two identical hops and three hops with embedded information. A total of rounded 4 bits can be conveyed in each hop with IM.

In addition, the impacts of information embedding on radar performance should also be evaluated, such as the ambiguity functions in the time domain [34] and in the angular domain [35].

2.6.2 Signal Reception and Processing for Communications

Since IM in all the three systems involves FH, and their receiver processing methods are similar in many aspects, we use FH-MIMO DFRC as an example to illustrate the signal reception and processing for its relative simplicity. Overall, the research on the receiver design for these systems is still limited. Our overview here is mainly based on [34, 36, 37] and also incorporates works on MIMO-OFDM and CAESAR DFRCs [32, 35] into the framework of FH-MIMO DFRC.

Consider a receiver with M_R antennas. The signal received from each antenna is passed to a mixer with local oscillator frequency f_c, which is generally the central frequency of the N subbands. Assume narrowband communications and the difference between multipath delays $|\tau_\ell - \tau_{\ell'}| \ll T$. Referring to (2.1) and (2.4) and ignoring the variation of Doppler frequencies, the noise-free baseband received signal can be approximated as follows:

$$\mathbf{y}(t) = \sum_{m=1}^{M_T} \sum_{\ell=1}^{L} c_\ell e^{j2\pi(f_{k,m}-f_c)(t-\tau-\tau_o(t))} \mathbf{a}(M_R, \phi_\ell)\mathbf{a}^T(M_T, \theta_\ell)\beta_{k,m} g(t - \tau - \tau_o(t) - kT),$$

where $\beta_{k,m}$ is the m-th column of B_k, $c_\ell \in \mathbb{C}$ is the equivalent path coefficient, subsuming multiple terms, and $\tau_\ell \approx \tau$ is used. The baseband signal is then sampled at $T_s = 1/B$, generating $L_p = \lfloor T/T_s \rfloor$ samples per hop. Let

$$\mathbf{H} \triangleq \sum_{\ell=1}^{L} c_\ell \mathbf{a}(M_R, \phi_\ell)\mathbf{a}^T(M_T, \theta_\ell). \tag{2.31}$$

For the simplicity of presentation, assume that $g(t)$ is a rectangular windowing function. Assume that synchronization is done perfectly. We can stack L_p

measurements from all M_R antennas to a matrix $\mathbf{Y}_k$, which is given by

$$\mathbf{Y}_k = \mathbf{H}\boldsymbol{B}_k\boldsymbol{\Phi}^T = \sum_{m=1}^{M_T} \mathbf{H}\beta_{k,m}\boldsymbol{\psi}_{k,m}^T, \tag{2.32}$$

where $\boldsymbol{\Phi} = [\boldsymbol{\psi}_{k,1}, \dots, \boldsymbol{\psi}_{k,M_T}]$, $\beta_{k,m}$ is the m-th column of B_k, and $\boldsymbol{\psi}_{k,m} = \{e^{j2\pi(f_{k,m}-f_c)\ell_p}\}$, $\ell_p = 0, \dots, L_p - 1$.

2.6.2.1 Demodulation

The task of demodulation is to retrieve information bits from $\mathbf{Y}_k$. We assume perfect synchronization and channel estimation here. It will become clear that channel estimation may not be necessary if only frequency combination needs to be identified.

An optimal formulation of the demodulator can be based on the maximum likelihood principle [32]. But it has very high-computational complexity and is infeasible for practical implementation. Alternatively, we can apply suboptimal methods such as CS techniques.

The patterns of both frequency combination and antenna permutation can be identified by formulating a sparse recovery problem. The basic idea is to construct an $L_p \times N$ dictionary matrix $\boldsymbol{\Phi}_d$ by expanding $\boldsymbol{\Phi}$ to cover all N subband frequencies. Each column of $\boldsymbol{\Phi}_d$ has the similar expression with $\boldsymbol{\psi}_{k,m}$ for each subband frequency. Then we can get a 1-D MMV-CS formulation as

$$\mathbf{Y}_k^T = \boldsymbol{\Phi}_d\mathbf{A}_{\text{MMV}}, \tag{2.33}$$

where only M_T out of N rows in $\mathbf{A}_{\text{MMV}}$ are nonzero, corresponding to $(\mathbf{H}\boldsymbol{B}_k)^T$. The M_T frequency estimates, which correspond to the frequency combination pattern, can then be found by using one of the well-known MMV-CS recovery algorithms. The estimate of the M_T nonzero rows of $\mathbf{A}_{\text{MMV}}$, obtained in the recovery process, can be used to find the antenna permutation pattern, by matching them with $(\mathbf{H}\boldsymbol{B}_k)^T$. This matching process can be realized in either a simpler rowwise way or a more complicated process jointly across all rows.

Another simpler suboptimal method is to exploit the orthogonality of frequencies across antennas and apply a discrete Fourier transform (DFT) matrix $\mathbf{F}$:

$$\mathbf{Y}_k\mathbf{F} = \mathbf{H}\boldsymbol{B}_k\boldsymbol{\Phi}_k^T\mathbf{F}, \tag{2.34}$$

where each row of $\boldsymbol{\Phi}_k^T\mathbf{F}$ is the windowed DFT output of a single-tone signal and its waveform has the shape of an impulse with a single peak. Thus, each row of $\mathbf{Y}_k\mathbf{F}$ represents the weighted sum of these impulses. Therefore, the frequency combination pattern may be identified via locating the peaks. This is particularly effective when either the inverse of $\mathbf{H}$ exists or when the LOS path is dominating in $\mathbf{H}$. In the former, we can compute $\mathbf{H}^{-1}\mathbf{Y}_k$ and obtain $\boldsymbol{B}_k\boldsymbol{\Phi}_k^T\mathbf{F}$. This leads to simple identification of both frequency combination and antenna permutation patterns, as $\boldsymbol{B}_k$ is a permutation matrix. In the latter, the frequency combination pattern can be found via the peaks and the antenna permutation pattern is determined via exhaustive searching, even in a single-antenna receiver [34, 36].

2.6.2.2 Channel Estimation

An accurate estimate of $\mathbf{H}$ is critical for demodulation. However, it is challenging to design and incorporate long training sequences, which is essential for estimating $\mathbf{H}$, in FH-MIMO DFRC systems. This is because training sequence requires certainty, which will affect the randomness of FH radar operation.

There are very limited results on channel estimation for FH-DFRC systems with IM. In [34], both synchronization and channel estimation are investigated for a single-antenna receiver, with the consideration of packet communications. For channels with a dominating LOS-path, which could be a typical operating condition for radar-centric DFRC, a frame structure is proposed with two identical hops serving as preamble followed by hops with embedded information. The two identical hops are designed to enable effective estimation of timing offset, carrier frequency offset, and channel. To simplify synchronization and channel estimation, reordered hopping frequencies are used, which slightly reduces the information-embedding capability in terms of antenna permutation. Timing offset and channel estimators are proposed by exploiting the signal differences between two neighboring antennas. The work is also extended to NLOS channels by using incomplete sampled hops and judiciously designed hopping frequencies to combat interhop and interantenna interference.

2.6.3 Codebook Design

The codebook determines how the patterns of frequency combination and antenna permutation are selected and mapped to information bits. As was disclosed in [32], the achievable communication rates are largely constrained by antenna permutation, as its demodulation performance is sensitive to the differences between different columns of $\mathbf{H}$.

The design criterion can hence be formulated based on the distance between two codewords of antenna permutation:

$$\lambda(m, m') = \|\mathbf{H}\beta_{k,m} - \mathbf{H}\beta_{k,m'}\|_2^2. \tag{2.35}$$

Maximizing the minimal distance among all $\lambda(m, m')$ is a typical design criterion. Since directly searching via (2.35) is computationally complicated, a method of projection into a lower-dimensional plane is proposed in [32]. However, given that the design needs to be updated once $\mathbf{H}$ is changed, the complexity is still very high.

Such a complicated design may be avoided by using precompensation. For example, for LOS-path dominating channels, the channel differences between antenna permutations will be small when the AoD is small. In [36], an elementwise phase compensation method is proposed to remove the AoD dependence of demodulating antenna permutation. Thus, the distances between different codewords become identical.

In addition to its impact on communication performance, codebook may also affect the radar performance, for example the ambiguity function as evaluated in [34, 37]. More specifically, it is demonstrated in [37] that the probability of radar waveform degeneration can be reduced by spreading the available frequency hops between waveforms as evenly as possible, and in [34], it is shown that by constraining the codewords, the receiver processing can be largely simplified, with negligible impact on the radar ambiguity function.

2.7 Conclusions

In this chapter, we introduce signal-processing basics for communications and sensing that are essential to the JCAS technologies to be presented in Parts II and III. The majority of the illustration of this chapter is based on the OFDM waveform given its wide applicability in modern communications and radar systems. This will greatly help understand Chapters 6 and 5 in Part II, as the sensing techniques presented therein are particularly for OFDM and its variant waveforms, e.g. OTFS. Part III of this book will be focused on the FH-MIMO DFRC, as briefly illustrated in Section 2.6.1.

References

1 K. Wu, W. Ni, T. Su, R. P. Liu, and Y. J. Guo. Exploiting spatial-wideband effect for fast AoA estimation at lens antenna array. *IEEE Journal on Selected Topics in Signal Processing*, 13(5):902–917, September 2019. ISSN 1941-0484. doi: 10.1109/JSTSP.2019.2937691.

2 B. Friedlander. Waveform design for MIMO radars. *IEEE Transactions on Aerospace and Electronic Systems*, 43(3):1227–1238, July 2007. ISSN 0018-9251. doi: 10.1109/TAES.2007.4383615.

3 J. Li and P. Stoica. *MIMO Radar Signal Processing*. John Wiley & Sons, Ltd., 2008. ISBN 9780470391488. doi: 10.1002/9780470391488.fmatter.

4 Z. Lin and Z. Wang. Interleaved OFDM signals for MIMO radar. *IEEE Sensors Journal*, 15(11):6294–6305, 2015. doi: 10.1109/JSEN.2015.2458178.

5 G. Hakobyan, M. Ulrich, and B. Yang. OFDM-MIMO radar with optimized nonequidistant subcarrier interleaving. *IEEE Transactions on Aerospace and Electronic Systems*, 56(1):572–584, 2020. doi: 10.1109/TAES.2019.2920044.

6 C. Chen and P. P. Vaidyanathan. MIMO radar ambiguity properties and optimization using frequency-hopping waveforms. *IEEE Transactions on Signal Processing*, 56(12):5926–5936, 2008. doi: 10.1109/TSP.2008.929658.

7 S. R. J. Axelsson. Analysis of random step frequency radar and comparison with experiments. *IEEE Transactions on Geoscience and Remote Sensing*, 45(4):890–904, 2007. doi: 10.1109/TGRS.2006.888865.

8 T. Huang, N. Shlezinger, X. Xu, D. Ma, Y. Liu, and Y. C. Eldar. Multi-carrier agile phased array radar. *IEEE Transactions on Signal Processing*, 68:5706–5721, 2020. doi: 10.1109/TSP.2020.3026186.

9 W. Baxter, E. Aboutanios, and A. Hassanien. Dual-function MIMO radar-communications via frequency-hopping code selection. In *2018 52nd Asilomar Conference on Signals, Systems, and Computers*, pages 1126–1130, October 2018. doi: 10.1109/ACSSC.2018.8645212.

10 I. P. Eedara, A. Hassanien, M. G. Amin, and B. D. Rigling. Ambiguity function analysis for dual-function radar communications using PSK signaling. In *52nd Asilomar Conference on Signals, Systems, and Computers*, pages 900–904, October 2018. doi: 10.1109/ACSSC.2018.8645328.

11 C. Chen and P. P. Vaidyanathan. MIMO radar ambiguity properties and optimization using frequency-hopping waveforms. *IEEE Transactions on Signal Processing*, 56(12):5926–5936, December 2008. ISSN 1941-0476. doi: 10.1109/TSP.2008.929658.

12 M. A. Richards. *Fundamentals of Radar Signal Processing*. Tata McGraw-Hill Education, 2005.

13 J. Li and P. Stoica. *MIMO Radar Signal Processing*. John Wiley & Sons, 2008.

14 M. A. Richards, J. Scheer, W. A. Holm, and W. L. Melvin. *Principles of Modern Radar*. CiteSeer, 2010.

15 P. Kumari, S. A. Vorobyov, and R. W. Heath. Adaptive virtual waveform design for millimeter-wave joint communication-radar. *IEEE Transactions on Signal Processing*, 68:715–730, 2020. doi: 10.1109/TSP.2019.2956689.

16 A. Ali, N. Gonzalez-Prelcic, R. W. Heath, and A. Ghosh. Leveraging sensing at the infrastructure for mmWave communication. *IEEE Communications Magazine*, 58(7):84–89, 2020. doi: 10.1109/MCOM.001.1900700.

17 G. Duggal, S. Vishwakarma, K. V. Mishra, and S. S. Ram. Doppler-resilient 802.11ad-based ultrashort range automotive joint radar-communications system. *IEEE Transactions on Aerospace and Electronic Systems*, 56(5):4035–4048, 2020. doi: 10.1109/TAES.2020.2990393.

18 M. L. Rahman, J. A. Zhang, X. Huang, Y. J. Guo, and R. W. Heath Jr. Framework for a perceptive mobile network using joint communication and radar sensing. *IEEE Transactions on Aerospace and Electronic Systems*, 56(3):1926–1941, 2020.

19 J. A. Zhang, M. L. Rahman, X. Huang, Y. J. Guo, S. Chen, and R. W. Heath. Perceptive mobile network: Cellular networks with radio vision via joint communication and radar sensing. *IEEE Vehicular Technology Magazine*, 1–11, 2020. doi: 10.1109/MVT.2020.3037430.

20 P. Kumari, J. Choi, N. G. Prelcic, and R. W. Heath. IEEE 802.11ad-based radar: An approach to joint vehicular communication-radar system. *IEEE Transactions on Vehicular Technology*, PP(99):1, 2017. ISSN 0018-9545. doi: 10.1109/TVT.2017.2774762.

21 E. Grossi, M. Lops, L. Venturino, and A. Zappone. Opportunistic radar in IEEE 802.11ad networks. *IEEE Transactions on Signal Processing*, 66(9):2441–2454, 2018. doi: 10.1109/TSP.2018.2813300.

22 Z. Zhang and B. D. Rao. Extension of SBL algorithms for the recovery of block sparse signals with intra-block correlation. *IEEE Transactions on Signal Processing*, 61(8):2009–2015, April 2013. ISSN 1053-587X. doi: 10.1109/TSP.2013.2241055.

23 D. Nion and N. D. Sidiropoulos. Tensor algebra and multidimensional harmonic retrieval in signal processing for MIMO radar. *IEEE Transactions on Signal Processing*, 58(11):5693–5705, 2010. doi: 10.1109/TSP.2010.2058802.

24 C. W. Rossler, E. Ertin, and R. L. Moses. A software defined radar system for joint communication and sensing. In *IEEE Radar Conference*, pages 1050–1055. IEEE, 2011.

25 A. R. Chiriyath, B. Paul, and D. W. Bliss. Radar-communications convergence: Coexistence, cooperation, and co-design. *IEEE Transactions on Cognitive Communications and Networking*, 3 (1):1–12, 2017.

26 A. Hassanien, M. G. Amin, Y. D. Zhang, and F. Ahmad. Signaling strategies for dual-function radar communications: An overview. *IEEE Aerospace and Electronic Systems Magazine*, 31(10):36–45, 2016. doi: 10.1109/MAES.2016.150225.

27 A. Hassanien, M. G. Amin, E. Aboutanios, and B. Himed. Dual-function radar communication systems: A solution to the spectrum congestion problem. *IEEE Signal Processing Magazine*, 36(5):115–126, September 2019.

28 L. Zheng, M. Lops, Y. C. Eldar, and X. Wang. Radar and communication coexistence: An overview: A review of recent methods. *IEEE Signal Processing Magazine*, 36(5):85–99, September 2019. ISSN 1558-0792. doi: 10.1109/MSP.2019.2907329.

29 Z. Xu, A. Petropulu, and S. Sun. A joint design of MIMO-OFDM dual-function radar communication system using generalized spatial modulation. In *2020 IEEE Radar Conference*, pages 1–6, 2020. doi: 10.1109/RadarConf2043947.2020.9266486.

30 D. Ma, N. Shlezinger, T. Huang, Y. Liu, and Y. C. Eldar. Joint radar-communication strategies for autonomous vehicles: Combining two key automotive technologies. *IEEE Signal Processing Magazine*, 37(4):85–97, 2020. doi: 10.1109/MSP.2020.2983832.

31 E. BouDaher, A. Hassanien, E. Aboutanios, and M. G. Amin. Towards a dual-function MIMO radar-communication system. In *2016 IEEE Radar Conference*, pages 1–6, 2016. doi: 10.1109/RADAR.2016.7485316.

32 T. Huang, N. Shlezinger, X. Xu, Y. Liu, and Y. C. Eldar. MAJoRCom: A dual-function radar communication system using index modulation. *IEEE Transactions on Signal Processing*, 68:3423–3438, 2020. doi: 10.1109/TSP.2020.2994394.

33 W. Baxter, E. Aboutanios, and A. Hassanien. Dual-function MIMO radar-communications via frequency-hopping code selection. In *Asilomar Conference on Signals, Systems, and Computers*, pages 1126–1130, 2018. doi: 10.1109/ACSSC.2018.8645212.

34 K. Wu, J. Andrew Zhang, X. Huang, Y. J. Guo, and R. W. Heath. Waveform design and accurate channel estimation for frequency-hopping MIMO radar-based communications. *IEEE Transactions on Communications*, 1–16, 2020. doi: 10.1109/TCOMM.2020.3034357.

35 X. Wang, A. Hassanien, and M. G. Amin. Dual-function MIMO radar communications system design via sparse array optimization. *IEEE Transactions on Aerospace and Electronic Systems*, 55(3):1213–1226, 2019.

36 K. Wu, J. A. Zhang, X. Huang, and Y. J. Guo. Integrating secure and high-speed communications into frequency hopping MIMO radar, 2020. arXiv: 2009.13750.

37 W. Baxter, H. Nosrati, and E. Aboutanios. A study on the performance of symbol dictionary selection for the frequency hopped DFRC scheme. In *2020 IEEE Radar Conference*, pages 1–6, 2020. doi: 10.1109/RadarConf2043947.2020.9266476.

3

Efficient Parameter Estimation

We have shown in Chapter 2 that one of the most common problems encountered in both wireless communications and sensing is the parameter estimation. Radio sensing, traditionally, is mainly for detecting targets and estimating their parameters, including ranges, velocities, and angles. Although modern-sensing systems, such as those targeted at gesture recognition, may not intend to extract these physical parameters, it is sometimes still better to have these parameters jointly used to reduce the system complexity and enhance accuracy of recognition. In communication systems, channels, along with other parameters, e.g. sampling timing and carrier frequency offsets, are generally estimated and compensated before information can be effectively extracted. With communications marching toward millimeter-wave and terahertz frequency bands, line of sight propagation and therefore, sparse channel becomes predominant. Consequently, channel estimation, which used to estimate the complex entries of a channel matrix, can now be performed via estimating physical parameters of dominant scatters, such as impinging directions. Recently, a new waveform, called the orthogonal time-frequency space modulation (OTFS), is introduced for future 6G communications. OTFS modulates the information in the so-called "delay-Doppler domain," making the estimations of ranges (corresponding to delay) and velocities (corresponding to Doppler) of dominant scatters also interesting to sensing.

The parameter estimation in communications and sensing, with dimension reduction, can be unitedly expressed as a frequency estimation problem, namely, estimating the parameter f of the following single-tone signal,

$$s(n) = Ae^{j\left(\frac{2\pi f n}{f_s}+\phi\right)} + z(n),\ \ n = 0, 1, \dots, N-1, \tag{3.1}$$

where A is the signal amplitude, f_s is the sampling rate, ϕ is the initial phase, N is the sample number, and $z(n)$ is an additive white Gaussian noise (AWGN) with the noise variance σ^2. Here, f can be the following:

1. The equivalent angle of arrival (AoA) or angle of departure (AoD) in terms of a communications/sensing antenna array. The n-th entry of the steering vector

Joint Communications and Sensing: From Fundamentals to Advanced Techniques, First Edition.
Kai Wu, J. Andrew Zhang, and Y. Jay Guo.

of an N-antenna uniform linear array can be written as $2\pi nd\sin\theta/\lambda$, where d is antenna spacing, θ is the AoA or the AoD, and λ is the wavelength. Comparing with the exponential term in (3.1), we can treat $d\sin\theta$ as the central frequency to be estimated and λ as the sampling frequency (of the array), which is actually a popular way of interpreting steering vectors in antenna array signal processing.

2. The delay caused by wave propagation between a pair of transceivers. In a single-node OFDM sensing with a colocated transceiver, f becomes $2R/\mathsf{C}$, where R denotes the propagation distance between the transceiver and the target, and C denotes the light speed. Reflecting this in (3.1), n becomes the subcarrier index and $1/f_s$ becomes the subcarrier frequency interval, i.e. B/N (with B denoting the bandwidth).
3. The Doppler frequency over the so-called "slow time" in pulse-Doppler radar sensing. In this case, f in (3.1) becomes the Doppler frequency, n is pulse index, and f_s becomes the pulse repetition frequency (PRF), the reciprocal of pulse repetition time.

For clarity, we base our discussions in this chapter on the single-tone signal model in (3.1). When multiple frequency components exist, they are generally superimposed additively in radar and communications. Thus, the techniques introduced in this chapter can be employed for estimating each component. This is often done through the estimate-and-subtract strategy. The strategy is to first estimate each frequency coarsely as a single tone, as if there were no other tones; then, from the second round, each tone is refined by subtracting the recovered signals of other tones. With more rounds of refinement performed, the estimates of all tones can be increasingly accurate. Single-tone estimation forms the basis of multitone estimation. Using the single-tone model is also convenient for investigating the analytical performance of an estimator. As such, we only focus on the single-tone frequency estimation in this chapter.

Numerous conventional methods are available for single-tone frequency estimation, including the maximum likelihood estimator, the super-resolution spectrum analysis methods, and compressive sensing methods, etc. While the above methods have been well documented in the literature and numerous textbooks [1, 2], this chapter introduces several low-complexity yet efficient estimators based on interpolating DFT coefficients. Such estimators are promising for real-time communications and sensing.

3.1 Q-Shifted Estimator (QSE)

We start with reviewing a recently developed estimator, called the q-shift estimator (QSE) [3]. As will be seen shortly, the method iteratively interpolates the DFT coefficients at frequencies that are $\pm q$ times the frequency interval plus the current frequency estimate, hence the name q-shift estimator. Taking the DFT of $s(n)$ given in (3.1) and identifying the peak of the DFT result, we can obtain an estimate of f yet only as an integer multiples of $2\pi/N$. Let $k^\star$ denote the integer. Then f can be expressed as follows:

$$f = \frac{k^\star + \delta}{N} f_s,\ \delta \in [-0.5, 0.5], \tag{3.2}$$

where δ denotes the frequency residual. QSE assumes that $k^\star$ is known and estimates δ only [3].

Given $k^\star$, the iteration number Q and the initial residual estimate $\hat{\delta}_0 = 0$, QSE updates the i-th ($i = 1, \ldots, Q$) frequency residual by [3, eq. 7]

$$\hat{\delta}_i = \frac{1}{c(q)} \times \mathcal{R}e\left\{\beta_i\right\} + \hat{\delta}_{i-1},\ |q| < 0.5. \tag{3.3}$$

The intermediate expressions $c(q)$ and β_i are given by

$$c(q) = \frac{1 - \pi q \cot(\pi q)}{q \cos^2(\pi q)};\ \beta_i = \frac{S_{+q} - S_{-q}}{S_{+q} + S_{-q}}. \tag{3.4}$$

Here, $S_{\pm q}$ is the $\frac{1}{\pm q}$-interpolated DFT coefficients around $k^\star$ and defined as follows:

$$S_{\pm q} = \sum_{n=0}^{N-1} s(n) e^{-j\frac{2\pi(k^\star + \hat{\delta}_{i-1} \pm q)n}{N}}. \tag{3.5}$$

Equation 3.3 is obtained based on the following linear approximation [3, eq. 19]:

$$\mathcal{R}e\{\beta_i\} = c(q)\Delta_i + \mathcal{O}(\Delta_i^3),\ |q| < 0.5, \tag{3.6}$$

where $\Delta_i = \hat{\delta}_i - \delta$ is the estimation error in the i-th iteration. After Q iterative updates in (3.3), QSE produces the final frequency estimation as $\hat{f} = \frac{k^\star + \hat{\delta}_Q}{N} f_s$.

Note that the linear approximation (3.6) can be invalidated by some values of q. To illustrate this, Figure 3.1 plots $\mathcal{R}e\{\beta_i\}$ w.r.t. $\Delta_i \in [-0.5, 0.5]$ by taking different values of q. We see that not only the linearity of $\mathcal{R}e\{\beta_i\}$ w.r.t. Δ_i degrades as q increases, but the monotonicity of $\mathcal{R}e\{\beta_i\}$ over Δ_i can also change twice for large

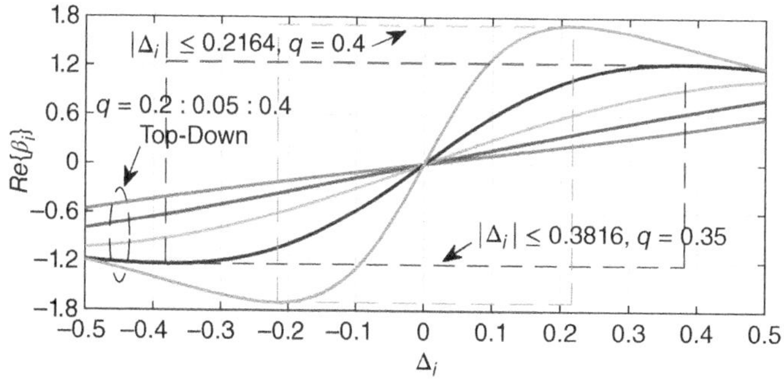

Figure 3.1 Illustration of the monotonicity of $\mathcal{R}e\{\beta_i\}$ w.r.t. Δ_i, affected by q.

q's, e.g. $q = 0.4$. Obviously, the invalidation of (3.6) fails the core step of QSE, as given in (3.3). Thus, if q is not properly selected, the frequency residual δ fall out of the monotonic region of $\mathcal{R}e\{\beta_i\}$ w.r.t. Δ_i, potentially leading to the divergence of QSE from the CRLB. The divergence of QSE was observed in [3, figure 2], where $N = 16$, $\delta = 0.25$ and $q = \frac{1}{\sqrt[3]{N}} \approx 0.4$ were taken. According to Figure 3.1, $q = 0.4$ makes $\delta = 0.25$ fall out of the monotonic region of $\mathcal{R}e\{\beta_i\}$ w.r.t. Δ_i which is $|\Delta_i| \leq 0.2164$.

3.2 Refined QSE (QSEr)

3.2.1 Impact of q

As revealed at the end of Section 3.1, q is vital to preserving the monotonicity of $\mathcal{R}e\{\beta_i\}$ and hence, the validity of the core step of QSE, i.e. (3.3). Thus, it is worth determining a feasible region of q through analyzing the monotonicity of $\mathcal{R}e\{\beta_i\}$ w.r.t. $\Delta_i (\in [-0.5, 0.5])$. However, it is mathematically intractable to carry out the analysis based on β_i in (3.3) due to the strong coupling of linear and complex exponential functions of Δ_i. To facilitate finding the feasible region of q, let us reformulate β_i by collecting the common terms of Δ_i and q into an auxiliary function, denoted by $\mathcal{Q}(\Delta_i, q)$. Based on (3.3), β_i in (3.3) can be reformulated as

$$\beta_i = \frac{1 - \mathcal{Q}(\Delta_i, q)e^{j2\pi q}}{1 + \mathcal{Q}(\Delta_i, q)e^{j2\pi q}}, \tag{3.7}$$

where $\mathcal{Q}(\Delta_i, q)$ is given by

$$\mathcal{Q}(\Delta_i, q) = \frac{\Delta_i - q}{\Delta_i + q} \times \frac{\sin[\pi(\Delta_i + q)]}{\sin[\pi(\Delta_i - q)]}. \tag{3.8}$$

Taking the real part of β_i, we have

$$\beta_i^{\mathcal{R}} = \mathcal{R}e\{\beta_i\} = \frac{1 - \mathcal{Q}^2(\Delta_i, q)}{1 + 2\mathcal{Q}(\Delta_i, q)\cos(2\pi q) + \mathcal{Q}^2(\Delta_i, q)}. \tag{3.9}$$

Next, we apply the chain rule of nested functions [4] to separately investigate the monotonicity of $\mathcal{R}e\{\beta_i\}$ w.r.t. $\mathcal{Q}(\Delta_i, q)$ and also $\mathcal{Q}(\Delta_i, q)$ w.r.t. Δ_i. By applying the Chain Rule of nested functions [4], the first derivative of $\beta_i^{\mathcal{R}}$ w.r.t. Δ_i is given by

$$(\beta_i^{\mathcal{R}})'_{\Delta_i} = (\beta_i^{\mathcal{R}})'_{\mathcal{Q}} \times (\mathcal{Q})'_{\Delta_i}, \tag{3.10}$$

where $(\beta_i^{\mathcal{R}})'_{\mathcal{Q}}$ is the first derivative of $\beta_i^{\mathcal{R}}$ w.r.t. $\mathcal{Q}(\Delta_i, q)$, and $(\mathcal{Q})'_{\Delta_i}$ is the first derivative of $\mathcal{Q}(\Delta_i, q)$ w.r.t. Δ_i. In the following, we examine the signs of $(\mathcal{Q})'_{\Delta_i}$ and $(\beta_i^{\mathcal{R}})'_{\mathcal{Q}}$ first. Then the region of q can be determined based on the overall monotonicity of $\mathcal{R}e\{\beta_i\}$ w.r.t. Δ_i.

According to (3.8), $(\mathcal{Q})'_{\Delta_i}$ can be given by

$$(\mathcal{Q})'_{\Delta_i} = \mathcal{P}(\Delta_i, q)/(\Delta_i + q)^2 \sin^2[\pi(\Delta_i - q)] \tag{3.11}$$

where the denominator is always nonnegative. Therefore, the sign of its numerator, as given by $\mathcal{P}(\Delta_i, q) = 2q\sin[\pi(\Delta_i + q)]\sin[\pi(\Delta_i - q)] - \pi(\Delta_i^2 - q^2)\sin(2\pi q)$, determines the sign of $(\mathcal{Q})'_{\Delta_i}$, and hence, is analyzed below. Let $(\mathcal{P})'_{\Delta_i}$ and $(\mathcal{P})''_{\Delta_i}$ denote the first and second derivatives of $\mathcal{P}(\Delta_i, q)$ w.r.t. Δ_i, respectively. They can be calculated as follows:

$$(\mathcal{P})'_{\Delta_i} = 2\pi q \sin(2\pi\Delta_i) - 2\pi\Delta_i \sin(2\pi q); \tag{3.12}$$

$$(\mathcal{P})''_{\Delta_i} = 4\pi^2 q \cos(2\pi\Delta_i) - 2\pi \sin(2\pi q). \tag{3.13}$$

By setting $(\mathcal{P})'_{\Delta_i} = 0$, we obtain $\Delta_i = \pm q$ at the local optima of $\mathcal{P}(\Delta_i, q)$. By taking $\Delta_i = \pm q$ in (3.13), $(\mathcal{P})''_{\pm q} = 4\pi^2 q \left[\cos(2\pi q) - \frac{\sin(2\pi q)}{2\pi q}\right]$, where $\cos(2\pi q) - \frac{\sin(2\pi q)}{2\pi q} < 0$ holds for $\forall |q| < 0.5$. Therefore, $(\mathcal{P})''_{\pm q}$ satisfies

$$(\mathcal{P})''_{\pm q} \begin{cases} < 0, & \text{if } q > 0 \\ > 0, & \text{if } q < 0 \end{cases}. \tag{3.14}$$

Equation (3.14) indicates that both local optima $\Delta_i = \pm q$ of $\mathcal{P}(\Delta_i, q)$ are local minima when $q < 0$ or local maxima when $q > 0$. By taking $\Delta_i = \pm q$ in $\mathcal{P}(\Delta_i, q)$, we obtain $\mathcal{P}(\pm q, q) = 0$. This indicates that $\mathcal{P}(\Delta_i, q) \geq 0$ when $q < 0$ and $\mathcal{P}(\Delta_i, q) \leq 0$ when $q > 0$. However, $\mathcal{P}(\pm q, q) = 0$ does not guarantee $(\mathcal{Q})'_{\Delta_i} = 0$ due to the singular denominators of $(\mathcal{Q})'_{\Delta_i}$ at $\Delta_i = \pm q$; see (3.11). By taking $\Delta_i = \pm q$ in (3.11), we achieve

$$\lim_{\Delta_i \to \pm q} (\mathcal{Q})'_{\Delta_i} = \begin{cases} \frac{(\mathcal{P})''_q}{8\pi^2 q^2} & \text{when } \Delta_i \to q \\ \frac{(\mathcal{P})''_q}{2\sin^2[\pi(\Delta_i - q)]} & \text{when } \Delta_i \to -q \end{cases}. \tag{3.15}$$

Combining (3.14) and (3.15), we can conclude that the local minima of $(\mathcal{Q})'_{\Delta_i}$ are always positive, when $q < 0$; and the local maxima of $(\mathcal{Q})'_{\Delta_i}$ are always negative, when $q > 0$. This finally leads to

$$(\mathcal{Q})'_{\Delta_i} \begin{cases} < 0, & \text{if } q > 0 \\ > 0, & \text{if } q < 0 \end{cases}, \quad \forall |\Delta_i| \leq 0.5. \tag{3.16}$$

We proceed to investigate the sign of $(\beta_i^{\mathcal{R}})'_{\mathcal{Q}}$ which, based on $\beta_i^{\mathcal{R}}$ in (3.9), can be given by

$$(\beta_i^{\mathcal{R}})'_{\mathcal{Q}} = \frac{\mathcal{N}(\mathcal{Q})}{[1 + 2\mathcal{Q}(\Delta_i, q)\cos(2\pi q) + \mathcal{Q}^2(\Delta_i, q)]^2}, \tag{3.17}$$

where the numerator function is

$$\mathcal{N}(\mathcal{Q}) = -[2\mathcal{Q}^2(\Delta_i, q) + 2]\cos(2\pi q) - 4\mathcal{Q}(\Delta_i, q). \tag{3.18}$$

Since the denominator of $(\beta_i^R)'_Q$ is nonnegative, we focus on identifying the sign of its $\mathcal{N}(Q)$. It is noted that $\mathcal{N}(Q)$ is a quadratic function of $Q(\Delta_i, q)$, and hence, the sign of $\mathcal{N}(Q)$ only changes if the solutions to $\mathcal{N}(Q) = 0$ are valid. Based on (3.18), the solutions to $\mathcal{N}(Q) = 0$ can be solved as follows:

$$Q_\pm^* = (-1 \pm |\sin(2\pi q)|)/\cos(2\pi q). \tag{3.19}$$

In the following, we examine the validity of (3.19) to analyze the sign of $\mathcal{N}(Q)$.

***Case 1**: In this case of* $|q| < 0.25$, $\cos(2\pi q) > 0$; and $(-1 \pm |\sin(2\pi q)|) < 0$ due to $|\sin(2\pi q)| < 1$. Thus, we have

$$Q_\pm^* < 0, \ \forall |q| < 0.25. \tag{3.20}$$

However, (3.8) indicates that

$$Q(\Delta_i, q) > 0, \ \forall |\Delta_i| \le 0.5, \ \forall |q| < 0.5, \tag{3.21}$$

since $\frac{\sin[\pi(\Delta_i \pm q)]}{\Delta_i \pm q} > 0$ given $|\Delta_i \pm q| < 1$. Clearly, (3.20) contradicts with (3.21), and hence, $Q_\pm^*$ is invalid given $\forall |q| < 0.25$.

***Case 2**: In the case of* $|q| = 0.25$, $\cos(2\pi q) = 0$, and hence, $\lim_{|q| \to 0.25} Q_-^* \to -\infty$, contradicting with (3.21). Similarly, substituting $|q| = 0.25$ in Q_+^* leads to

$$\lim_{q \to \pm 0.25} Q_+^* = \lim_{q \to \pm 0.25} \frac{-1 + \sin(2\pi q)}{\cos(2\pi q)} \overset{(a)}{=} \lim_{q \to \pm 0.25} \frac{2\pi \cos(2\pi q)}{-2\pi \sin(2\pi q)} = 0, \tag{3.22}$$

where the *L'Höpital's* rule [4] is applied to achieve the equality (a). However, by substituting $q = 0.25$ in (3.8), we notice that $Q = 0$ does not happen.[1] Therefore, we can conclude that both $Q_\pm^*$ are invalid solutions in the case of $q = 0.25$. By substituting $|q| = 0.25$ into (3.17), we have $\mathcal{N}(Q) = -4Q(\Delta_i, q) < 0$ based on (3.21). Accordingly, by substituting (3.10) into (3.16), we can obtain

$$(\beta_i^R)'_{\Delta_i} \begin{cases} \ge 0, & q = 0.25 \\ \le 0, & q = -0.25 \end{cases}, \ \forall |\Delta_i| \le 0.5. \tag{3.23}$$

***Case 3**: In the case of* $0.25 < |q| < 0.5$, $\cos(2\pi q) < 0$, and hence, $Q_\pm^* > 0$. To further examine the validity of $Q_\pm^* > 0$, we can check whether the solutions to (3.24), i.e. $\Delta_{i\pm}^*$, are valid by falling in $[-0.5, 0.5]$.

$$Q(\Delta_{i\pm}^*, q) = Q_\pm^* \tag{3.24}$$

Remark 3.1 ***(Monotonicity of** $Q(\Delta_{i\pm}^*, q)$ **w.r.t.** q)* Consider the positive value of q first, i.e. taking $0.25 < q < 0.5$. $Q_\pm^*$ in (3.19) becomes $\frac{-1 \pm \sin(2\pi q)}{\cos(2\pi q)}$ which, based on the basic properties of trigonometric functions, can be simplified into

---

1 According to (3.8), $Q = 0$ can only happen when $\Delta_i \pm q = 1$; however, we have $|\Delta_i \pm q| < 1$ given $|\Delta_i| \le 0.5$ and $|q| < 0.5$.

(3.25). Given $0.25 < q < 0.5$, we have $\cos(\pi q) > -\cos(\pi q)$, and hence, $\cos(\pi q) + \sin(\pi q) > \sin(\pi q) - \cos(\pi q)$.

$$Q_+^* = \frac{\sin(\pi q) - \cos(\pi q)}{\sin(\pi q) + \cos(\pi q)}, \quad Q_-^* = \frac{\sin(\pi q) + \cos(\pi q)}{\sin(\pi q) - \cos(\pi q)}. \tag{3.25}$$

Substituting this into (3.25) leads to

$$Q_+^* < 1 \text{ and } Q_-^* > 1. \tag{3.26}$$

By letting $Q(\Delta_i, q)$ in (3.8) equal to 1, $\Delta_i = 0$ is the solution. Combining this with (3.16) and (3.26), we obtain

$$\Delta_{i+}^* > 0 \text{ and } \Delta_{i-}^* < 0, \tag{3.27}$$

where $\Delta_{i\pm}^*$ ($\in [-0.5, 0.5]$) is the valid solution to equation (3.24). From (3.25), we have $Q_+^* Q_-^* = 1$. Replacing $Q_\pm^*$ with $Q(\Delta_{i\pm}^*, q)$ and then employing (3.8), we have

$$\Delta_{i+}^* = -\Delta_{i-}^*. \tag{3.28}$$

By substituting (3.26)~ 3.28 into (3.17), the sign of $\mathcal{N}(Q)$ satisfies (3.31), where $\Delta_i^* = |\Delta_{i\pm}^*|$. Given (3.19), the first derivative of $Q_\pm^*$ w.r.t. q can be examined, leading to

$$\begin{cases} (Q_+^*)'_q > 0 \\ (Q_-^*)'_q < 0 \end{cases}. \tag{3.29}$$

Combining (3.16) and (3.29), (3.30) can be achieved.

Remark 3.1 indicates

$$q \uparrow (\downarrow) \implies \begin{cases} Q_+^* \uparrow (\downarrow) \implies \Delta_{i+}^* \downarrow (\uparrow) \\ Q_-^* \downarrow (\uparrow) \implies \Delta_{i-}^* \uparrow (\downarrow) \end{cases}, \tag{3.30}$$

where "↑," "↓," and " ⟹" denote "increasing," "decreasing," and "leading to," respectively. From (3.30), we can assert that *there exists a q_L^* such that $\Delta_i^* = 0.5$ if* $|q| = q_L^*$; otherwise, $\Delta_i^* > 0.5$ for $|q| < q_L^*$, and $\Delta_i^* < 0.5$ for $|q| > q_L^*$, where $\Delta_i^* = |\Delta_{i+}^*| = |\Delta_{i-}^*|$, as proved in Remark 3.1. To this end, when $|q| \leq q_L^*$, (3.24) has feasible solutions, and hence, $Q_\pm^*$ is valid, which, according to (3.17), gives

$$\mathcal{N}(Q) \begin{cases} < 0, & \text{if } |\Delta_i| \underset{(>)}{<} \Delta_i^* \\ (>) & \\ = 0, & \text{if } |\Delta_i| = \Delta_i^* \end{cases}, \quad |q| \leq q_L^*. \tag{3.31}$$

On the contrary, when $|q| > q_L^*$, (3.24) does not have feasible solutions, hence invalidating $Q_\pm^*$. By examining the coefficient of $Q^2(\Delta_i, q)$ in $\mathcal{N}(Q)$ (see (3.17), we obtain $-2\cos(2\pi q) > 0$ for $|q| > q_L^*$ and, in turn,

$$\mathcal{N}(Q) > 0, \text{ for } |\Delta_i| \leq 0.5, \ |q| > q_L^*. \tag{3.32}$$

Finally, by substituting (3.16), (3.31), and (3.32) into (3.10), the monotonicity of $\beta_i^{\mathcal{R}}$ w.r.t. Δ_i can be revealed as follows:

$$(\beta_i^{\mathcal{R}})'_{\Delta_i} \begin{cases} \geq 0, & \text{for } |\Delta_i| \leq 0.5,\ 0.25 < q \leq q_{\text{L}}^* \approx 0.32 \\ \leq 0, & \text{for } |\Delta_i| \leq 0.5, -q_{\text{L}}^* \leq q < -0.25 \end{cases}; \tag{3.33a}$$

$$(\beta_i^{\mathcal{R}})'_{\Delta_i} \begin{cases} < 0, & \text{for } |\Delta_i| \in (\Delta_i^*, 0.5] \\ \geq 0, & \text{for } |\Delta_i| \leq \Delta_i^* \end{cases},\ q \in (q_{\text{L}}^*, 0.5); \tag{3.33b}$$

$$(\beta_i^{\mathcal{R}})'_{\Delta_i} \begin{cases} > 0, & \text{for } |\Delta_i| \in (\Delta_i^*, 0.5] \\ \leq 0, & \text{for } |\Delta_i| \leq \Delta_i^* \end{cases},\ \underset{(-0.5,-q_{\text{L}}^*)}{q\ \in}, \tag{3.33c}$$

where the critical point $q_{\text{L}}^* \approx 0.32$ is derived in the following remark.

Remark 3.2 ***(Derivation of*** $q_L^* \approx$ ***0.32)*** By substituting $\Delta_i = \Delta_i^* = 0.5$ into (3.8) and applying the basic manipulations, we obtain $\mathcal{Q}(0.5, q_{\text{L}}^*) = \frac{0.5-q_{\text{L}}^*}{0.5+q_{\text{L}}^*} \times \frac{\sin[\pi(0.5+q_{\text{L}}^*)]}{\sin[\pi(0.5-q_{\text{L}}^*)]} = \frac{0.5-q_{\text{L}}^*}{0.5+q_{\text{L}}^*}$. Similarly, $\mathcal{Q}_+^*$ can be simplified into $\mathcal{Q}_+^* = \frac{-1+\sin(2\pi q_{\text{L}}^*)}{\cos(2\pi q_{\text{L}}^*)} = \frac{\sin(\pi q_{\text{L}}^*)-\cos(\pi q_{\text{L}}^*)}{\cos(\pi q_{\text{L}}^*)+\sin(\pi q_{\text{L}}^*)}$. Setting $\mathcal{Q}(0.5, q_{\text{L}}^*) = \mathcal{Q}_+^*$ and collecting terms, we have

$$\cot(\pi q_{\text{L}}^*) = 2q_{\text{L}}^*. \tag{3.34}$$

An accurate analytical solution is intractable mathematically. To solve (3.34), we replace $\cot(\pi q_{\text{L}}^*)$ with its Taylor series, i.e. $\cot(\pi q_{\text{L}}^*) = \frac{1}{\pi q_{\text{L}}^*} - \frac{\pi q_{\text{L}}^*}{3} + \mathcal{O}\{(q_{\text{L}}^*)^3\}$, and obtain $\frac{1}{\pi q_{\text{L}}^*} - \frac{\pi q_{\text{L}}^*}{3} \approx 2q_{\text{L}}^*$. This finally yields $q_{\text{L}}^* \approx \left(\frac{\pi^2}{3} \pm 2\pi\right)^{-\frac{1}{2}} = 0.3232$, where the invalid solution is suppressed. Similarly, we can prove that the solution to $\mathcal{Q}(-0.5, q^*) = \mathcal{Q}_-^*$ is $q^* = -q_{\text{L}}^*$.

3.2.2 Refined Optimal *q*

From *Cases 1~3*, we conclude that QSE only works for the following region of q:

$$|q| \leq q_{\text{L}}^* \approx 0.32. \tag{3.35}$$

Without considering the above feasible region of q, the optimal q specified in [3], i.e. $q_{\text{opt}}^{\text{o}} = 1/\sqrt[3]{N}$, can invalidate QSE. A simple way to refine the optimal q is given as follows:

$$q_{\text{opt}} = \min\{q_{\text{opt}}^{\text{o}}, q_{\text{L}}^*\},\ \text{s.t. } q_{\text{opt}}^{\text{o}} = 1/\sqrt[3]{N}. \tag{3.36}$$

For convenience, we refer to QSE based on (3.36) as the refined QSE (QSEr). The asymptotic convergence of QSEr toward the CRLB is analyzed as follows. As evident from (3.36), $q_{\text{opt}} \leq q_{\text{opt}}^{\text{o}}$ and therefore, $\text{var}\left(\hat{\delta}_Q\right)|_{q=q_{\text{opt}}} \leq \text{var}\left(\hat{\delta}_Q\right)|_{q=q_{\text{opt}}^{\text{o}}}$. Here, $\text{var}\left(\hat{\delta}_Q\right) = \frac{6}{2\pi^2 N\gamma} + \mathcal{O}(q^4) + o(N^{-1})$ is the estimation variance of QSE after

Q iterations [3, eq. (38)]. Note that $\gamma = \frac{A^2}{\sigma^2}$ is the received SNR. Since the convergence of $\mathrm{var}\left(\hat{\delta}_Q\right)|_{q=q^{\mathrm{o}}_{\mathrm{opt}}}$ has been confirmed by [3, Thrm. 3], we conclude that $\mathrm{var}\left(\hat{\delta}_Q\right)|_{q=q_{\mathrm{opt}}}$ asymptotically converges toward the CRLB.

The advantages of QSEr over the original QSE [3] are summarized as follows: *First*, QSEr is able to converge to the CRLB even for small values of N, e.g. 8 and 16, while QSE diverges from the CRLB for small N. *Second*, validated by the convergence analysis given earlier, QSEr converges faster than QSE if $N \leq 30$. *Third*, QSEr has more uniform convergence performance across the whole region of δ ($|\delta| \leq 0.5$), as compared to QSE.

3.2.3 Numerical Illustration of QSEr

In this section, we exploit the optimal q refined in (3.36) to reevaluate QSE [3]. The following parameters $\delta = 0.25$, $k^\star = 2$, $f_s = N$, and $N = 8$ are taken here to examine the small-sample estimation performance. The simulation codes provided by the authors of [3] on the web page [5] are modified to simulate QSE and its variants. By taking $N = 8$ in (3.36), the value $q^{\mathrm{o}}_{\mathrm{opt}} = 0.5$ would be used by the original QSE [3]. However, $q = q^{\mathrm{o}}_{\mathrm{opt}} = 0.5$ leads to the singularity of $c(q)$; see (3.3). Thus, $q = (0.5 - 10^{-8})$ is taken to simulate the original QSE [3].

Figure 3.2 plots the MSE of frequency estimates against the estimation SNR, where a hybrid algorithm of A&M [8] and QSE, referred to as HAQSE [3], is simulated as a benchmark, and the phase-corrected Quinn estimator (PCQ) [6] and the estimator using three DFT points (TDP) [7] are also simulated. In the figure,

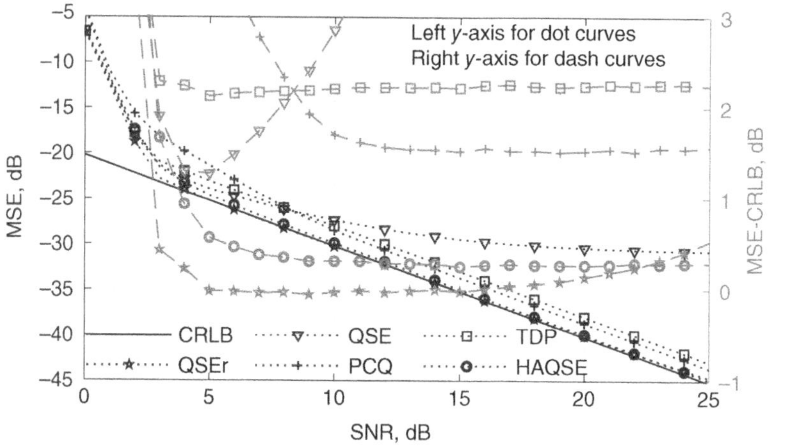

Figure 3.2 MSE of frequency estimates vs. SNR, where QSE and HAQSE [3], PCQ [6] and TDP [7] are simulated as benchmarks.

we see that only QSEr and HAQSE [3] can approach CRLB asymptotically. From the right y-axis, QSEr outperforms HAQSE mostly with the MSE improvement up to 0.5 dB. We also see that, while QSEr can reach the CRLB in a wide SNR region, PCQ [6] and TDP [7] remain 1.5 and 2.2 dB away from the CRLB, respectively, even after convergence.

Figure 3.3 compares the MSEs of QSEr, QSE, and HAQSE, as the estimation of SNR increases. We see that the refined optimal q can improve the estimation accuracy of QSEr substantially at $N = 8$ and 16, as compared to the original QSE [3]. Specifically, the MSE improvement of QSEr over QSE can be as large as 13.14 dB. The reason is because the refined q_{opt} can guarantee the monotonicity of $\beta_i^{\mathcal{R}}$ w.r.t. Δ_i, whereas the original $q_{\text{opt}}^{\text{o}}$ cannot. We also see that QSEr and QSE have the same performance at $N = 64$, which is because $q_{\text{opt}} = q_{\text{opt}}^{\text{o}}$ according to (3.36).

In Figure 3.3, we also see that QSEr outperforms HAQSE for a large SNR range at $N = 8$ and 16, with an MSE improvement of up to 1.897 dB. This is different to what was shown in [3, figure 2], where HAQSE outperformed QSE substantially. The reason underlying the superiority of HAQSE over the original QSE [3] is that HAQSE runs A&M estimator [8] prior to QSE. Unlike QSE, A&M does not suffer from the monotonicity issue as QSE does. Thus, the initial frequency bias for QSE can be reduced after running A&M and is very likely to fall within the monotonic region of $\beta_i^{\mathcal{R}}$ w.r.t. Δ_i even at $q_{\text{opt}}^{\text{o}} = N^{-1/3} > q_{\text{L}}^*$. To this end, running A&M makes QSE valid. However, it cannot be guaranteed that A&M always reduces the

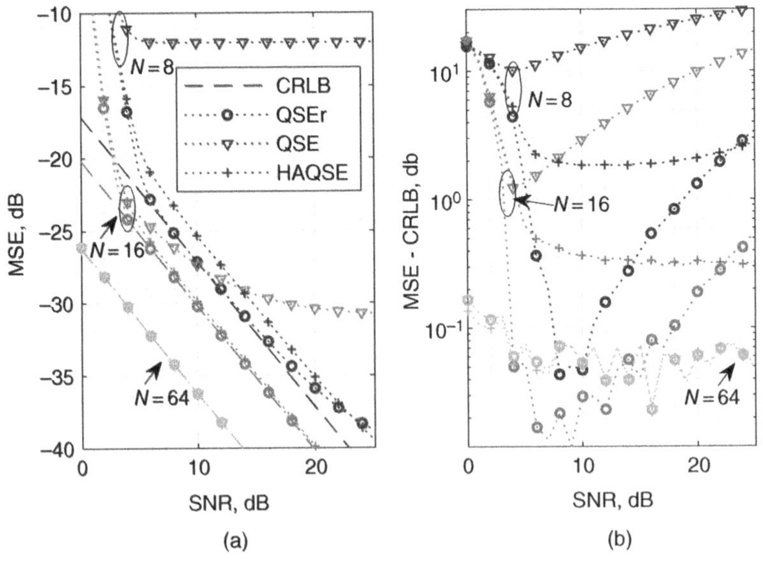

Figure 3.3 MSE of frequency estimates against SNR.

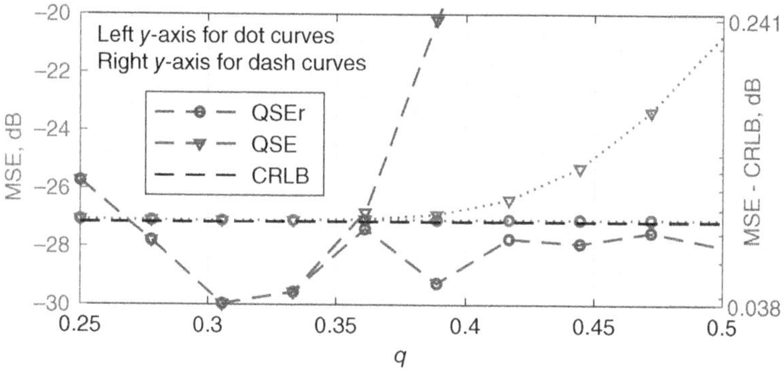

Figure 3.4 MSE of frequency estimates vs. q, where $N = 8$ and SNR = 10 dB.

frequency bias to within the monotonic region of $\beta_i^{\mathcal{R}}$ w.r.t. Δ_i. In contrast, the refined q_{opt} guarantees that $\beta_i^{\mathcal{R}}$ is always monotonic against Δ_i.

Figure 3.4 compares the MSEs of QSEr and QSE [3] as q varies. We see that a smaller q produces a better frequency estimation with the MSEs of both QSEr and QSE overlapping with the CRLB. However, we see that the MSE of QSE starts to increase, as q becomes larger than $q_{\mathrm{L}}^* = 0.32$. In contrast, QSEr is able to converge to the CRLB in the whole region of q, since the new optimal q in (3.36) is always confined below 0.32. It is noteworthy that the asymptotic performance of QSE is supposed to be consistent for any $|q| < q_{\mathrm{opt}} = 0.5$; see [3, Sec. II-G]. However, QSE diverges from the CRLB due to the discussed impact of q, which is adequately addressed in QSEr.

Figure 3.5 compares the MSEs of QSEr and QSE [3] as N increases, where, for fair comparison, q_{opt} and $q_{\mathrm{opt}}^{\mathrm{o}}$ are applied for QSEr and QSE, respectively. We see that QSEr based on the new optimal q ensures the convergence of the MSE to the

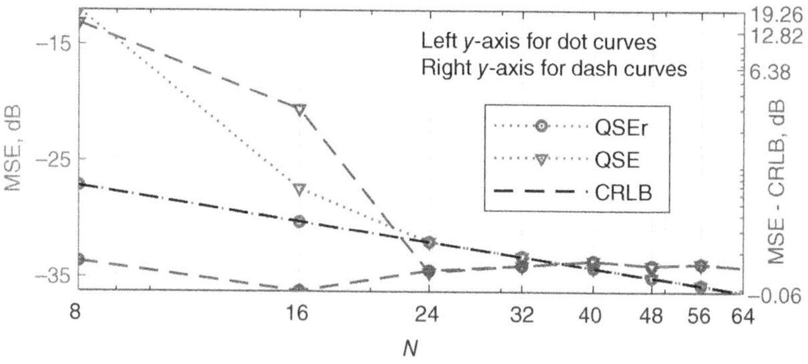

Figure 3.5 MSE of frequency estimates vs. N, where SNR = 10 dB.

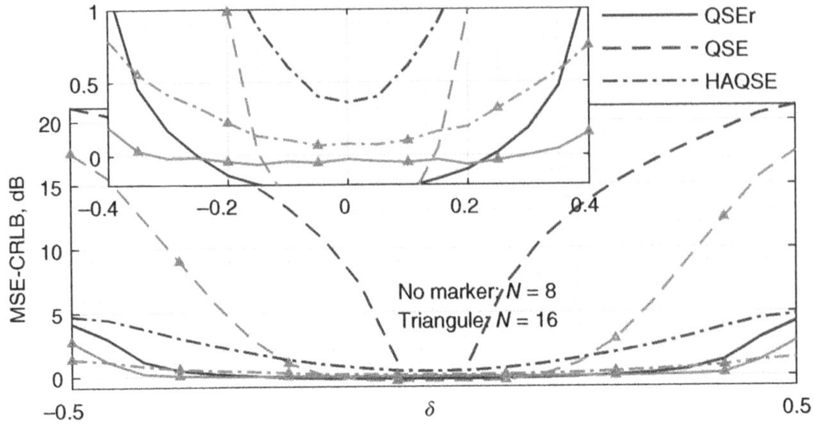

Figure 3.6 MSE of frequency estimates vs. δ, where SNR = 10 dB.

CRLB even for small values of N, e.g. 8 and 16. From Figure 3.5, we also see that the MSE improvement of QSEr is about 15 dB and 3 dB for $N = 8$ and 16, respectively, as compared to QSE. Moreover, Figure 3.5 also validates the convergence analysis of QSEr.

Figure 3.6 plots MSE against δ. We see that only QSEr with the optimal q can achieve a relatively uniform estimation performance (by approaching the CRLB), across the whole region of δ. We also see that, as expected, the original QSE diverges from the CRLB substantially for large values of δ, which is caused by overlooking the impact of q on the monotonicity of $\beta_i^{\mathcal{R}}$ w.r.t. Δ_i. From the zoomed in subfigure, we see that QSEr also outperforms HAQSE obviously, with an MSE improvement of 0.5 dB at $\delta = \pm 0.4$. The superiority of QSEr over QSE and HAQSE validates the significance of our finding on the overlooked impact of q to the convergence of the state-of-the-art QSE [3] and also validates the efficacy of the optimal q in (3.36).

3.3 Padé Approximation-Enabled Estimator

We proceed to introduce another frequency estimator [9] that employs the Padé approximation for better accuracy and also less interpolations. For convenience, let us briefly recap the signal model and problem formulation. Consider the estimation of the unknown frequency f from the following N samples of a complex single-tone exponential signal,

$$s(n) = Ae^{j\left(\frac{2\pi f n}{f_s}+\phi\right)} + z(n), \quad n = 0, 1, \ldots, N-1, \tag{3.37}$$

where A, f_s, ϕ, and N denote the signal amplitude, the sampling rate, the initial phase, and the sample number, respectively. The AWGN is denoted by $z(n)$. The frequency f can be decomposed into the sum of an integer and a fractional multiples of f_s/N, i.e.

$$f = (k^\star + \delta)f_s/N, \ \ \delta \in [-0.5, 0.5], \tag{3.38}$$

where $k^\star$ denotes an integer and δ a noninteger. Taking the N-point DFT of $s(n)$, i.e. $\sum_{n=0}^{N-1} s(n)e^{-j\frac{2\pi kn}{N}}$, yields,

$$S(k) = \sum_{n=0}^{N-1} \tilde{A}e^{j\frac{2\pi(k^\star+\delta)n}{N}} e^{-j\frac{2\pi kn}{N}} + Z(k), \tag{3.39}$$

where $\tilde{A} = Ae^{j\phi}$, f in (3.37) is replaced by its expression given in (3.38), and $Z(k)$ denotes the DFT of $z(n)$. The integer $k^\star$ can be estimated by identifying the maximum of $|S(k)|^2$, i.e.

$$k^\star \ : \ \max{}_{k \in [0, N-1]} |S(k)|^2. \tag{3.40}$$

In general, $k^\star$ can be accurately estimated [3, 6, 8, 10, 11]. Thus, we only focus on refining the frequency estimate by estimating δ, as done in Section 3.2.

3.3.1 Core Updating Function

Consider the i-th iteration, where the estimate of δ from the previous iteration, as denoted by δ_{i-1}, is available. With reference to [3], we can interpolate the DFT coefficients at

$$k_{i,\pm} = k^\star + \delta_{i-1} \pm q_i, \tag{3.41}$$

where $k^\star$ is given in (3.38), and q_i is an extra controlling parameter used to shift the interpolation positions. Plugging $k = k_{i,\pm}$ into (3.39), the interpolated DFT coefficients are

$$S(k_{i,\pm}) = \breve{A}\sin(\pi(\xi_i \mp q_i)) / \sin(\pi(\xi_i \mp q_i)/N), \tag{3.42}$$

where $\breve{A} = \tilde{A}e^{j\frac{(N-1)\pi}{N}(\xi_i \mp q_i)}$, $\tilde{A}$ given in (3.39), $\xi_i (= \delta - \delta_{i-1})$ denotes the estimation error in the $(i-1)$-th iteration, and the noise term is dropped for brevity. Denoting $S_{i,\pm} = S(k_{i,\pm})$, we can construct the following ratio:

$$\rho_i = (|S_{i,+}|^2 - |S_{i,-}|^2)/(|S_{i,+}|^2 + |S_{i,-}|^2) \tag{3.43}$$

The purpose of doing so is to estimate ξ_i from ρ_i. Let $\hat{\xi}_i$ denote the estimate of ξ_i. We can use $\hat{\xi}_i$ to refine the δ estimation as follows:

$$\delta_i = \delta_{i-1} + \hat{\xi}_i, \ \text{s.t.} \ \hat{\xi}_i \approx \xi_i = \delta - \delta_{i-1}. \tag{3.44}$$

Next, we illustrate how to estimate $\hat{\xi}_i$ from ρ_i.

Introducing the function $S(\xi_i, \pm q_i) = \frac{\sin^2(\pi(\xi_i \mp q_i))}{\sin^2(\frac{\pi(\xi_i \mp q_i)}{N})}$, the right-hand side (RHS) of (3.43) can be written into

$$f(\xi_i) = \frac{S(\xi_i, q_i) - S(\xi_i, -q_i)}{S(\xi_i, q_i) + S(\xi_i, -q_i)}. \tag{3.45}$$

Jointly observing (3.43) and (3.45), we see that estimating $\hat{\xi}_i$ from ρ_i is equivalent to solving the equation $f(\xi_i) = \rho_i$. Some features of $f(\xi_i)$, which are useful for deriving the solution, are provided in the following lemma.

Lemma 3.1 *For $|\xi_i| \leq q_i$, $f(\xi_i)$ monotonically increases with ξ_i, and presents odd symmetry around the origin, i.e. $f(-\xi_i) = -f(\xi_i)$.*

Proof: Since $0 < q_i \leq 0.5$ is satisfied, $S(\xi_i, \pm q_i)$ approximates the square of a sinc function [12]. Hence, we know that $S(\xi_i, q_i)$ is a monotonically increasing function from $-q_i$ to q_i, while $S(\xi_i, -q_i)$ monotonically decreases in the same region. These can be translated into $S'_+ > 0$, $S'_- < 0$, where S'_+ denotes the first derivative of $S(\xi_i, q_i)$ w.r.t. ξ_i and S'_- is similarly defined for $S(\xi_i, -q_i)$. Accordingly, the monotonicity of $f(\xi_i)$ can be deduced from its first derivative, i.e. $f' = 2(S'_+ S_- - S'_- S_+)/[S_+ + S_-]^2 > 0$, where S_+ denotes $S(\xi_i, q_i)$, and S_- denotes $S(\xi_i, -q_i)$.

From the expression given in (3.45), we know that $S(\xi_i, \pm q_i)$ is symmetric about $\xi_i = \pm q_i$. Thus, they are also symmetric about the axis of $\xi_i = 0$ in the region of $|\xi_i| \leq q_i$, i.e.

$$S(\xi_i, q_i) = S(-\xi_i, -q_i), \quad S(-\xi_i, q_i) = S(\xi_i, -q_i). \tag{3.46}$$

Applying this symmetry in (3.45) leads to

$$\begin{aligned} f(-\xi_i) &= \frac{S(-\xi_i, q_i) - S(-\xi_i, -q_i)}{S(-\xi_i, q_i) + S(-\xi_i, -q_i)} \\ &= \frac{S(\xi_i, -q_i) - S(\xi_i, q_i)}{S(\xi_i, q_i) + S(\xi_i, -q_i)} = -f(\xi_i), \end{aligned}$$

which shows the symmetry of $f(\xi_i)$ about the origin.

From (3.45), we see that it can be difficult to directly solve the equation $f(\xi_i) = \rho_i$, due to the existence of the squared sine functions. To simplify the equation, the first-order Taylor series of $f(\xi_i)$ has been widely employed [3, 6, 8, 10, 11, 13], which, however, is valid for large N's. Here, we use the Padé approximation to approximate the following Taylor series of $f(\xi_i)$ (of degree six), i.e.

$$\tilde{f}(\xi_i) = \sum_{l=0}^{6} c_l \xi_i^l = c_1 \xi_i + c_3 \xi_i^3 + c_5 \xi_i^5, \tag{3.47}$$

where the even powers of ξ_i are suppressed since $f(\xi_i)$ is an odd function of ξ_i, as illustrated in Lemma 3.1. The Padé approximation of the above Taylor series can be given by

$$\hat{f}(\xi_i) = \left(\sum_{p=0}^{P} a_p \xi_i^p\right) \Big/ \left(\sum_{r=0}^{R} b_r \xi_i^r\right), \quad P = R = 3, \tag{3.48}$$
$$\text{s.t. } \hat{f}(0) = \tilde{f}(0);\ \hat{f}^{(k)}(0) = \tilde{f}^{(k)}(0),\ k = 1, \ldots, P+R,$$

where $h^{(k)}(x)$ denotes the k-th derivative of $h(x)$ (h can be $\hat{f}$ or $\tilde{f}$) and $h^{(k)}(0)$ is value of $h^{(k)}(x)$ at $x = 0$. The rationale for setting $P = R = 3$ is illustrated in Remark 3.3.

Also note that the constraints in (3.48) constitute $(P + R + 1)$ equations, which can be used to express the coefficients of the Padé approximation, i.e. a_p and b_r, in terms of those of Taylor series, i.e. c_l. We underline that the properties of $f(\xi_i)$ unveiled in Lemma 3.1 can be used to suppress some high-order terms in $\hat{f}(\xi_i)$. As illustrated in Remark 3.3, we have $a_0 = a_2 = 0$ and $b_1 = b_3 = 0$. Plugging these constraints into the $(P + R + 1)$ equations, the Padé approximation coefficients can be solved, with the solution achieved in 3.2.

Remark 3.3 Based on (3.47), the Taylor series $\tilde{f}(\xi_i)$ is of degree six. Thus, according to [14, Sec. 5.12], we have $P + R = 6$ for the Padé approximation $\hat{f}(\xi_i)$ given in (3.48). Lemma 3.1 shows $f(0) = 0$. Solving $\hat{f}(0) = 0$ yields $a_0 = 0$, which then indicates $P \geq 1$. Lemma 3.1 also states $f(-\xi_i) = -f(\xi_i)$. To preserve the odd symmetry, the numerator and denominator of $\hat{f}(\xi_i)$ can only have odd and even powers of ξ_i, respectively, since there is a nonzero constant b_0 in the denominator. Based on the above analysis, $(P, R) = (1, 5)$ or $(2, 4)$ leads to the same Padé approximation with the degree of the denominator polynomial up to four; $(P, R) = (3, 3)$ or $(4, 2)$ leads to the same Padé approximation with the degree of the denominator polynomial up to three; and $(P, R) = (5, 1)$ or $(6, 0)$ makes the Padé approximation degenerated into Taylor series. Given that a cubic polynomial can be more tractable than a quartic one, we employ the Padé approximation with $(P, R) = (3, 3)$.

Lemma 3.2 *The function $f(\xi_i)$ can be approximated by $\hat{f}(\xi_i)$ with the approximation error in the order of $\mathcal{O}(\xi_i^7)$, where*

$$\hat{f}(\xi_i) = (a_1\xi_i + a_3\xi_i^3)/(1 + b_2\xi_i^2), \text{ with } a_1 = c_1, a_3 = c_3 - c_1c_5/c_3, b_2 = c_5/c_3. \tag{3.49}$$

Equating $\hat{f}(\xi_i)$ obtained in Lemma 3.2 to ρ_i calculated in (3.43), we obtain that $a_1\xi_i + a_3\xi_i^3 = \rho(1 + b_2\xi_i^2)$, which can be further turned into a cubic equation of ξ_i,

$$\xi_i^3 + k_2\xi_i^2 + k_1\xi_i + k_0 = 0, \text{ s.t. } k_2 = -\rho b_2/a_3, k_1 = a_1/a_3, k_0 = -\rho/a_3, \tag{3.50}$$

Using the cubic formula [15], the three roots of the above equation can be solved as follows:

$$z_1 = -k_2/3 + 2B,\ z_2 = -k_2/3 - B + D,\ z_3 = -k_2/3 - B - D, \tag{3.51}$$

with the intermediate variables given by

$$B = (S+T)/2,\ D = \sqrt{3}(S-T)\mathrm{j}/2;$$
$$S = \sqrt[3]{R + \sqrt{D}},\ T = \sqrt[3]{R - \sqrt{D}},\ D = Q^3 + R^2;$$
$$R = \frac{(9k_1k_2 - 27k_0 - 2k_2^3)}{54},\ Q = \frac{(3k_1 - k_2^2)}{9},$$

where k_0, k_1 and k_2 are given in (3.50). Among the three roots, only one is the final estimate of ξ_i, as dictated below.

Lemma 3.3 *The estimate of ξ_i is given by*

$$\hat{\xi}_i = z_{i^*},\ s.t.\ i^* = \mathrm{argmin}_i\ |z_i|. \tag{3.52}$$

Proof: As proved in Lemma 3.1, $f(\xi_i)$ is monotonic against ξ_i for $|\xi_i| < q_i$. Thus, we can only have one solution to $f(\xi_i) = \rho_i$ in the region of $|\xi_i| < q_i$. Since the continuous region covers $\xi_i = 0$ (the smallest value that can be taken), the solution to the equation $\hat{f}(\xi_i) = \rho_i$ in the region is the smallest root given in (3.51). This leads to the (3.52).

Based on the above analyses and derivations, the steps of estimating δ_i from δ_{i-1} are summarized in Algorithm 3.1.

Algorithm 3.1 Estimating δ_i

Given $a_1, a_3, b_2, \delta_{i-1}$ from iteration $(i-1)$ and q_i, and provided that $|\xi_i| \leq q_i$, δ_i can be estimated as follows:

1. Interpolate the DFT coefficients at $k_{i,\pm}$ given in (3.41), leading to $S_{i,\pm}$ given in (3.43);
2. Construct ρ_i, as illustrated in (3.43);
3. Compute the coefficients k_0, k_1, and k_2 based on (3.50);
4. Compute the three roots in (3.51);
5. Obtain the estimate of ξ_i, as given in Lemma 3.3;
6. Update δ_i as done in (3.44).

3.3.2 Initialization and Overall Estimation Procedure

A high-quality initialization can speed up the asymptotic convergence of an iterative frequency estimator. Next, we provide a high-quality initialization of the estimator with a single interpolation; c.f., two or more interpolations in many previous designs [3, 6, 8, 10, 11, 13].

Assume that $\delta \in [0, 0.5]$ holds for the moment. If we set $\delta_0 = 0.25$, then the estimation error ξ_1 satisfies that $\xi_1 = \delta - \delta_0 \in [-0.25, 0.25]$. Accordingly, we can set $q_1 = 0.25$ and run Algorithm 3.1 to estimate δ_1. Moreover, we notice that $\delta_0 - q_1 = 0$. This indicates that one of the interpolated DFT coefficients is at $k_{1,-} = k^\star + \delta_0 - q_1 = k^\star$. This DFT coefficient has been computed when identifying $k^\star$; see (3.40) in Section 3.1. Thus, we only need to interpolate one DFT coefficient at $k_{1,+} = k^\star + \delta_0 + q_1 = 0.5$. The above analysis also applies for the case of $\delta \in [-0.5, 0]$. Then, the next question is how to determine the initial region of δ. To answer that, the following sign test [10] can be performed by reusing the DFT results for identifying $k^\star$,

$$\alpha = \text{sign}\{[S(k^\star - 1) - S(k^\star + 1)]S^\dagger(k^\star)\}, \tag{3.53}$$

where $k^\star$ is given in (3.38), and $()^\dagger$ takes the complex conjugate. Using α, we have

$$\delta \in [0, 0.5], \text{ if } \alpha > 0; \text{ or } \delta \in [-0.5, 0], \text{ if } \alpha < 0. \tag{3.54}$$

As will be illustrated in Section 3.3.3, the sign test has a high accuracy in the sense that the estimators with or without using the sign test approach the CRLB from the same SNR.

Algorithm 3.2 Overall Estimation Procedure

Input: N, $\delta_0 = 0.25\alpha$, $q_1 = 0.25$, q_i $(i = 2, \dots, I)$, *the coefficient set* $C_1 = \{a_1, a_3, b_2\}$ *for* δ_1, *and the sets* $C_i = \{a_1^{(i)}, a_3^{(i)}, b_2^{(i)}\}$ *for* δ_i $(i \geq 2)$. *The estimator performs as follows:*

1. Estimate δ_1 by running Algorithm 3.1 once based on N, δ_0, q_1 and C_1;
2. For each $i = 2, \dots, I$, run Algorithm 3.1 iteratively based on N, δ_{i-1}, q_i and C_i.

The final frequency estimate is given by $\hat{f} = \frac{f_s(k^\star + \hat{\delta}_I)}{N}$.

Based on the above initialization and Algorithm 3.1, we summarize the overall processing of the proposed frequency estimator in Algorithm 3.2. The overall computational complexity of the estimator is analyzed next. We first evaluate the computational complexity of Algorithm 3.1. It is dominated by that of interpolating DFT coefficients in Step (1). According to (3.39), one interpolation needs N complex multiplications. As illustrated above (3.53), a single interpolation is required for the first iteration, while for iteration $i(\geq 2)$, we need $2N$ complex multiplications to interpolate the DFT coefficients twice. Algorithm 3.2 runs Algorithm 3.1 for I times, and thus, its overall computational complexity is dominated by $N + (I - 1) \cdot 2N = (2I - 1)N$ complex multiplications. As will be illustrated in Section 3.3.3, the estimator is able to approach the CRLB after only two iterations, i.e. $I = 2$.

We recommend taking $q_i = q_1 = 0.25$ ($\forall i$) given two reasons. *First*, the estimator is robust against the value of q_i, or in other words, the estimation performance remains almost the same across a wide range of q_i's. This will be illustrated in Figure 3.11 of Section 3.3.3. *Second*, a benefit of taking of $q_i = q_1 = 0.25$ ($\forall i$) is that only one set of Padé approximation coefficients, i.e., C_1, are required to be stored onboard, saving storage and time for indexing different sets (if used) in practical systems.

3.3.3 Numerical Illustrations

In this section, we provide simulation results to validate the performance of the estimator introduced in this section. Several related estimators are simulated as benchmarks which can be implemented in the framework of Algorithm 3.2. Below, we only highlight their differences from our estimator.

(1) A&M [8]: This estimator always interpolates the DFT coefficients at $k_{i,\pm} = k^\star + \delta_{i-1} \pm 0.5$ ($\forall i \geq 1$) with $\delta_0 = 0$ taken. The ratio $\rho_i = \frac{|S(k_{i,+})|-|S(k_{i,-})|}{|S(k_{i,+})|+|S(k_{i,-})|}$ is constructed in each iteration, and $\xi_i = \rho_i/2$. A&M also has a different construction of ρ_i which leads to the same asymptotic performance as the one given above and hence is not considered here.

(2) Generalized A&M (GAM) [10]: This estimator iterates as A&M but starts from a different initial δ_0. In particular, GAM takes $\delta_0 = \alpha\beta$, where α is given in (3.53) and $\beta = 0, 0.25$ and 0.5 are considered and compared in the work. Here, for a fair comparison with our estimator, we only consider $\beta = 0.25$.

(3) Hybrid A&M and q-Shift Estimator (HAQSE) [3]: This estimator applies A&M for δ_1. Then, starting from $i = 2$, HAQSE interpolates the DFT coefficients at $k_{i,\pm} = k^\star + \delta_{i-1} \pm q_{\text{H}}$, where $q_{\text{H}} = N^{-1/3}$ is proven to be sufficient for the estimator to converge to the CRLB for large N's and $q_{\text{H}} \leq 0.32$ is suggested in [16] to ensure the validity of HAQSE also for small N's. HAQSE constructs ρ_i as $\rho_i = \Re\left\{\frac{S(k_{i,+})-S(k_{i,-})}{S(k_{i,+})+S(k_{i,-})}\right\}$ and updates $\xi_i = \frac{q_{\text{H}}\cos^2(\pi q_{\text{H}})}{1-\pi q_{\text{H}}\cot(\pi q_{\text{H}})}\rho_i$.

Unless otherwise specified, the following parameters are used for all the estimators: $k^\star = 2, f_s = 1, I = 2, \delta \in \mathcal{U}_{[-0.5,0.5]}, q_1 = q_2 = 0.25$ (for the estimator), and $q_{\text{H}} = N^{-1/3}$ (for HAQSE). Note that $\mathcal{U}_{[-0.5,0.5]}$ denotes the uniform distribution in $[-0.5, 0.5]$. All the results to be presented are averaged over 5×10^4 independent trials. Moreover, the CRLB [8], given by $\frac{6f_s^2}{4\pi^2\gamma N^3}$, is provided in most simulation results, where $\gamma = \frac{|A|^2}{\sigma_0^2}$ is the SNR of the single-tone signal given in (3.37), and σ_0^2 denotes the noise variance of $z(n)$ therein. As interpreted in Table 3.1, different estimators in the simulation results are differentiated by markers, while different values of N are distinguished by line styles.

Figure 3.7 plots the MSE of $\hat{f}$ against γ, where $N = 8, 16$, and 32 are simulated. We see that the MSEs of $\hat{f}$ converge to the CRLB for all the estimators. Comparing

Table 3.1 Marker and line style definitions in simulation results.

Marker	∘	+	×	▽	**None**
Estimator	AM	GAM	HAQSE	Proposed	CRLB

Line style	**Dash**	**Dash-dotted**	**Solid**
N	8	16	32

the two subfigures, it is obvious that the estimation performance of all the estimators is further improved (closer to the CRLB) after the second iteration. Figure 3.8 zooms in the differences among the estimators by normalizing the MSEs plotted in Figure 3.7 against their respective CRLBs. We see from Figure 3.8a that, after the first iteration, the proposed estimator already achieves the MSE as low as 1.079 times the CRLB for a small $N = 8$ and reduces the MSE to 1.063 times the CRLB for $N = 32$, which is notably based on a single interpolation. We see from Figure 3.8b that, after the second iteration, the proposed estimator persistently outperforms the benchmark estimators across the whole region of γ and approaches the CRLB more tightly.

We see from Figure 3.8b that the simulated MSE can be smaller than the CRLB, yet with a considerably small difference. Two reasons may cause this phenomenon. *First*, CRLB is derived for a deterministic parameter that is under estimation, while the frequency taken for the simulations is random over independent trials. We remark that this random configuration is necessary for a fair

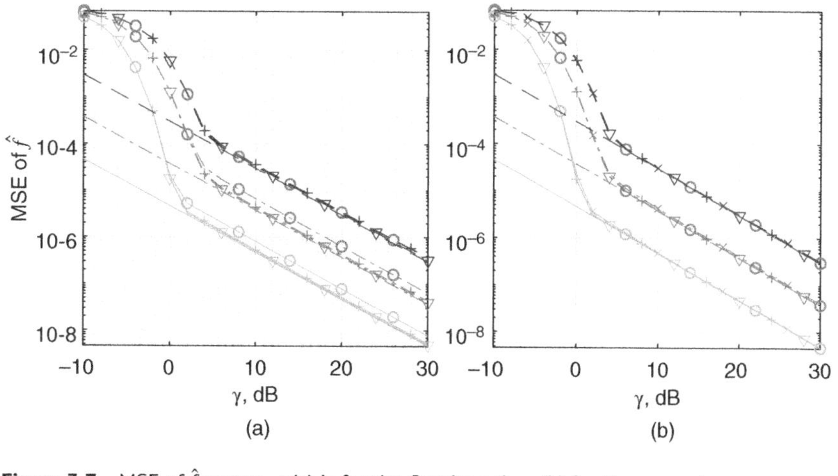

Figure 3.7 MSE of $\hat{f}$ versus γ: (a) is for the first iteration, (b) for the second.

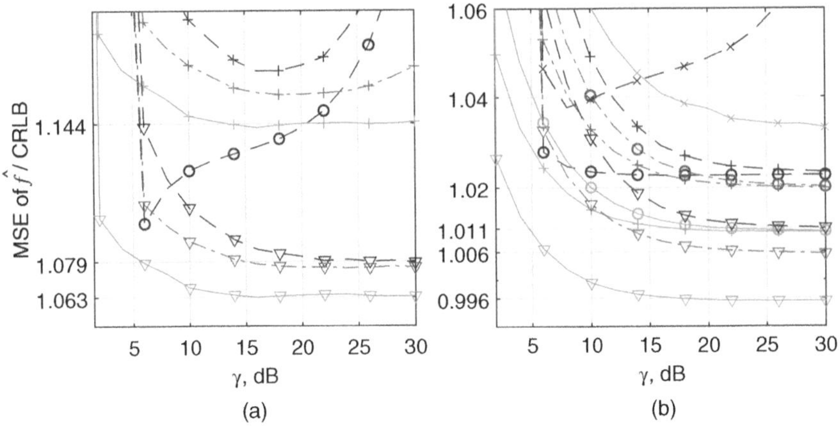

Figure 3.8 Illustration of the ratio between MSE and CRLB to better compare different estimators: (a) for the first iteration, (b) for the second. For better clarity, range limits are imposed on the y-axes, and hence the estimators with larger MSEs become invisible in the figure.

comparison of different estimators, since they can have dramatically distinct estimation performance over frequencies, as will be illustrated in Figure 3.10. *Second*, this phenomenon can be caused by the finite number of independent (Monte-Carlo) trials. Refer to [3, Sec. V] for a detailed analysis of this aspect.

Figure 3.9 illustrates the MSE performance w.r.t. N. From Figure 3.9a, we see that, after the first iteration, the normalized MSEs of the proposed estimator is as low as 1.065 which is improved by about 6.74% over A&M and GAM (both achieve

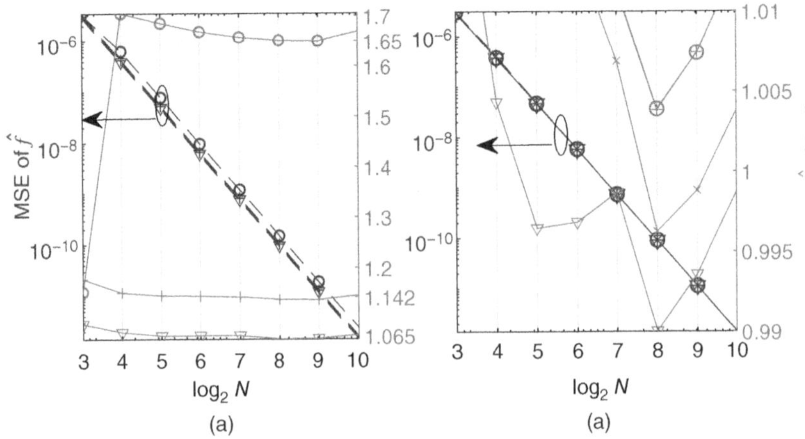

Figure 3.9 MSE performance against N, where $\gamma = 20$ dB, (a) for the first iteration, and (b) for the second.

the minimum about 1.142). We also see that the proposed estimator present a more stable asymptotic performance than A&M and GAM, as N increases. From Figure 3.9b, we see that, after the second iteration, the proposed estimator is able to approach the CRLB for almost all values of N, while HAQSE can only achieve this for large N's. This is expected since HAQSE is designed for large N's. On the other hand, this highlights the benefit of introducing the Padé approximation.

Figure 3.10 plots the MSE of different estimators against the whole region $\delta \in [-0.5, 0.5]$. We see from Figure 3.10a that the proposed estimator already has a close-to-CRLB performance at some δ after the first iteration. We also see that the proposed estimator provides a performance lower bound for GAM. This is expected, since both estimators take $\delta_0 = 0.25$, while the proposed one achieves a more accurate approximation between ρ_i and ξ_i. We see from Figure 3.10b that, after the second iteration, the proposed estimator achieves the best flatness in the whole region of δ.

Figure 3.11 compares HAQSE and the proposed estimator in terms of q_2 (for the proposed) or q_{H} (for HAQSE). We see that both estimators show the tight convergence to the CRLB in the case of $N = 16$ and 32 and show the robustness against q. This is consistent with the analysis in [3] that the asymptotic performance of HAQSE shall remain the same for $q_{\mathrm{H}} \leq N^{-1/3}$. We also see that our estimators always provides a performance lower bound for the HAQSE across the whole region of q_2. Notably, we see that, for the small $N = 8$, our estimator still presents a stable MSE performance over q_2's, while HAQSE shows an increasingly worse performance as q_{H} increases. The visible improvement achieved by the

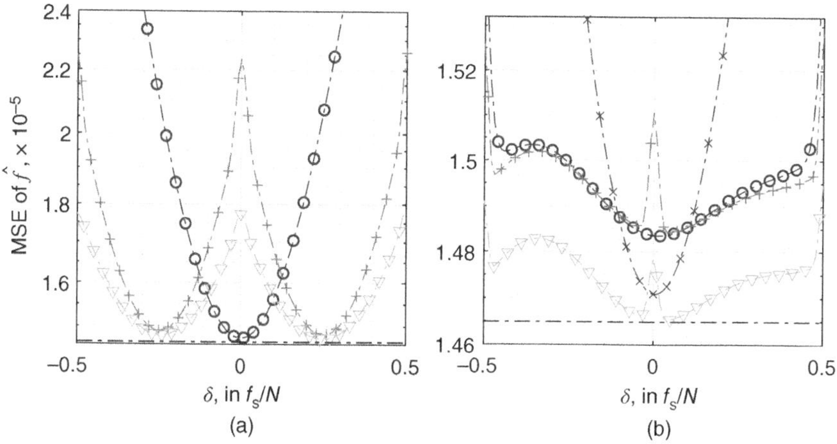

Figure 3.10 MSE performance versus δ, where (a) for the first iteration, and (b) for the second. Note that $N = 16$, $\gamma = 20$ dB, and 100 discrete values are evenly taken in the region of $\delta \in [-0.5, 0.5]$.

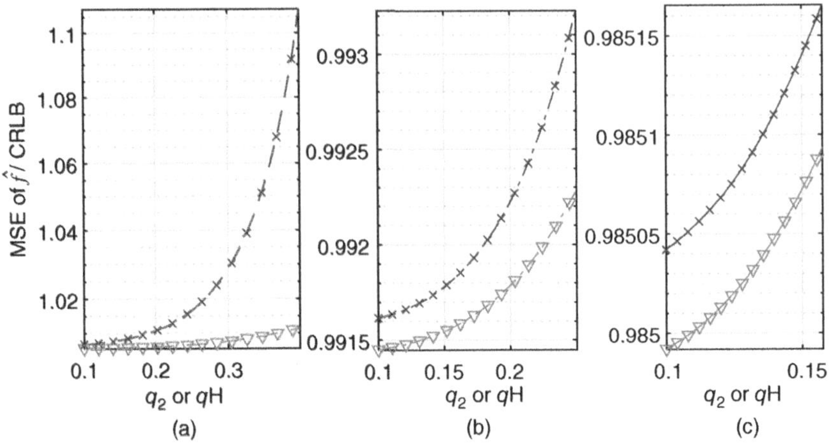

Figure 3.11 MSE performance versus q_2, where $\gamma = 20$ dB. The subfigures (a), (b), and (c) are for $N = 8, 16$, and 32, respectively. For fair comparison with HAQSE, we set $q_{\mathrm{H}} = q_2$ which is evenly taken from $[0.1, N^{-1/3}]$.

proposed estimator over the state-of-the-art HAQSE is *on the one hand*, due to the more accurate initial estimate in the first iteration, as has been demonstrated in Figures 3.7–3.10; and *on the other hand*, ensured by the newly proposed much more accurate approximation between ρ_i and ξ_i.

3.4 Conclusions

Two accurate frequency estimators are presented in this chapter, both performed by interpolating the DFT coefficients. Since only elementary math operations are involved, the computational complexity of the two estimators are low, making them particularly suitable for real-time communication and sensing systems. In fact, these estimators will be frequently used in later chapters to perform the estimations of key parameters in the two radio frequency systems.

References

1 H. L. Van Trees. *Detection, Estimation, and Modulation Theory, Part I: Detection, Estimation, and Linear Modulation Theory*. John Wiley & Sons, 2004.

2 S. Haykin. *Array Signal Processing*. Prentice-Hall, Inc., Englewood Cliffs, NJ, 1985.

3 A. Serbes. Fast and efficient sinusoidal frequency estimation by using the DFT coefficients. *IEEE Transactions on Communications*, 67(3):2333–2342, March 2019. ISSN 0090-6778. doi: 10.1109/TCOMM.2018.2886355.
4 S. G. Krantz. *A Handbook of Real Variables: With Applications to Differential Equations and Fourier Analysis*. Springer Science Business Media, 2011.
5 Codes for "Fast and efficient sinusoidal frequency estimation by using the DFT coefficients", December 2018. URL https://codeocean.com/capsule/0939761/tree.
6 J.-R. Liao and C.-M. Chen. Phase correction of discrete Fourier transform coefficients to reduce frequency estimation bias of single tone complex sinusoid. *Signal Processing*, 94:108–117, 2014. ISSN 0165-1684.
7 X. Liang et al. A new and accurate estimator with analytical expression for frequency estimation. *IEEE Communications Letters*, 20(1):105–108, January 2016. ISSN 2373-7891. doi: 10.1109/LCOMM.2015.2496149.
8 E. Aboutanios and B. Mulgrew. Iterative frequency estimation by interpolation on Fourier coefficients. *IEEE Transactions on Signal Processing*, 53(4): 1237–1242, April 2005. ISSN 1053-587X. doi: 10.1109/TSP.2005.843719.
9 K. Wu, J. A. Zhang, X. Huang, and Y. J. Guo. Accurate frequency estimation with fewer DFT interpolations based on PADÉ approximation. *IEEE Transactions on Vehicular Technology*, 70(7):7267–7271, 2021. doi: 10.1109/TVT.2021.3087869.
10 Y. Liu, Z. Nie, Z. Zhao, and Q. H. Liu. Generalization of iterative Fourier interpolation algorithms for single frequency estimation. *Digital Signal Processing*, 21(1):141–149, 2011.
11 S. Ye, J. Sun, and E. Aboutanios. On the estimation of the parameters of a real sinusoid in noise. *IEEE Signal Processing Letters*, 24(5):638–642, May 2017. ISSN 1070-9908. doi: 10.1109/LSP.2017.2684223.
12 A. V. Oppenheim. *Discrete-time Signal Processing*. Pearson Education India, 1999.
13 B. G. Quinn. Estimating frequency by interpolation using Fourier coefficients. *IEEE Transactions on Signal Processing*, 42(5):1264–1268, 1994.
14 W. H. Press, S. A. Teukolsky, W. T. Vetterling, and B. P. Flannery. *Numerical Recipes 3rd Edition: The Art of Scientific Computing*. Cambridge University Press, 2007.
15 E. W. Weisstein. Cubic formula, 2002. URL https://mathworld.wolfram.com/.
16 K. Wu, W. Ni, J. A. Zhang, R. P. Liu, and Y. J. Guo. Refinement of optimal interpolation factor for DFT interpolated frequency estimator. *IEEE Communications Letters*, 24(4):782–786, 2020. doi: 10.1109/LCOMM.2019.2963871.

Part II

Communication-Centric JCAS

4

Perceptive Mobile Network (PMN)

Owing to the global deployment and coverage of mobile communications networks, it is expected that the most dominant joint communications and sensing (JCAS) networks will be communications-centric, i.e. an enhancement of the current mobile communications networks to include sensing capabilities. We referred to such JCAS networks as perceptive mobile networks (PMNs). This chapter provides a comprehensive overview of PMNs that realizes JCAS technology in mobile networks with a focus on signal processing techniques. It is intended to provide a clear picture on what the PMN will look like and how it may evolve from the current communications-only network from the viewpoints of both infrastructure and signal-processing techniques. The receiver designed in PMN to perform JCAS tasks will be elaborated. More specifically, we consider mobile-network-specific JCAS challenges and solutions associated with the heterogeneous network architecture and components, sophisticated mobile signal formats, and complicated signal propagation environment. We discuss the required changes to system infrastructure for the paradigm shift from communications-only mobile networks to PMNs with integrated communications and sensing and provide a comprehensive discussion on existing technologies and open research problems.

4.1 Framework for PMN

To start with, we present a framework of PMN that integrates radio sensing into the current communications-only mobile network. Within this framework, we describe the optional system architectures, introduce three types of unified sensing techniques and discuss what communication signals can be effectively used for sensing.

Joint Communications and Sensing: From Fundamentals to Advanced Techniques, First Edition.
Kai Wu, J. Andrew Zhang, and Y. Jay Guo.

4.1.1 System Platform and Infrastructure

By modifying and enhancing the hardware, algorithms, and software, the current mobile networks can evolve into PMNs. In principle, sensing can be realized in either the user equipment (UE) or base stations (BS). One advantage of sensing in the UE is that it may motivate wider end-user applications. Compared to the UE, however, the BS possesses the advantages of having networked connections, flexible cooperation, large antenna array, powerful computation capability, and known and fixed locations to enable more stable sensing. Therefore, in the following, we mainly consider BS-based sensing.

It should be noted that the evolution to PMN is not limited to a particular cellular standard. Hence, instead of limiting to any specific standard, the following discussions are generic in nature, focusing on key components and technologies in modern mobile networks, such as antenna array, broadband, multiuser multiple-input and multiple-output (MIMO) and orthogonal frequency-division multiple access (OFDMA). When necessary, we will relate our discussions to the 5G new radio (NR) standard.

Depending on the network setup, there are two types of topologies in which JCAS can be implemented, that is, a cloud radio access network (CRAN) and a standalone BS. The realization of sensing in a PMN based on these two topologies is illustrated in Figure 4.1. Below we elaborate on the system and network setups

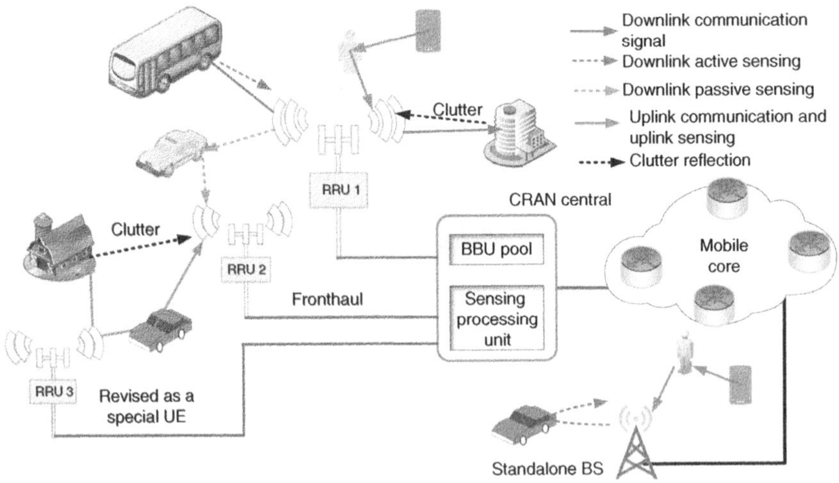

Figure 4.1 Illustration of sensing in a PMN with both standalone BS and CRAN topologies. RRU1 is a node supposed to have full-duplexing capability or equivalent. RRU3 is modified to be a special UE, transmitting uplink signals for uplink sensing in RRU2, with clock synchronization between them. RRU2 can also be modified as a receiver only to do both uplink and downlink sensing as well as communications (receiver only).

for the two topologies. We will then discuss the three types of sensing operations based on the topologies in Section 4.1.2. Methods that modify the setup to enable sensing will be discussed in Section 4.2.

4.1.1.1 CRAN

A typical CRAN consists of a central unit and multiple distributed antenna units, which are called remote radio units (RRUs) [1]. The RRUs are typically connected to the CRAN center via optical fiber. Either quantized radio frequency (RF) signals or baseband signals can be transmitted between RRUs and the central unit. As shown in Figure 4.1, in a CRAN-based PMN, the densely distributed RRUs that are coordinated by the central unit provide communications to the UEs. Their received signals from themselves, other RRUs, and UEs, are collected and processed by the CRAN central, for both communications and sensing. The CRAN central unit hosts the original baseband unit (BBU) pool for communications and the new processing unit for sensing. This setup is actually aligned with the topologies for distributed radar systems [2].

A typical communication scenario can be described as follows: several RRUs work cooperatively to provide connections to the UEs, using multiuser MIMO techniques over the same resource blocks (same time and frequency slots). In CRAN communication networks, power control is typically applied such that signals from one RRU may not reach other RRUs. While it is not necessary, we can relax this constraint and assume that cooperative RRUs are within the signal coverage area of each other. This assumption is reasonable when dense RRUs are deployed and used to support surrounding UEs via coordination. Admittedly, this may not be necessary for certain types of sensing activities as we are going to discuss in Section 4.1.1.2, but it does increase the options of sensing [3]. Technically, it is feasible at the cost of increased transmission power, even if only for supporting sensing, as the downlink signals do not cause mutual communication interference to RRUs.

Note that, in this configuration, all RRUs are typically synchronized using the timing clock from the GPS signals. This forms an excellent network with distributed nodes for sensing applications [4]. Such CRAN-based PMNs are reported in [3, 5, 6], together with the sensing algorithms.

4.1.1.2 Standalone BS

As an alternative to CRAN-based PMN, a standalone BS can also perform sensing using the received signals either from its own transmitted signals or from UEs. This is actually the typical and simpler setup that has been widely considered in the literature. This setup includes the small BS that may be deployed within a household, which pushes for the concepts of edge computing and sensing. Like Wi-Fi sensing [7], such a small BS can be used to support indoor-sensing applications such as fall detection and house surveillance. It also includes a roadside unit (RSU) that is part of the mobile network but specifically deployed to support vehicular communications.

In what follows, our discussions will be referred to the CRAN topology, but most of the results are applicable to standalone BS based PMNs. In the case there is no confusion, we will use CRAN and BS interchangeably.

4.1.2 Three Types of Sensing Operations

There are three types of sensing that can be unified and implemented in PMNs, defined as *uplink and downlink sensing*, to be consistent with uplink and downlink communications. In uplink sensing, signals received from UEs are used for sensing, while in downlink sensing, the sensing signals are from BSs. The downlink sensing is further classified as *Downlink Active Sensing* and *Downlink Passive Sensing*. The former refers to the cases when an RRU collects the echoes from its own, whereas in the latter an RRU collects echoes from other RRU-transmitted signals. In other words, the terms active and passive are used to differentiate the cases of sensing using self-transmitted signals and signals from other nodes. Below, we explain each sensing operation in details.

4.1.2.1 Downlink Active Sensing

In downlink active sensing, an RRU (or BS) uses the reflected/diffracted signals of its own transmitted downlink communication signals for sensing. This is the typical case considered in systems where the sensing receiver is colocated with the transmitter [8–10]; it functions like a monostatic radar. Downlink active sensing enables a BS to sense its surrounding environment. Since the transmitter and receiver are on the same platform, they can be readily synchronized at the clock-level, and the sensing results can be clearly interpreted by the node without external assistant. However, this setup would require certain full-duplexing capability or something equivalent [10].

4.1.2.2 Downlink Passive Sensing

Downlink passive sensing refers to the case where an RRU uses the received downlink communication signals from other RRUs for sensing. This is analogous to the setup of bistatic and multistatic radars, where the transmitter(s) and receiver are spatially separated, but their clocks may be synchronized [11, 12]. Downlink passive sensing signals can be available to a chosen RRU when the transmission power is sufficiently strong. In this case, they will always be there together with the downlink active sensing signals, the reflection and refraction of the RRU's own transmitted signal. They may arrive at the sensing receiver slightly later than the downlink active sensing signals due to longer propagation distances. When all RRUs cooperatively communicate with multiple UEs using spatial domain multiple access (SDMA), these two types of signals cannot be readily separated in time or frequency. Therefore, the corresponding sensing algorithms also need to consider downlink active sensing signals if downlink passive sensing is in operation. The setup and its sensing algorithms for this complicated scenario have been investigated in [3]. In general, downlink passive sensing enables the observation of the environments between RRUs.

4.1.2.3 Uplink Sensing

The uplink sensing conducted at the BS utilizes the received uplink communication signals from UE transmitters. It is similar to passive sensing [13] in the sense that the transmitter and receiver are spatially separated and nonsynchronized. The difference is that, in uplink sensing, the receiver is fully aware of the system protocol and signal structure. Uplink sensing can be directly implemented without requiring a change of hardware and network setup, and without requiring for full-duplex operation. However, it estimates the relative, instead of absolute, time delay and Doppler frequency since the clock/oscillator is typically not locked between spatially separated UE transmitters and BS receivers. This ambiguity may be resolved with special techniques, as will be discussed in Section 4.3.6. Uplink sensing senses UEs and the environment between UEs and RRUs, and has been studied in, e.g. [14].

4.1.2.4 Comparison

Downlink sensing can potentially achieve more accurate sensing results than uplink sensing. This is because, in the downlink sensing case, RRUs generally have more advanced transmitters such as more antennas and higher transmission power, and the whole transmitted signals are centrally known. Additionally, as the sensing results in downlink sensing are not directly linked to any UEs, the privacy issue is largely not a problem. Comparatively, uplink sensing may disclose the information of UE, potentially causing privacy concerns.

Both downlink and uplink sensing in PMNs are feasible for practical applications in terms of sensing capabilities. According to the results in [5] and [6], the downlink and uplink sensing with practical transmission power values (lower than 25 dBm) can reliably detect objects more than 50 and 150 m away, respectively, in a dense multipath propagation environment. Additionally, a distance resolution at a few meters can be achieved for a signal bandwidth of 100 MHz, an angle resolution of about 10° for a uniform linear array of 16 antennas, and a resolution of 5 m/s moving speed within channel coherence period.

A comparison of the three types of sensing is provided in Table 4.1.

Table 4.1 Comparison of three types of sensing operations.

Types	Signals	Action	Advantages	Disadvantages
Downlink active sensing	Reflects from a RRU/BS's own transmitted downlink communication signal	Sense surrounding environment of the RRU/BS	All data symbols in the received signals can be used and are centrally known	Generally require full-duplex operation and other network modifications. Devices can be specially deployed to resolve this problem
Downlink passive sensing	Received downlink communication signals from other RRUs	Sense environment between RRUs	RRUs are synchronized. Privacy is less an issue because sensing results not directly linked to any UEs	
Uplink sensing	Uplink communication signals from UE transmitters	Sense UEs and the environment between UEs and RRU	Require minimum modification to communication infrastructure. Does not require full-duplexing	Timing and Doppler frequency measurement could be relative. Transmitted information signals are not directly known. Rapid channel variation when UEs are moving

4.1.3 Signals Usable from 5G NR for Radio Sensing

To help understand what signals in mobile communications networks can be used for sensing, we examine the 5G NR case in details. For 5G NR, we can exploit the following signals for sensing: reference signals used for channel estimation, synchronization signal blocks (SSBs), and data payload. The properties of these signals in terms of sensing are summarized in Table 4.2. These communication signals may be further jointly optimized for communications and sensing, using methods reported in, e.g. [15].

4.1.3.1 Reference Signals Used for Channel Estimation

Deterministic signals specifically designed for channel estimation are available in many systems. The 5G NR includes the demodulation reference signals (DMRS) for both uplink (physical uplink shared channel [PUSCH]) and downlink (physical downlink shared channel [PDSCH]), sounding reference signals (SRS) for uplink, and channel state information-reference signals (CSI-RS) for downlink [16]. Most of them are comb-type pilot signals, circularly shifted across orthogonal frequency-division multiplexing (OFDM) symbols, and are orthogonal between different spatial streams. Especially, DMRS signals accompanying the shared

Table 4.2 A summary of the properties of the signals that can be used for sensing in PMN with reference to 5G NR.

Type of signals	Signal pattern in time domain	Signal pattern in frequency domain	Signal correlation	Signal values
Reference signals	Irregular and varying in length	May be on a regular comb structure	Typically orthogonal	Known and fixed
Synchronization signal blocks (SSBs)	Short and less frequent (every 20 ms)	Sparsely distributed, using 20 resource blocks	Orthogonal over smaller spatial layers	Known and fixed
Data payload	UE specific, irregular, and long	Allocation dependent, resource block based	Statistically independent	Known in downlink sensing, unknown in uplink sensing (become known after demodulation)

channel are always transmitted with data payload and exhibit user-specific features. Therefore, DMRS signals are random and irregular over time, which requires sensing algorithms that can deal with such irregularity. Comparatively, signals used for beam management in connected modes, like SRS and CSI-RS, can be either periodic or aperiodic, and hence, they are more suitable for sensing algorithms based on conventional spectrum estimation techniques such as ESPRIT. Such training signals for channel estimation are the most widely exploited ones for sensing in the JCAS literature [14, 17].

The number and position of DMRS OFDM symbols are known to BSs, and they can be adjusted and optimized across the resource grid, including slots and subcarriers (resource blocks). This implies good prospects for both channel estimation and sensing in different channel conditions. The allocation of resource grids can be optimized by considering requirements from both communications and sensing. With a given subcarrier spacing, the available radio resources in a subframe are treated as a resource grid composed of subcarriers in frequency and OFDM symbols in time. Accordingly, each resource element in the resource grid occupies one subcarrier in frequency and one OFDM symbol in time. A resource block consists of 12 consecutive subcarriers in the frequency domain. A single NR carrier is limited to 3300 active subcarriers as defined in Sections 7.3. and 7.4 of TS 38.211 in [16]. The number and pattern of the subcarriers that DMRS signals occupy have a significant impact on the sensing performance, as we will see in Section 4.3.5.

In [18], some simulation results for both uplink and downlink sensing using DMRS are provided. The signal is generated according to the Gold sequence as defined in *3GPP TS 38.211*, for both *PDSCH* and *PUSCH*. The generated physical resource-block (PRB) is over a 3-D grid comprising a 14-symbol slot for the full subcarriers across the DMRS layers or ports. The interleaved DMRS subcarriers of PDSCH are used in downlink sensing, while groups of noninterleaved DMRS subcarriers of PUSCH are used in uplink sensing. The results demonstrate the feasibility of achieving excellent sensing performance with the use of the DMRS signals. However, a major problem of sensing ambiguity is also noted due to the interleaved pattern of the subcarriers.

4.1.3.2 Nonchannel Estimation Signals

Several deterministic nonchannel estimation signals such as the synchronization signal (SS) and the physical broadcast channel (PBCH), also called the SSB, can also be used for sensing. Such signals typically have regular patterns with a periodic appearance at an interval of several to tens of milliseconds. However, they only occupy a limited number of subcarriers, which may lead to limited identification of multipath delay values. Little research on using this class of signals for JCAS has been reported.

4.1.3.3 Data Payload Signals

Additionally, one can also exploit the data payload signals in both the PDSCH and PUSCH for sensing. In downlink sensing, the data symbols are known to the sensing receiver and hence can be directly used. In uplink sensing, symbols need to be used in a decision-directed mode. Since these data symbols are random and signals in different spatial streams are nonorthogonal, they are not ideal for sensing. If it is used for uplink sensing, the signals need to be demodulated first, which could also introduce demodulation error. However, they can significantly increase the number of available sensing signals, and hence improve the overall sensing performance at the cost of increased complexity. Precoders for these signals can be optimized by jointly considering the requirements from communications and sensing.

4.2 System Modifications to Enable Sensing

Communications and sensing in a PMN can share a number of processing modules in a widely employed MIMO-OFDM transceiver, including the whole transmitter and many receiver modules, as can be seen from Figure 2.4.

Despite the numerous modules shareable by communications and sensing, some modifications at hardware and network levels to existing mobile networks are necessary for realizing PMNs. Communication signals can generally be directly used for estimating sensing parameters, but the communication system platform is not directly ready for sensing. On the one hand, a communication node does not have the full-duplex capability at the moment, that is, transmitting and receiving signals of the same frequency at the same time. This makes monostatic radar sensing infeasible without modifying current communication infrastructure. On the other hand, for transmitter and receiver in two nodes are spatially separated, and there is typically no clock synchronization between them. This can cause ambiguity in ranging estimation and makes processing signals across packets difficult. Thus, bistatic radar techniques cannot be directly applied in this case. These are fundamental problems that need to be solved at the system level in order to make sensing in communications-centric systems feasible.

We now describe the modifications of current hardware and systems that are required for current communications-only mobile networks to evolve to PMNs. The depicted changes focus on the fundamental ones that allow the current mobile network to do radio sensing simultaneously with communications. For clarity, we do not consider low-level changes such as joint waveform optimization, joint antenna placement and sparsity optimization and power optimization, and leave them to Section 4.3. For the three types of sensing integrated in PMN, the realization of uplink sensing is relatively easy; the major challenges are with downlink sensing, where the leakage and reflected signals from the transmitter can cause significant interference to the received signals for both sensing and communications. We now review four options of system modifications to enable sensing in PMN.

4.2.1 Dedicated Transmitter for Uplink Sensing

Conventional uplink sensing can be realized in a similar way to passive sensing [19], with the difference that the receivers in PMN are part of the network and process signals for both communications and sensing. Uplink sensing confronts the same sensing ambiguity problem with passive sensing. Conventionally, the phase clock between UEs and BSs is not synchronized; hence, sensing ambiguity in time and Doppler frequency is present in uplink sensing. If the ambiguity can be tolerated, no change to hardware and system architectures of current mobile systems is required. Such sensing ambiguity may also be resolved using signal processing techniques under certain special situations [14], as will be detailed in Section 4.3.6.

To eliminate the ambiguity, dedicated (static) UEs that are clock-synchronized to BSs can be used. In terms of the required system modification, uplink sensing with the use of such static UEs would be the most convenient way for achieving nonambiguity sensing in the PMNs. This is shown as RRU3 in Figure 4.1 for a CRAN, where RRU3 can be modified to operate as a UE, transmitting uplink signals. This option can also enable distributed (single-input and multiple-output (SIMO)) sensing, where the transmitted signals can be processed jointly at multiple spatially separated BSs for collaborative sensing.

4.2.2 Dedicated Receiver for Downlink (and Uplink) Sensing

For downlink sensing without requiring full-duplexing capability, one near-term option is to deploy a BS that only works on the receiving mode [20]. It can be configured as a receiver either for downlink sensing only or for both communications and downlink sensing.

To implement this near-term downlink sensing, changes to the hardware may be required. This is because the receiver in current BSs is conventionally designed to receive uplink communication signals only, and downlink sensing requires receiving downlink communication signals. The required change is insignificant for time-division duplexing (TDD) systems since a TDD transceiver generally uses a switch to control the connection of antennas to the transmitter or receiver. Thus, the change is only the adjustment of the transmitting and receiving period so that the switch is equivalently always connected to the receiver. For frequency-division duplexing (FDD) systems, the BS receivers may be incapable of working on downlink frequency bands, and modification to the hardware is required. Therefore, it is more cost-effective to implement downlink sensing in TDD than in FDD systems.

Alternatively, we can also deploy a dedicated receiving-only node for both downlink and uplink sensing, as well as communications if desired. This is particularly feasible for TDD systems. In TDD systems, downlink and uplink sensing signals can then be largely separated in time at the receiver. Naturally, to remove the ambiguity in delay estimation, clock synchronization is required between the transmitters and this node. An example is shown as RRU2 in Figure 4.1 for a CRAN, which can perform downlink and uplink sensing using received signals from RRU1 and RRU3, respectively.

4.2.3 Full-Duplex Radios for Downlink Sensing

Resolving the signal leakage problem in downlink sensing would ideally require *full-duplex* technologies [21]. Although full-duplex communications techniques have been widely investigated, it is still not very practical to be applied in an environment involving mobility and dynamics, particularly for TDD systems. In full-duplex communications, a device needs to be able to receive signals from other communication devices, while transmitting its signals using the same frequency channels. The receiver needs to recover the weak received communication signals, while suppressing the onboard leakage and echoes from the nearby environment. The signal to be recovered can be several orders lower than the interference signal. Full duplex communications generally use a combination of antenna separation, RF suppression, and baseband suppression to mitigate the interference from the transmitter at the receiver.

We note that it may be easier to realize full-duplex for JCAS compared to for full-duplex communications, if we only require colocated sensing while transmitting, but not a simultaneous transmission from two communication nodes. This is because we only need to remove the impact of the directly leaked signal from the transmitter while keeping the echoes of the transmitted signals for sensing, without the presence of communication signals from other devices. There is generally no need to consider the interference between sensing and received communication signals. As illustrated in Figure 4.2, a sufficient guard interval between communications transmission (for sensing reception) and communication reception, existing in current TDD systems, should be able to prevent mutual interference between communications and sensing. The guard interval following transmission in Node A, GI_t, may only need to be slightly increased compared to conventional communications-only TDD systems, to accommodate possible longer echo sensing signals.

The above discussions indicate that full-duplex operation is a potentially long-term solution to enable seamless integration of downlink sensing with communications. In [22], it is shown that moving targets are more robust to the leaked self-interference, whereas limited transmitter-receiver isolation is primarily a concern for detecting static objects. In [10], multibeam [23, 24] techniques are investigated for full-duplex JCAS systems. Using the multibeam technology, where separated beams can be generated and optimized for communications and sensing, it is demonstrated that leakage and clutter signals can be significantly reduced.

Nevertheless, it is still very challenging to implement full-duplex JCAS, particularly for MIMO systems. The main reason is that in MIMO systems, a large number of leakage signals between pairs of transmitter and receiver antennas need to be handled simultaneously. Overall, full-duplex JCAS would be an ideal future

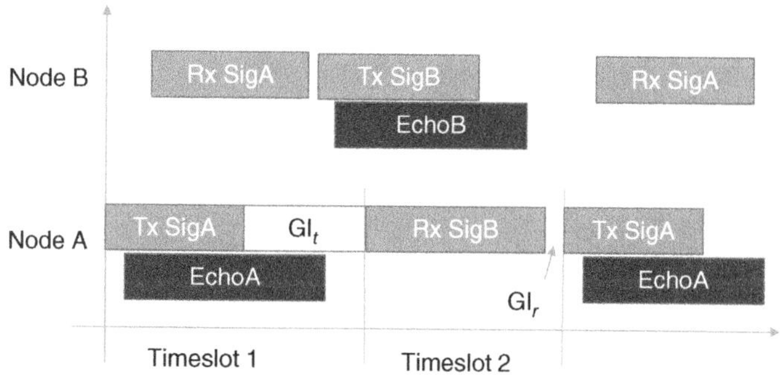

Figure 4.2 Illustration of the timeslot allocation in a full-duplex TDD JCAS system from the viewpoint of Node A. Full-duplex operation is only required during Timeslot 1. GI stands for guarding interval, "Sig" is a shortened form of signal.

technology for PMN but, unfortunately, it is immature and impractical for real implementations in the near future.

4.2.4 Base Stations with Widely Separated Transmitting and Receiving Antennas

One possible solution for downlink sensing is to use well-separated transmitting and receiving antennas. The large separation can significantly reduce the leakage from transmitted signals. The baseband receiver also accepts feedback from the baseband transmitter so that a baseband self-interference cancellation may be further applied. However, this spatially well-separated antenna structure requires extra antenna installation space and can increase the overall cost. One option to minimize the cost is to *add a dedicated single antenna that is spatially separated from the existing antennas for receiving sensing signals* in a conventional MIMO system.

Figure 4.3 shows an example of this option in TDD systems. The system has a normal transceiver for communications with four antennas. A fifth antenna is installed at a position well separated from the four antennas, and it is connected to the receiver via a cable. Signals from the fifth antenna are used for downlink active sensing. Figure 4.3a depicts the general concept, and Figure 4.3b shows a potential implementation in existing TDD systems. The switches (SPDT1–4) operate normally for a TDD communication system. For the fifth antenna, it is always connected to the fifth receiver. Given that the onboard circuit leakage is small and

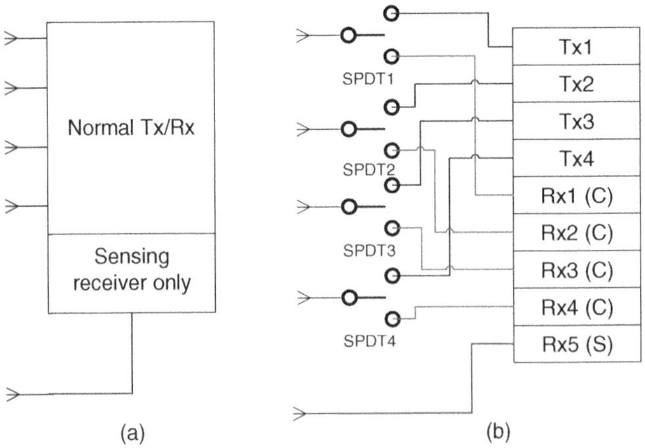

Figure 4.3 Simplified TDD transceiver model with a single receiving antenna dedicated to sensing. (a) A general concept and (b) possible realization on existing hardware platform.

Table 4.3 Comparison of options of enabling sensing in mobile networks, in terms of system deployment.

Options	Advantages	Disadvantages
Using full-duplex radios	An ideal future option with best system integration and flexibility in system design and optimization	Immature today; high system and signal processing complexity
Dedicated transmitter for uplink sensing	Easy to realize. Still support transmission for communications. Enable SIMO sensing (similar to bistatic and multistatic radar)	Slightly increased cost
Dedicated receiver for downlink sensing	Easy to realize in TDD systems. Still support communication reception. Enable multiple-input and single-output (MISO) sensing	Not suitable for current FDD systems. Slightly increased cost
Using one spatially widely separated receive antenna for sensing	Colocated transmitter and receiver enables monostatic sensing for environment directly surrounding the sensing node. Lowest-cost solution for TDD systems	Require separated control of RF chains. Slightly increased space requirement for installation

the TDD switches can be separately controlled, this option can be conveniently realized in an existing TDD system that supports 5×5 MIMO.

Sensing using a single receiving antenna in this case can be realized by exploring the multiple transmitted spatial streams [20]. Although the sensing performance may be slightly degraded due to the reduced number of receiving antennas for sensing, the degradation is mainly on the received signal energy. All the sensing parameters except for the angle-of-arrival (AoA) can still be correctly estimated.

If there are spaces for installing multiple dedicated receive antennas for sensing, a single sensing receiver (RF down-converter channel) can also be connected to these antennas in a time-interleaved (multiplexing) way so that more spatial diversity gain can be achieved and AoA can also be estimated. This is due to the relatively low variation speed of channel parameters in sensing. The feasibility has been demonstrated in [25], where a single receiver is connected to multiple antennas using fast antenna array multiplexing to identify the time-variant directional structure of mobile radio channel impulse responses. Practical measurement results show that the delay, AoA, and Doppler can be effectively estimated using this setup. The four types of modifications introduced in Sections 4.2.1 to 4.2.4 are summarized in Table 4.3.

Before we conclude this section, we note that, in addition to hardware modifications, hardware calibration is also important for sensing. In [25, 26], it is shown how an imperfect receiver and antenna array can impact high-resolution sensing parameter estimation in channel sounding experiments. Antenna array calibration techniques are further investigated in [25, 26] and shown to be able to mitigate the impact of hardware imperfection effectively.

4.3 System Issues

As a new platform, PMN is still in its early stage of research and development. As described in Section 4.2, there are a number of challenges to overcome to make PMN a reality, which also offer great research opportunities. In the following, we review existing technologies and algorithms that have been developed to address these challenges, organized under nine topics. We also discuss open research problems for each topic. Since the major issue in PMN is how to achieve radio sensing without compromising the performance of existing communications, we focus on the issues in realizing radio sensing, leveraging the existing cellular communication infrastructure, and on how communications may be affected and improved by integrating sensing functions.

4.3.1 Performance Bounds

There are two types of performance bounds that can be used to characterize the performance limits of sensing in PMNs. One is based on mutual information (MI), and the other is based on the estimation accuracy of sensing parameters, such as the Cramer–Rao lower bounds (CRLBs). A very recent work provides a comprehensive review of the performance limits of JCAS systems [27].

MI can be used as a tool to measure both the radar and communication performance [28]. To be specific, for communications, the MI between wireless channels and the received communication signals can be employed as the waveform optimization criterion, while for sensing, the conditional MI between sensing channels and the sensing signals is used [29, 30]. For sensing, MI is a measure of how much information about the channel, the propagation environment, is conveyed to the receiver. Maximizing MI hence is particularly useful for sensing applications that rely more on feature signal extraction than on sensing parameters estimation, for example for target recognition. The usage of MI and capacity is well known to the communications community. The usage of MI for radar waveform design can also be traced back to the 1990s [29]. MI has also been used to optimize the performance of coexisting radar communication systems [31].

Let us consider simplified signal models to illustrate the formulations of MI. Let $\mathbf{H}_S$, $\mathbf{H}_C$ denote the channels for sensing and communications, respectively; $\tilde{\mathbf{Y}}_S$ and $\tilde{\mathbf{Y}}_C$ be the received signals for sensing and communications, respectively; $\mathbf{X} = f(\mathbf{S})$ be the transmitted signals where $\mathbf{S}$ is the information symbols and $f(\mathbf{S})$ denotes the function converting $\mathbf{S}$ to $\mathbf{X}$. The function $f(\mathbf{S})$ can be linear or nonlinear, and can be across multiple domains, including time, frequency, and spatial domains. When a spatial precoding matrix $\mathbf{P}$ is used in place of $f(\cdot)$, the received signals are given by

$$\tilde{\mathbf{Y}}_s = \mathbf{H}_s\mathbf{PS} + \mathbf{Z}; \quad \tilde{\mathbf{Y}}_c = \mathbf{H}_c\mathbf{PS} + \mathbf{Z},$$

for sensing and communications, respectively. The MI expressions for communications and sensing can be represented as follows:

$$\begin{aligned} &I(\mathbf{H}_s; \tilde{\mathbf{Y}}_s|\mathbf{X}) = h(\tilde{\mathbf{Y}}_s|\mathbf{X}) - h(\mathbf{Z}), \text{ for sensing;} \\ &I(\mathbf{S}; \tilde{\mathbf{Y}}_c|\mathbf{H}_c) = h(\tilde{\mathbf{Y}}_c|\mathbf{H}_c) - h(\mathbf{Z}), \text{ for comm.,} \end{aligned} \tag{4.1}$$

where $I(\cdot)$ denotes the conditional MI, and $h(\cdot)$ denotes the entropy of a random variable. We can thus see that the MI for communications and sensing are defined as the metrics conditional on the channel and the transmitted signals, respectively. In each case, the signals $\mathbf{X}$, or more specifically the precoder $\mathbf{P}$, are optimized to maximize the MI.

MI for JCAS systems has been studied and reported in a few publications. They are typically conducted by jointly optimizing the two MI expressions and their variations in (4.1). The work in [32] formulates radar MI and the communication channel capacity for a JCAS system and provides preliminary numerical results. In [28], radar waveform optimization is studied for a JCAS system by maximizing MI expressions. In [33], the estimation rate, defined as the MI within a unit time, is used for analyzing the radar performance, together with the capacity metric for communications. A more general and flexible formulation can be based on a weighted sum of the two MIs [34]

$$F = \frac{w_R}{F_s} I(\mathbf{H}_s; \tilde{\mathbf{Y}}_s|\mathbf{X}) + \frac{1 - w_R}{F_c} I(\mathbf{S}; \tilde{\mathbf{Y}}_c|\mathbf{H}_c), \tag{4.2}$$

where F_c and F_s are the maximal MI for communications and sensing, respectively, and are treated as two known constants in the optimization, and $w_R \in [0, 1]$ is a weighting factor. The function in (4.2) is concave and can be maximized by using, e.g. the Karush–Kuhn–Tucker (KKT) conditions. The optimal $\mathbf{P}$ turns out to be a water-filling type of solution that jointly considers the distributions of the Eigen-values of $\mathbf{\Sigma}_X$ and $\mathbf{\Sigma}_{H_R}$.

The following available results serve as good bases for studying the MI for PMNs. Some specific problems to PMNs can be considered to make the results more practical.

- First, the MI formulations for uplink and downlink sensing are different due to the different knowledge on the transmitted signals and the channel differences. In the downlink sensing, the symbols are known to the receiver, and the channels for communications and sensing are different but could be correlated. For uplink sensing, the symbols are unknown to the receiver, and the channels are the same for communications and sensing. Hence, the optimization objective functions and results can be quite different for uplink and downlink sensing.

- Second, the specific packet and signal structures in cellular networks can have a significant impact on MI for both communications and sensing. For example, a packet signal may include training sequence and data symbols which will lead to different MI formulations and results, as their statistical properties are different. In [34], the MI is studied for PMN, considering the frame structure and estimation errors. The findings from [34] indicate that the optimal solution for one function (communications or sensing) is generally not optimal for the other, and some trade-off needs to be made, for example when the directions of sensing and communications deviate significantly. This implies the importance of sensing-motivated user scheduling, i.e. taking user scheduling into the joint optimization of communications and sensing.

CRLB is a more traditional metric that has been widely used in characterizing the lower bound of parameter estimation in radar [35]. For PMN, the closest work on CRLB is reported in [20], where a SDMA system is considered. The CRLB expressions are not always available in closed-form, particularly for MIMO-OFDM signals, primarily because the received signals are nonlinear functions of sensing parameters. Therefore, although they can be found and evaluated numerically, the CRLB metrics are not easy to be applied in analytical optimization. It is even harder to apply them in optimization, jointly with another cost function. So typically, iterative approaches are required to optimize CRLB based formulation in PMN [20].

MI has also been combined with other metrics to study the performance of radar systems. For example two criteria, namely, maximization of the conditional MI and minimization of the minimum mean-square error (MMSE), are studied in [36] to optimize the waveform design for MIMO radar by exploiting the covariance matrix of the extended target impulse response. In [37], the optimal waveform design for MIMO radar in colored noise is also investigated by considering two criteria: maximizing the MI and maximizing the relative entropy between two hypotheses that the target exists or does not exist in the echoes. In [20], waveform optimization is studied with the application and comparison of multiple sensing performance metrics including MI, MMSE, and CRLB. It has been shown that there are close connections between MI-based and CRLB-based optimizations, and the MI-based method is more efficient and less complicated compared to the CRLB-based method. Overall, research for JCAS and PMNs based on these combined criteria is still very limited.

4.3.2 Waveform Optimization

For JCAS, joint waveform optimization is a key research problem as the single transmitted signal is used for both functions, but the two functions have different requirements for the signal waveform. As discussed earlier in Part I, traditional radar and communication systems use very different waveforms, which are optimized for respective applications. For example, recall that radar uses orthogonal and unmodulated pulsed or frequency modulated continuous wave (FMCW) signals, while in PMNs, typically the signals are random, with multicarrier modulation and multiuser access. However, the waveform for one function may be modified to accommodate the requirements of the other by virtue of joint design and optimization. The work in [47] represents one of the earliest research activities that investigate waveform design for JCAS systems. The waveform design and signal parameters can have a significant impact on the overall performance of a JCAS system. For example, the numerical analysis in [48] demonstrates the close linkage between the sensing resolution capabilities and the signal parameters for both single carrier and multicarrier communication systems.

In Section 4.3.2, we discuss three classes of waveform optimization techniques for PMN, classified in the spatial, time, and frequency domains, as shown in Table 4.4. The first one is based on optimizing the spatial precoding matrix, and it typically does not require the change of existing signals and can be seamlessly realized in current cellular networks. The second and third ones optimize signal parameters and resource allocation in the time and frequency domain,

Table 4.4 Classification of waveform optimization techniques in PMN.

Waveform optimization (single carrier, OFDM, and new waveforms such as OTFS)	Spatial domain	Single spatial precoder	Minimize waveform difference [9, 15]
			Weighted-optimal and Pareto-optimal waveform design based on MI and/or CRLB [20, 34, 38]
			Multi-objective function optimization [39]
		Decomposed spatial precoder	Multibeam optimization where two pregenerated subbeams are optimally combined, to support fast varying scanning directions and multiple targets [23, 40–42]
			Additional spatial streams are introduced to either remove multiuser interference caused by scanning beam or enhance sensing performance [43]
	Time domain	Optimize the signal structure in terms of length of packet and interval between packets [44]	
		Optimize power distribution between different parts of packets [34]	
	Frequency domain	Optimize subcarrier allocations, e.g. nonuniform subcarriers to achieve better delay estimation [45]	
		Mimic traditional frequency-modulated radar signals in the preamble [17, 46]	

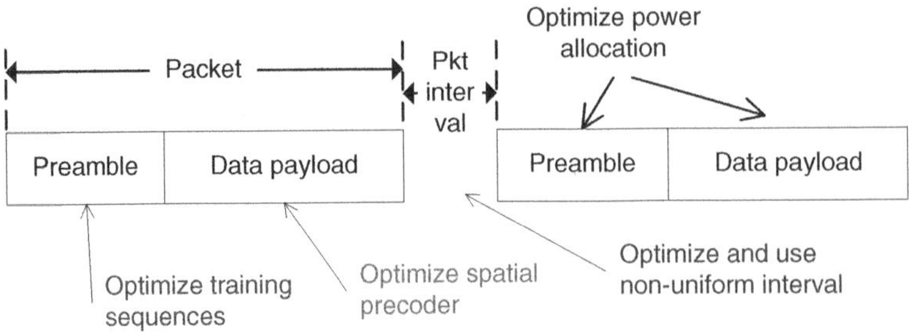

Figure 4.4 Waveform optimization depicted with respect to packets.

respectively, requiring slight modification to the signals. They are also depicted in Figure 4.4 with respect to signal packets.

4.3.2.1 Spatial Optimization

For spatial optimization with respect to the precoding matrix **P** in PMNs, apart from the MI-based waveform optimization as discussed in Section 4.3.1, there are two more practical methods. The first method is to optimize the signal conversion function $f(\cdot)$, or specifically the precoding matrices $\mathbf{P}_{q,n,t}$ when $f(\cdot) = \mathbf{P}$, to make the statistical properties of the transmitted signals **X** best suitable for both communications and sensing. The second method is to add sensing waveform to the underlying communication waveform, while considering a coherent combination of the two waveforms for destination nodes. The two methods have respective advantages and disadvantages, which are elaborated below.

In the first method, when optimization is with respect to the spatial precoding matrix, it is designed to alter the statistical properties of the transmitted signal. This method is particularly suitable for global optimization of cost functions jointly formulated for communications and sensing. The basic optimization formulation is as follows:

$$\arg\max_{\mathbf{P}} \lambda(\mathbf{P}), \quad \text{subject to Constraints } 1, 2, \ldots, \tag{4.3}$$

where $\lambda(\mathbf{P})$ is the objective function. There are various methods and combinations in defining the objective functions and the constraints. Each can be for either communications or sensing individually or a weighted joint function.

In the second method, the basic waveforms can be designed in advance for either or both communications and sensing, and the two waveforms are then added in a way to jointly optimize the performance of communications and sensing. The basic idea can be mathematically represented as follows:

$$\underset{\alpha,\mathbf{Pc},\text{or }\mathbf{Ps}}{\text{argmax}} \lambda(\mathbf{P}), \quad \text{subject to Constraints } 1, 2, \ldots, \tag{4.4}$$

where $\mathbf{P}_c$ and $\mathbf{P}_s$ are the precoding matrices primarily targeted for communications and sensing, respectively, and **P** is a function of $\mathbf{P}_c$ and $\mathbf{P}_s$. Examples are

$\mathbf{P} = \alpha\mathbf{P}_c + (1-\alpha)\mathbf{P}_s$ and $\mathbf{P} = [\alpha\mathbf{P}_c, (1-\alpha)\mathbf{P}_s]$, where α is a complex scalar. Here, among $\{\alpha, \mathbf{P}_c, \mathbf{P}_s\}$, either $\mathbf{P}_c$, $\mathbf{P}_s$, or both of them can be predesigned and fixed, while real-time optimization is with respect to the other parameters. Although the results may be suboptimal, this method provides great flexibility and can adapt quickly to changes on the requirements for communications and sensing.

Below, we use the example of multibeam optimization in an analog array to further elaborate the ideas. Steerable analog array, which is also the basic component of a hybrid array, could be widely applied in mmWave JCAS systems. A JCAS system may need to support communications and sensing at different directions, which is challenging to address given the limits on the beamforming (BF) capability of analog arrays. A good solution is the multibeam technology [23, 40, 49]. The multibeam consists of a fixed subbeam dedicated to communications and a scanning subbeam with directions varying over different communication packets. By optimizing the beam with multiple subbeams, communications and radar sensing at different directions can be supported simultaneously with a single signal. This can largely extend the field of view of sensing. The multibeam can be applied at both transmitter and receiver, while transmitter is considered here.

Two classes of methods have been proposed for the multibeam optimization: the subbeam-combination method [23, 40, 49] via constructively combining two pregenerated subbeams according to given criteria, corresponding to the second method in (4.4), and the *global optimization* [49] by jointly considering the communications and sensing BF requirements and optimizing a single BF vector directly, corresponding to the first method in (4.3).

In the subbeam-combination method, two basic beams for communications and sensing are separately generated according to the desired BF waveform, using, e.g. the iterative least squares method [23]. The two beams are further shifted to the desired directions by multiplying a sequence, and then combined by optimizing a power distribution factor ρ and a phase shifting coefficient φ.

The BF vector $\mathbf{w}_t$ in (4.5) can be represented as follows:

$$\mathbf{w}_T = \sqrt{\rho}\mathbf{w}_{T,F} + \sqrt{1-\rho}e^{j\varphi}\mathbf{w}_{T,S}, \tag{4.5}$$

where $\mathbf{w}_{T,F}$ and $\mathbf{w}_{T,S}$ are the BF vectors that are predetermined, corresponding to the fixed subbeam and scanning subbeam, respectively. The value of ρ can typically be determined via balancing communications and sensing distances [40], and the optimization is conducted mainly with respect to φ, which can ensure the pregenerated subbeams are coherently combined when generating the multibeams.

The multibeam optimization problem can then be formulated with desired objective functions and constraints. Consider one example of maximizing the received signal power for communications with constrained BF gain in given scanning directions. Let the threshold of the BF gain in the ith sensing direction ϕ_i be $C_i^2(1-\rho)M_T$, where $C_i \in [0,1]$ is a scaling coefficient, representing the ratio between the gain of the scanning subbeam in the interested direction and the maximum gain that the array can achieve, i.e. $(1-\rho)M_T$. We can formulate the constrained optimization problem as follows:

$$\varphi_{\text{opt}}^{(1)} = \arg\max_{\varphi} \frac{\mathbf{w}_T^H \mathbf{H}_C{}^H \mathbf{H}_C \mathbf{w}_T}{\|\mathbf{w}_T\|^2}, \tag{4.6}$$

$$\text{s.t.} \quad \frac{|\mathbf{a}^T(M_T, \phi_i)\mathbf{w}_T|^2}{\|\mathbf{w}_T\|^2} \geq C_i^2(1-\rho)M_T, \quad i = 1, 2, \dots, N_s,$$

where N_s is the number of the total constraints, ϕ_is are the angles of interest where the BF gain is constrained, and a maximal ratio combiner is assumed to be applied at the communication receiver.

With $\mathbf{w}_{T,F}$, $\mathbf{w}_{T,S}$, and ρ being predetermined, the optimization problem can be solved by first finding the unconstrained optimal solution for the objective function and then looking for its intersection with the intervals determined by the constraints. Closed-form solutions can then be obtained as detailed in [49].

The subbeam-combination method enables simple and flexible multibeam generation and optimization, and hence is promising for practical applications that require fast adaptation to changing BF requirements. It is particularly useful for mmwave systems where directional BF is used.

The subbeam-combination method is an efficient low-complexity solution, but it is suboptimal. A globally optimal solution can be obtained by solving a problem formulated directly with respect to $\mathbf{w}_T$. Considering a similar example of maximizing the received signal power for communications with constraints on BF waveform, the formulation can be represented as follows:

$$\mathbf{w}_{t,\text{opt}} = \underset{\mathbf{w}_T, \mathbf{w}_T^H \mathbf{w}_T = 1}{\text{argmax}} \; \mathbf{w}_T^H \mathbf{H}_C{}^H \mathbf{H}_C \mathbf{w}_T, \tag{4.7a}$$

$$\text{s.t.} \quad \|\mathbf{A}(M_T, \phi_i)\mathbf{w}_T - \mathbf{d}_v\|^2 \leq \varepsilon_w, \quad \text{and/or} \tag{4.7b}$$

$$|\mathbf{a}^T(M_T, \phi_i)^T \mathbf{w}_T|^2 \geq \varepsilon_i, \; i = 1, 2, \dots, N_s, \tag{4.7c}$$

where (4.7b) bounds the mismatches between the generated BF waveform and the desired one $\mathbf{d}_v$, (4.7c) constrains the gain of the scanning subbeam in N_s directions, and ε_w and ε_i are the thresholds for these constraints. These constraints can be applied individually or jointly.

The optimization in (4.7) is generally a nonconvex NP-hard problem. In [49], this problem is converted to a homogeneous quadratically constrained quadratic program (QCQP), which is then solved by the software-defined radio (SDR) technique. Although it is not easy to obtain a closed-form optimal solution, the global optimization method provides a benchmark for suboptimal multibeam optimization schemes.

For waveform optimization in PMNs, the following specific problem associated with multiuser access is yet to be considered, particularly for uplink sensing.

For downlink sensing, multiuser access and multiuser interference do not need to be considered for communications, because the transmitted signals are known to the sensing receiver and the environment to be sensed is common to multiuser signals. Thus, waveform optimization only needs to consider the multiuser aspect for communications, as studied in [15, 20]. However, for uplink, signals need to be specific to each user for both communications and sensing, because the signal propagation environments between different users and the BS could be different. But these environments could also be correlated. Thus, waveform optimization in the uplink is a more challenging task.

4.3.2.2 Optimization in Time and Frequency Domains

In addition to spatial optimization, communication signals can also be optimized across the time and frequency domains, by jointly considering the communications and sensing requirements. The optimization can be with respect to the frame structure, subcarrier occupation, power allocation, and pilot design and typically requires some slight changes of the mobile signals.

The preamble part in a data frame can be optimized with respect to both the signal format and resource allocation. A typical communication frame consists of preamble and data payload, and in cellular networks, they are structured via logical channels. The preamble typically contains unmodulated and orthogonal signals, which can be directly used for sensing, as described in Section 4.1.3. For a MIMO-OFDM signal, the format of the preamble signals can be designed to mimic the traditional radar waveform, while maintaining the properties required. For example, in [46], orthogonal linear frequency modulation (LFM) signals, which are commonly used in MIMO radar, are generated based on MIMO-OFDM JCAS. A similar LFM signal is also generated in [17], for a JCAS system using a hybrid antenna array. The subcarrier occupation of the preamble can also be designed by incorporating the idea of nonequidistant subcarriers in MIMO-OFDM radar [45] to balance the performance of communications and sensing.

The spatiotemporal power optimization between preamble and data payload is investigated for JCAS in [34]. It is shown that the length of the preamble and its power allocation have a much larger impact on communications than on sensing. Using a cost function as the weighted sum of the MIs for communications and sensing, a closed-form solution is obtained for optimal power allocation.

The interval of preambles or pilots can also be optimized to improve the sensing performance, while maintaining communication efficiency. This can also be realized via varying the length of the data payload over packets, if packets are continuously transmitted. In [44], for a single data-stream single-carrier joint communications and radar (JCR) system based on 802.11ad, nonuniformly placed preambles are proposed to enhance velocity estimation accuracy. It is found that when preambles/pilots are equally spaced, the performance of radar or communications cannot be effectively improved without affecting the other. Comparatively, nonuniform preambles/pilots are found to achieve a better performance trade-off between communications and sensing, particularly at large radar distances. Although developed for a single carrier system, an extension to MIMO-OFDM in PMN is possible with the usage of nonuniform pilots in both frequency and time domains.

4.3.2.3 Optimization with Next-Generation Signaling Formats

Most of communications-centric JCAS systems have been formulated on either signal carrier or OFDM(A) systems, which are consistent with those waveforms used in radar. Joint JCAS waveform design may also be applied to next-generation communication signals currently considered for 6G, such as orthogonal time-frequency space (OTFS) signaling and fast-than-Nyquist (FTN) modulation.

The OTFS signaling is developed to address the signal reception problem in both frequency and time selective channels and is believed to be more effective than OFDM in such channels. The signal may be directly modulated in the so-called "delay-Doppler domain," where the channels are shown to be sparse. Equalization of OTFS signals requires channel estimation, which can be efficiently represented by sparse parameters of delay, angles, and Doppler frequency. Thus, the received signal processing of OTFS can be naturally linked to sensing parameter estimation. In [50], the effectiveness of using OTFS for JCAS is investigated. It is shown that using OTFS signals, like using OFDM, can generate as accurate sensing parameter estimation as using conventional radar waveform such as FMCW. In [51], an OTFS JCAS system is studied by explicitly taking into account the intersymbol interference (ISI) and intercarrier interference (ICI) effects. It is shown that ISI and ICI can be exploited to extend the maximum unambiguous detection limits in range and velocity. A generalized likelihood ratio test-based detector/estimator considering the ISI and ICI effects is developed. To reduce the conventionally high complexity of OTFS sensing, an efficient Bayesian learning scheme is proposed in [52], together with the reduction of the measurement matrix's dimension by incorporating the prior knowledge on the motion parameter limit of the true targets. These works demonstrate the feasibility and potential efficiency of OTFS JCAS systems. Although results on waveform optimization for OTFS JCAS is not available, we surmise that it can be conducted in a way similar to OFDM. In particular, the precoding may be more efficiently applied beyond the spatial domain, such as to the delay-Doppler domain.

4.3.3 Antenna Array Design

For radio sensing, each antenna with an independent RF chain is like a pixel in the camera. But a radio system allows more flexible controlling and processing

Table 4.5 Classification of antenna array design techniques in PMN.

Techniques	Key ideas	Examples
Virtual MIMO and antenna grouping	Group antennas and form virtual subarrays to achieve a balance between spatial multiplexing and diversity for communications, and spatial resolution and beamforming gain for sensing	[17, 23, 53, 54]
Sparse array design	Optimize the number and placement of antenna elements in an array by jointly considering the sparsity in sensing and communication channels	[55, 56]
Spatial modulation	Use the indexes and/or order of antennas to convey information bits. The randomness added to JCAS signals is also shown to improve sensing performance in some cases	[57–59]

of both transmitted and received signals. Therefore, there are more techniques which we can explore for antenna arrays in PMNs beyond the MIMO precoding for waveform optimization as discussed in Section 4.3.2. A classification of these techniques is presented in Table 4.5, including virtual array design, sparse array design, and spatial modulation. These techniques are elaborated on below.

4.3.3.1 Virtual MIMO and Antenna Grouping

There are many contradictory requirements for antenna array design between communications and sensing. Beamforming and antenna placement are two good examples. For beamforming, an array with dynamic beamforming and narrow beamwidth is typically required for sensing; however, during at least a packet period, communications require fixed and accurately pointed beams to obtain nontime-varying channels, and multibeams to support SDMA. For antenna placement, MIMO radar typically requires special antenna intervals to achieve increased virtual antenna aperture [53], while MIMO communications focuses on beamforming gain, spatial diversity, and spatial multiplexing; therefore, low correlation among antennas is more important. These different and contradicting requirements demand new antenna design methods.

One potential solution is to introduce the concept of antenna grouping and virtual subarrays [53, 54]. By dividing existing antennas into two or more groups/virtual subarrays, we can designate tasks of communications and sensing and optimize the design across groups of antennas. There could be an overlap between different groups of antennas, as shown in Figure 4.5. Virtual subarray introduces beamforming capability. Using orthogonal signals across virtual subarrays, we can maintain the orthogonality desired by MIMO radar, in order to achieve a larger aperture of an equivalent virtual array. Using overlapped antennas across neighboring virtual subarrays can increase the spatial degree of freedom of the MIMO radar. Using virtual subarrays, we can also conveniently generate multibeams [23]

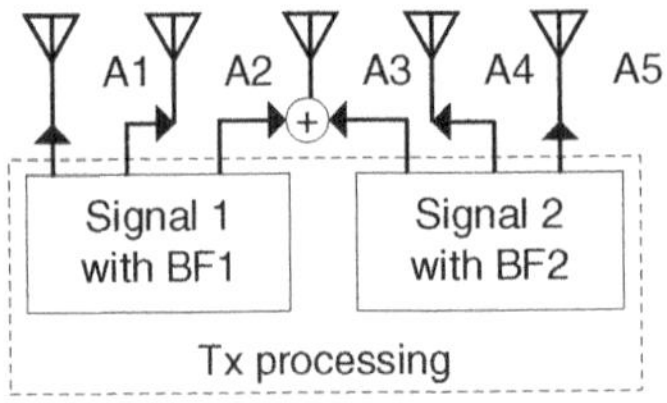

Figure 4.5 Two virtual subarrays with one overlapped antenna are formed: virtual subarray 1 with antennas A1, A2, and A3; and subarray 2 with A3, A4, and A5. Each virtual subarray will transmit one data signal, and the signals between two virtual subarrays are orthogonal over time. Beamforming is applied in each virtual subarray. BF for beamforming.

satisfying different beamforming requirements from communications and sensing. We can also virtually optimize the antenna placement, by antenna selection and grouping. Similar to the diversity and multiplexing trade-off in communications, there is a trade-off between processing gain and resolution in sensing, related to the number of independent spatial streams.

The array structure as shown in Figure 4.5 reminds us of the hybrid antenna array that has been widely studied in mmWave communications. Considering the benefits of antenna grouping for both communications and sensing, using hybrid antenna arrays [60, 61] will be an attractive low-cost option. This is particularly true for mmWave systems where propagation loss is high and beamforming gain is essential for achieving sufficiently high signal-to-noise ratio (SNR) for both communications and sensing. The research on hybrid array JCAS systems is still in its very early stage, with some limited results being reported in [17].

4.3.3.2 Sparse Array Design

Besides antenna grouping, sparse array design is another method to exploit the degrees of freedom that can be achieved via configuring the locations of antennas when the total number of antennas is fixed.

The design of sparse arrays or thinned arrays [62], such as coprime arrays [63], is often cast as optimally placing a given number of antennas on a subset of a large number of (uniform) grid points [55]. In this way, a small number of antennas can span a large array aperture with a high-spatial resolution and low sidelobes. So far, the sparse array design-based JCAS has mainly been studied in integrating communications to radar systems, i.e. embedding information into radar waveforms to perform data communications [55, 56]. In [55], antenna position and beamforming weights are optimized to design beams with mainlobe performing radar detection and sidelobe for communications through modulations such as amplitude shift keying (ASK) or phase shift keying (PSK). In [56], the MIMO waveform orthogonality is further exploited to permute the waveform across selected antenna grids and hence convey extra information bits.

Sparse array design is particularly suitable for massive MIMO arrays with tens to hundreds of antennas, but a limited number of RF chains, i.e. switched arrays or hybrid arrays. This setup can provide more degrees of freedom and potential performance enhancement, with reduced cost, in PMNs. For example, the sparse array design can add index modulation to the communication part, while the sparse array design can provide better spatial resolution for radar detection. To this end, some interesting problems remain to be solved, such as how to formulate the problems with two goals and new trade-offs between communications and sensing.

4.3.3.3 Spatial Modulation

Spatial modulation uses the set of antenna indexes to modulate information bits and has been extensively investigated for communication systems [64]. For multiantenna JCAS systems, spatial modulation can also be potentially applied. In [57, 58], a concept similar to spatial modulation is exploited to increase communication data rate in a frequency-hopping MIMO dual-function(al) radar communications (DFRC) system. In [59], spatial modulation is applied to JCAS by allocating antenna elements based on the transmitted message, achieving increased communication rates by embedding additional data bits in the antenna selection. A prototype is developed in [59] and demonstrates that the proposed scheme can improve the angular resolution and reduce the sidelobe level in the transmit beam pattern compared to using fixed antenna allocations.

Although these works are based on pulsed and continuous-wave radars, they can potentially be extended to PMN, by adding antenna selection to existing space–time modulations. In particular, the rich scattering environment in PMN provides a lower correlation between spatial channels, leading to potentially better performance.

4.3.3.4 Reconfigurable Intelligent Surface-Assisted JCAS

Reconfigurable intelligent surface (RIS) [65], also known as reconfigurable intelligent meta-surfaces or intelligent reflective surface, can be treated as a special type of relaying "antenna" arrays that are deployed to influence the signal propagation. The environmental objects are coated with artificial thin films of electromagnetic and reconfigurable meta-surfaces, which can be controlled to shape radio propagation. RIS provides a large volume of spatial degrees of freedom, which can be typically modeled as adjustable phase shifts. Using RIS can significantly improve communication performance by increasing beamforming gain, reducing interference, and reducing fading; it can also have a notable impact on sensing via generating location-dependent radio fingerprints and directed sensing. Therefore, RIS has been extensively studied for improving the performance of both radar

[66, 67] and communications [68], separately. Although limited, the research on RIS-assisted JCAS is emerging.

One direct application of RIS to JCAS is to treat the phase shifts in RIS as increased degrees-of-freedom in signal optimization. In [69], the precoding matrix of the transmitter and the phase shift matrix of the RIS are jointly optimized, with a problem formulated for maximizing the SNR at the radar receiver under the constrained SNR for communications. In [70], RIS is introduced as additional configurable channel parameters to assist the realization of a JCAS system based on a novel sparse code multiple access scheme.

Another important application of RIS to JCAS is to reduce the potential interference that may be caused by accommodating the sensing requirement, particularly for SDMA communications. In [71], the RIS technology is introduced to mitigate the multiuser interference, which may be increased when optimizing the beams by jointly considering the beamforming requirements for communications and sensing in a JCAS system. The JCAS waveform and the RIS phase shift matrix are jointly optimized, and the trade-off between communications and sensing performance is investigated. The work demonstrates that RIS can significantly improve the system throughput of communications, while without distorting the desired beampattern for sensing. In [72], it is also exploited to deal with the interference in a coexisting MIMO radar and SDMA communication system.

Given the extensive research of RIS in communications and sensing separately, we can expect rapidly growing research outputs on RIS-assisted JCAS. It would be interesting to see more diverse applications of RIS in JCAS, for example, how to use RIS to generate location-dependent radio fingerprints to enable simpler and more accurate sensing, while simultaneously improving the communication performance? Such work can better explore RIS's potential in JCAS, making it a game-changer in JCAS design, rather than just offering more adjustable phase values in an optimization problem.

4.3.4 Clutter Suppression Techniques

Rich multipath in mobile networks creates another challenge for sensing parameter estimation in PMNs. In a typical environment, BSs receive many multipath signals that are originated from permanent or long-period static objects. These signals are useful for communications, but for a fixed BS, they are generally not of interest for continuous sensing because they bear little new information. Such undesirable multipath signals are known as *clutter* in the traditional radar literature. In PMNs, we treat multipath signals as clutter if they remain largely unchanged and have near-zero Doppler frequencies over a period of interest. A lot of clutter could be present in the received signals because of the rich multipath environment of mobile networks. Clutter contains little information and is better

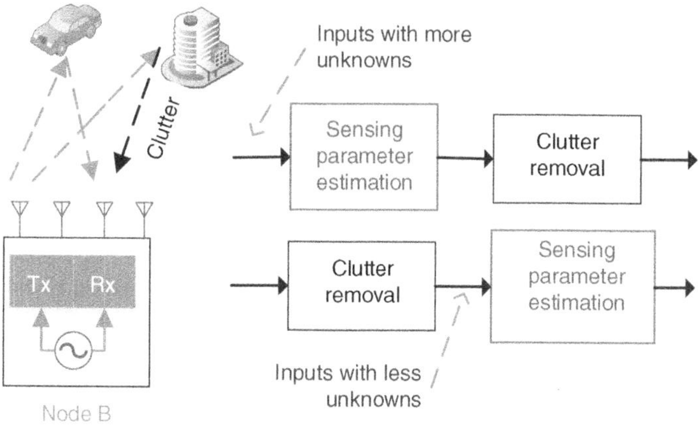

Figure 4.6 Clutter and two ways of clutter suppression.

to be removed from the signals sent to the sensing parameter estimator. In [73], the inner bounds of the impact of clutter on the performance of JCAS is evaluated. It is shown that clutter originating from motion objects can significantly degrade the inner bounds of the performance.

There may be two ways of performing clutter suppression, as shown in Figure 4.6: doing suppression after or before estimating the sensing parameters. The former does not introduce signal distortion for sensing parameter estimation, and the latter can reduce the unknown sensing parameters to be estimated. High-end military/domestic radar can simultaneously detect and track hundreds of objects, and the capability is built on advanced hardware such as huge antenna arrays of hundreds and thousands of antenna elements; thus, both ways can be applied. For a PMN BS with only tens of antennas and limited signal bandwidth, the sensing capability largely depends on the sensing algorithm, which is closely related to the number of unknown parameters. Most existing parameter estimation algorithms require more measurements than unknown parameters and the estimation performance typically degrades with the number of unknown parameters increasing. Therefore, it is crucial to identify and remove noninformation-bearing clutter signals before applying sensing parameter estimation.

Clutter suppression techniques for conventional radars are not directly applicable here because the signals and working environment for the two systems are very different. Typical radar systems are optimized for sensing a limited number of objects in open spaces using narrow beamforming, and clutter is typically from ground, sea, rain, etc., and has notable distinct features. The well-known algorithms in radar systems, such as space-time adaptive processing, independent component analysis, singular value decomposition, and Doppler focusing, are adapted to such scenarios. These techniques also need to exploit different features of desired and unwanted echoes, such as low correlation between them. These different features may not always be available in mobile networks because the desired multipath and clutter can come from the same classes of reflectors. For communications, narrow beamforming may occur in emerging millimeter wave systems, but not in more general microwave radio systems due to the limited number of antennas and the use of multibeam technology to support multiuser MIMO. The signal propagation environment in PMNs can also be very complex and different from a typical radar working environment. Therefore, existing clutter suppression methods developed for radar systems may not be directly suitable for clutter reduction in PMNs.

Alternative approaches exploit the correlation in time, frequency, and space domains, and use recursive averaging or differential operation to construct or remove clutter signals. These approaches could be more viable for PMNs. They have similarities to *background subtraction* in image processing [74]. However, there are two major differences:

- In image processing, the difference between two images is exhibited via pixel variation. In radio sensing, both Doppler shifts and variation in sensing parameters cause differences in received sensing signals at different time.
- In an image, the background is overlapped/covered by foreground. In radio sensing, clutter and desired multipath signals are typically additive, and coexist in the received signals.

Nevertheless, many background subtraction methods developed for image processing can be revised and applied for radio sensing in PMNs. Below we review two types of typical background subtraction techniques that can be used in PMNs: *recursive moving averaging (RMA) and Gaussian mixture model (GMM).*

4.3.4.1 Recursive Moving Averaging (RMA)

Assume sensing parameters are fixed over the coherence time period, then ideally the received signals for each path at two different times will only have a phase difference caused by the Doppler phase shift. If the Doppler frequency is near zero, then the two signals are nearly identical. Based on this assumption, we can use an RMA method [3] to estimate the clutter and then remove it from the received signal.

The received signals as shown in, e.g. (4.9), cannot be directly used in RMA, due to the transmitted signals $\mathbf{x}$ and possibly the unknown timing offset τ_0 which introduce a time-varying phase shift across discontinuous packets. They need to be removed or compensated before RMA can be applied. To demonstrate the idea of RMA, we refer to the signal stripping approach [3]. We assume that $\tau_0 = 0$ and an estimate of the channel matrix $\mathbf{H}_n$ is available, given by $\hat{\mathbf{H}}_n = \mathbf{H}_n + \mathbf{\Delta}_n$, where $\mathbf{\Delta}_n$ is the reconstruction error.

The RMA method uses a small forgetting vector to recursively average the received signal over a window, with a length sufficiently large to allow suppressing time-varying signals of nonstatic paths, but smaller than the coherent time. Mathematically, it can be represented as a recursive equation:

$$\bar{\mathbf{H}}_n(i) = \alpha\bar{\mathbf{H}}_n(i-1) + (1-\alpha)\hat{\mathbf{H}}_n(i), \tag{4.8}$$

where $\bar{\mathbf{H}}_n(i)$ is the output at the ith iteration, α is the forgetting factor (learning rate), and the initial $\hat{\mathbf{H}}_n(1)$ can be either zero or computed as the average of several initial $\hat{\mathbf{H}}_n(i)$s. In RMA, the window length can be adapted to the variation speed of the channels. The time interval between the inputs to the averaging determines how signals with different Doppler frequencies are added, either constructively or destructively. Hence, it has a significant impact on suppressing signals of different Doppler frequencies. The forgetting factor and the window length determine the suppression power ratio. Although experimental results have been reported in [3] for the relationship between these parameters and the effect of clutter estimation and suppression, optimal combinations of these parameters, in consideration of channel statistical properties, are yet to be studied.

Although the RMA method works well in principle, it may become inefficient due to practical issues, such as timing and frequency offset commonly existing in actual systems. These signal imperfectness needs to be well compensated before the RMA method can work effectively.

4.3.4.2 Gaussian Mixture Model (GMM)

GMM has been widely used for analyzing and separating moving objects from the background in image and video analysis, target identification and classification in radar system, and positioning solutions. The statistical learning of the GMM model with respect to the mean and variance in background subtraction is used to determine the state of each pixel whether a pixel is background or foreground. It has also been applied recently to extract static channel state information (CSI) from channel measurement in [75]. Different from GMM in video analysis, where background and foreground overlap each other, clutter and multipath of interest in PMNs are additive and can coexist. Therefore, it is infeasible for PMNs to place foreground (dynamic signals) and background (static signals) into two different sets by classical clustering approaches that happened in image or video signal processing.

GMM's working principle for clutter suppression in PMNs is as follows: Wireless channels can be modeled and estimated by a mixture of Gaussian distributions since each density represents the distribution of paths in the channel [75, 76]. Static and dynamic paths can be represented by Gaussian distributions with very different parameters over the time domain. This is because over a short time period, static paths change little and dynamic paths may vary significantly. It is also quite common that static paths typically have larger mean power than dynamic ones. Hence, in terms of their distributions, static paths have near-zero variances, which are much smaller than those of the dynamic ones. Therefore, by learning the mean values of the distribution, static paths can be identified and separated via comparing the variance.

The main advantage of GMM for clutter estimation in PMN is that much less samples are required to achieve a given accuracy, compared to the matched filtering and RMA methods. However, the estimation usually needs to be realized by high-complexity algorithms such as expectation maximization. Low-complexity estimation based on the GMM formulation is a key research problem here.

4.3.5 Sensing Parameter Estimation

Sensing tasks in PMNs include both explicit estimation of sensing parameters for locating objects and estimating their moving speeds, and application-oriented pattern recognition such as object and behavior recognition and classification. In this section, we review research on sensing parameter estimation, considering typical multiuser-MIMO OFDM signals used in modern mobile networks. We will review work on pattern recognition in Section 4.3.7.

Referring to the CRAN-based PMN presented in Section 4.1.2, let us formulate the sensing parameter estimation problem first. For downlink sensing, each RRU sees reflected downlink signals from itself and the other $Q-1$ RRUs. For any RRU, a general expression of its received signal at the nth subcarrier and the kth OFDM block is given by

$$\tilde{\mathbf{y}}_{n,k} = \sum_{q=1}^{Q}\sum_{\ell=1}^{L_q} b_{q,\ell} e^{-j2\pi n(\tau_{q,\ell}+\tau_{q,o})f_0} e^{j2\pi k f_{D,q,\ell} T_s} \cdot \mathbf{a}(M,\phi_{q,\ell})\mathbf{a}^T(M,\theta_{q,\ell})\tilde{\mathbf{x}}_{q,n,k} + \mathbf{z}_{n,k}, \tag{4.9}$$

where variables with subscript q are for the qth RRU, $\tilde{\mathbf{x}}_{q,n,k}$ are the transmitted signals at subcarrier n from the qth RRU, and $\mathbf{z}_{n,k}$ is the noise vector. There is typically a common clock between the transmitter and the receiver in one RRU, and between RRUs, hence the timing offset $\tau_{q,o}=0$. When RRUs' signals do not reach each other, then $Q=1$ in (4.9), and each RRU only sees its own reflected transmitted signals.

According to (4.9), we can see that packing $\tilde{\mathbf{y}}_{n,k}$ from multiple RRUs can increase its length, but the unknown parameters are similarly increased. Hence, sensing may not directly benefit from jointly processing signals from multiple RRUs, although there may be some correlation between sensing parameters for different RRUs. However, due to channel reciprocity, parameters for signal propagation between RRUs could be similar. Such a property can be exploited for joint processing across RRUs.

For uplink signal transmitted from up to U UEs, the received signal at a RRU has a similar form with (4.9), with Q being replaced by U. For uplink sensing, the received signals for sensing and communications are identical.

The task of sensing parameter estimation is to estimate the parameters $\{\tau_{q,\ell}, f_{D,q,\ell}, \phi_{q,\ell}, \theta_{q,\ell}\}$, and possibly $b_{q,\ell}$, from the received signals, in the presence of the signals $\tilde{\mathbf{x}}_{q,n,k}$. There are two methods, direct and indirect sensing that can be applied, as described in Chapter 2. The direct method applies sensing algorithms to the received signals, without removing $\tilde{\mathbf{x}}_{q,n,k}$ or separating channels for SDMA UEs. The indirect method, on the contrary, removes $\tilde{\mathbf{x}}_{q,n,k}$ and separate channels. They have respective advantages and disadvantages. Direct sensing has higher computational complexity and the applicable sensing algorithms are limited; while indirect sensing may introduce large noise during the removal of $\tilde{\mathbf{x}}_{q,n,k}$.

For either direct or indirect sensing, we may apply one or more sensing parameter estimation algorithms that can be classified as follows: periodogram such as 2D discrete Fourier transform (DFT), subspace-based spectrum analysis techniques, on-grid compressive sensing (CS) algorithms, off-grid CS algorithms and grid densification, and Tensor tools. Most of these techniques have higher complexity than classical channel estimation algorithms. Since the required sensing rate is typically in the order of milliseconds to seconds, such high-computational complexity is affordable at BSs. A comparison of these techniques for sensing parameter estimation in PMNs is summarized in Table 4.6. Details of some common techniques are elaborated below.

4.3.5.1 Periodogram such as 2D DFT

The classical 2D DFT method is a periodogram method being widely used in radar. It can be used to coarsely estimate sensing parameters by combining two of the following three transformations: converting the time-domain samples to the frequency domain, spatial-domain samples to angle domain, and phase shifting samples to Doppler frequency domain. A 3D DFT may also be used. But due to the complexity, it is generally replaced by two or three 2D DFTs. The resolution of this method is low because of the long tail of the inherent sinc function in the DFT. A windowing operation can be applied to slightly improve the resolution. This method typically requires a full set of continuous measurements in time or frequency domain, which can limit its application in PMNs due to the discontinuous samples.

4.3.5.2 Subspace-Based Spectrum Analysis Techniques

Classical subspace-based spectrum analysis techniques such as MUSIC and ESPRIT can estimate parameters of continuous values with high resolution.

Table 4.6 Classification and comparison of Sensing Parameter Estimation Algorithms.

Algorithms	Properties	Suitability and main limitation
Periodogram such as *2D DFT and other Fourier transform-based techniques*	Traditional techniques. Simple to implement. May be used as the starting point for other algorithms	Low resolution. Generally, requires a full set of continuous samples in all domains, which may not always be satisfied
Subspace methods such as MUSIC, ESPRIT, and Matrix Pencil	High resolution and can do off-grid estimation. High complexity. Work with a small number of measurements	Typically require a large segment of consecutive samples, with the exclusion of the MUSIC algorithm, which may not always be satisfied
Compressive sensing (On-grid)	Flexible. Does not require consecutive samples. Many options of problem formulations that can be tailored to signal models. Various recovery algorithms can be selected to adapt to complexity and performance requirements	Works well even for estimating a small amount of off-grid parameters. Performance can degrade significantly with many paths of continuous parameter values
Compressive Sensing (Off-grid) such as atomic norm minimization	Flexible and do not require consecutive samples. Capable of estimating off-grid values	Limitation in real time operation due to very high complexity. Still require sufficient separation between parameter values
Tensor-based algorithms	High-order formulation using the Tensor tools such as 3D Tensor CS simplifies computational complexity and provides capability in resolving multipath with repeated parameter values. Improve SNR by combining multiple measurements coherently	Need to be combined with other algorithms such as ESPRIT and CS. Thus, they face the inherent problems of these algorithms

However, the sensing accuracy of MUSIC depends on the searching granularity, and ESPRIT typically requires samples of equal intervals. Techniques that can deal with nonuniform sampling have been proposed, e.g. the coupled canonical polyadic decomposition approach in [77] and the generalized array manifold separation approach in [78], but they have very high-computational complexity. To achieve high resolutions, MUSIC and ESPRIT typically also require a large number of samples so that the signal subspace and noise subspace can be well separated via computing the signal correlation matrix. This may not always be available in some domains, such as the spatial domain, which would require a large number of antennas. However, it may be a good option to combine them with other techniques for sensing parameter estimation, by exploiting their capabilities of high resolution and estimating parameters of continuous values. One good example can be seen from [79], where MUSIC is first used to obtain angle estimates via the spatial correlation matrix, and the delay and Doppler are then estimated by forming a maximum likelihood estimation problem.

4.3.5.3 On-Grid Compressive Sensing Algorithms

CS techniques have been widely used in communication systems for channel estimation and in radar systems. CS techniques formulate parameter estimation as a sparse signal recovery problem, which can be solved by many algorithms such as l_1 recovery (convex relaxation), greedy algorithms and probabilistic inference. At the least, only twice the number of samples are required to accurately recover a certain number of unknown parameters, in the noise-free case. Typical CS techniques use on-grid quantized dictionaries, and hence, errors are caused due to quantization when the original parameters have continuous values. One main advantage of CS for sensing parameter estimation in PMNs is that it does not require consecutive samples. Actually, higher randomness of samples in time, frequency, and spatial domains can generally lead to better estimation performance.

We now demonstrate how a CS problem can be formulated from the received signals, referring to (4.9). Let N_u and S be the number and index set of available subcarriers for sensing, respectively. For downlink sensing, $N_u = N$. We assume that $N \gg L$ and N is large enough such that the quantization error of τ_ℓ is small and the delay estimation can be well approximated as an on-grid estimation problem. Let the delay term $e^{-j2\pi n\tau_\ell f_0}$ be quantized to $e^{-j2\pi n\ell/(gN)}$, where g is a small integer and its value depends on the method used for estimating τ_ℓ. The minimal delay quantization is then $1/(gB)$. We can now convert (4.9) to a generalized on-grid (delay only) sparse model, by representing it using $N_p \gg L, N_p \leq gN$ multipath signals where only L signals are nonzeros:

$$\tilde{\mathbf{y}}_{n,k} = \underbrace{\mathbf{A}(M, \boldsymbol{\phi})\mathbf{C}_n\Delta_k\mathbf{B}\mathbf{U}^T}_{\tilde{\mathbf{H}}_n}\tilde{\mathbf{x}}_{n,k} + \mathbf{z}_{n,k}, \tag{4.10}$$

with

$$\mathbf{A}(M, \boldsymbol{\phi}) = (\mathbf{A}_1(M, \boldsymbol{\phi}_1), \dots, \mathbf{A}_Q(M, \boldsymbol{\phi}_Q)),$$
$$\tilde{\mathbf{x}}_{n,k} = (\tilde{\mathbf{x}}_{1,n,k}^T, \dots, \tilde{\mathbf{x}}_{Q,n,k}^T)^T,$$
$$\mathbf{U} = \text{diag}\{\mathbf{A}_1(M, \theta_1), \mathbf{A}_2(M, \theta_2), \dots, \mathbf{A}_Q(M, \theta_Q)\}.$$

Here, $\mathbf{C}_n = \text{diag}\{e^{-j2\pi n/(gN)}, \dots, e^{-j2\pi nN_p/(gN)}\}$; $\mathbf{B}$ is a $N_p \times L$ rectangular permutation matrix that maps the signals from a user/RRU to its multipath signal, and has only one nonzero element of value 1 in each row; $\mathbf{U}$ is an $M_TQ \times L$ block diagonal radiation pattern matrix for M_T arrays; $\tilde{\mathbf{x}}_{n,k}$ is the $M_TQ \times 1$ symbol vector; the ℓth column in $\mathbf{A}_q(M, \boldsymbol{\phi}_q)$ (or $\mathbf{A}_q(M, \boldsymbol{\theta}_q)$) is $\mathbf{a}(M, \phi_{q,\ell})$ (or $\mathbf{a}(M, \theta_{q,\ell})$); and Δ_t is a diagonal matrix with the diagonal element being $b_\ell e^{j2\pi k f_{D,\ell} T_s}$. Note that the columns in $\mathbf{A}(M, \boldsymbol{\phi})$ of size $M \times N_p$ and the diagonal elements in Δ_k of size $N_p \times N_p$ are ordered and tied to the multipath delay values.

Based on (4.10), the signal samples from all available subcarriers and OFDM blocks can then be combined to formulate CS estimation problems. The sensing parameters to be estimated in PMNs include delay, AoA, and Doppler in three different domains. Sometimes, the angle-of-departure (AoD) and magnitude of a path are also of interest, which are not considered here. Since the signals are relatively independent in the three domains, they can be formulated in a high-dimension (3D here) vector Kronecker product form or even Tensor form. Therefore, we can apply 1D to 3D CS techniques to estimate these sensing parameters [80]. The following two problems need to be considered when selecting CS techniques of different dimensions.

- *Quantization error and number of available samples*: Although high-dimensional, on-grid CS algorithms such as the Kronecker CS could offer better performance, they require more samples than unknown variables in each dimension. In a typical BS, we can get a sufficient number of observations for the delay (linked to subcarriers), a reasonable number of samples in the Doppler frequency domain (linked to intermittent packets over a segment of channel coherent period), and a limited number of AoA observations (linked to antennas).
- *Complexity*: Exploiting the Kronecker CS property, the computational complexity is in the order of the product of the complexity in each domain, which is typically proportional to the cube of the number of samples.

Therefore, a high-dimensional CS algorithm is not always the viable option, particularly for the Doppler frequency and AoA estimation due to the limited number of samples. Comparatively, mobile signals generally have tens to thousands of subcarriers, which provide numerous samples for delay estimation. Thus, we can formulate two multimeasurement vector (MMV) CS problems, by stacking spatial-domain and Doppler-frequency domain signals, respectively, with frequency domain signals. From the MMV-CS amplitude estimates, we can then estimate the AoA and Doppler frequencies [3, 18]. The details of CS algorithms from 1D to 3D and their performance are presented in [18, 80]. One common problem associated with using lower-dimension CS is that parameters with overlapped values in one or more dimensions cannot be separately estimated. In this case, techniques such as the one proposed in [81] can be used, by taking advantage of the capability of model-based algorithms, for example modified matrix enhancement and matrix pencil.

Overall, on-grid CS algorithms are promising for sensing parameter estimation in PMNs. However, the quantization error is a major problem as true sensing parameters have continuous values. For parameters of continuous values, there exists a mismatch between the assumed and actual dictionaries, generally known as "dictionary mismatch," which can cause significant performance degradation [82]. The degradation is severer when the number of unknown variables is larger. Therefore, resolving the quantization error and dictionary mismatch is a major challenge here.

4.3.6 Resolution of Sensing Ambiguity

As discussed in Section 4.2, there is typically no clock-level synchronization between a sensing receiver and the transmitter in PMNs, particularly in uplink sensing. In this case, there exist both timing and carrier frequency offsets in the received signals. The timing offset is illustrated in Figure 4.7. Both of them, as shown in (2.2), are typically time-varying due to oscillator stability. In communications, timing offset can be absorbed into channel estimation and Carrier frequency offset (CFO) can be estimated and compensated. Their residuals become sufficiently small and can be ignored. Differently, in sensing they cause measurement ambiguity and accuracy degradation. Timing offset can directly cause timing ambiguity and then ranging ambiguity, and CFO can cause Doppler estimation ambiguity and then speed ambiguity. They also prevent aggregating signals from discontinuous packets for joint processing, as they cause unknown and random phase shifting across packets or CSI measurements. Thus, it is very important to resolve the clock timing offset problem. Should it be solved, uplink sensing can be efficiently realized, requiring little changes of network infrastructure.

There exist three categories of techniques to resolve the sensing ambiguity due to clock asynchronism: using a global reference clock, single-node-based, and

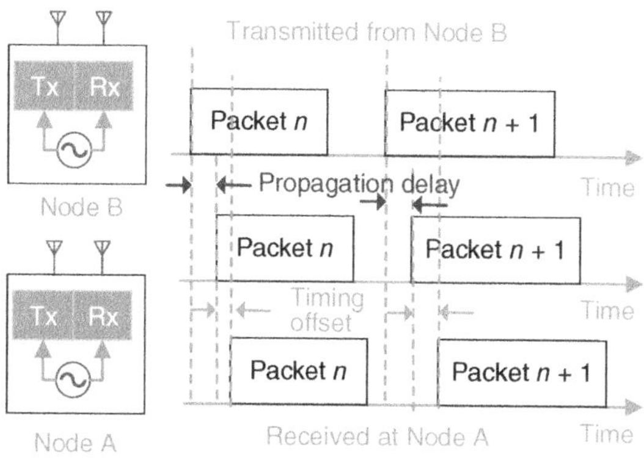

Figure 4.7 Illustration of the propagation delay τ_ℓ and timing offset $\tau_o(t)$. $\tau_o(t)$ is time varying due to instability of the oscillators' clocks.

network-based solutions. The first relies on an external accurate reference clock, such as the one extracted from GPS. The second can be implemented in a single receiver, exploiting the fact that multiple receiver antennas have a common local clock and, hence, their signals experience the same clock offset. The third exploits measurements from multiple cooperative nodes and relies on, e.g. triangulation technique to remove the offset. Below we mainly discuss the second technique which is relatively simpler to implement.

There have been a limited number of works that address this problem in passive sensing [83–85] based on the technique. A cross-antenna cross-correlation (CACC) method is applied to passive Wi-Fi-sensing to resolve the sensing ambiguity issue. The basic assumption is that timing offsets and CFO across multiple antennas in the receiver are the same because the common oscillator clock is used in the RF circuits for all antennas. Hence, they can be removed by computing the cross-correlation between signals from multiple receiving antennas. However, cross-correlation causes increased terms and unknown parameters. The sensing parameters also become relative ones.

Considering a single transmitter with a single antenna and referring to (4.9), the received noise-free signal at the mth antenna can be rewritten as follows:

$$\tilde{y}_{n,k,m} = \sum_{\ell=1}^{L} b_\ell e^{-j2\pi n(\tau_\ell+\tau_{o,k})f_0} e^{j2\pi k(f_{D,\ell}+f_{o,k})T_s} e^{jm\pi \sin(\phi_\ell)} \tilde{x}_{n,k}.$$

Let the m_0th antenna be the reference. Computing the cross-correlation between $\tilde{y}_{n,k,m}$ and $\tilde{y}_{n,k,m_0}$ yields

$$\begin{aligned} R(n,k,m) &= \tilde{y}_{n,k,m}\tilde{y}^*_{n,k,m_0} \\ &= \sum_{\ell_m=1}^{L}\sum_{\ell_{m_0}=1}^{L} b_{\ell_m} b^*_{\ell_{m_0}} e^{-j2\pi n(\tau_{\ell_m}-\tau_{\ell_{m_0}})f_0} e^{j2\pi k(f_{D,\ell_m}-f_{D,\ell_{m_0}})T_s} e^{jm\pi(\sin(\phi_{\ell_m})-\sin(\phi_{\ell_{m_0}}))} |\tilde{x}_{n,k}|^2. \end{aligned} \tag{4.11}$$

Note that in (4.11), $\tau_{o,k}$ and $f_{o,k}$ are removed. However, cross-correlation causes doubled terms and sensing parameters become relative.

To proceed with the estimation of sensing parameters, it is assumed that line-of-sight (LoS) path exists and has much larger magnitude than the non-line-of-sight (NLoS) paths. The UE and BS are assumed to be static and the relative location of the UE is also assumed to be known to the BS. In this case, terms multiplied with the LoS path are much larger than the others. The cross-product between LoS paths contain static multipath only and is invariant over the channel coherent time. Thus, it can be removed by passing the correlation output through a high-pass filter. The cross-terms between NLoS and LoS paths thus dominate in the output of the filter. The sensing parameters can then be estimated, with respect to the known parameters of the LoS path.

The CACC method has been successfully used in Wi-Fi-sensing. In [85], CACC is used to obtain estimates for ranges and velocities of targets. In [84], CACC is adopted to get the *AoA spectrum*, which represents the probabilities of the direction or angle of target. However, the outputs after CACC contain cross-product terms, which doubles the number of unknown parameters to be estimated. In particular, the relative delays and Doppler frequencies have values symmetric to zero. These increased symmetric terms are known as image components, which causes sensing ambiguity and degrades the performance of sensing algorithms. The authors in [84] proposed a method to suppress image signals, by adding one constant value to and subtracting another from to signals between the cross-correlation operation. The method works to some extent; however, it is found to be susceptible to the number and power distribution of static and dynamic signal propagation paths. Therefore, although the idea of CACC looks attractive in resolving the sensing ambiguity problem, more advanced techniques need to be developed to handle the output signals from CACC.

The required setup to enable CACC working is practical in PMNs. For example, the fixed nodes that receive fixed broadband service are ideal options. Those with LoS-path links can also be found, particularly when mmWave communications is deployed. It is generally more challenging to extend this method to more complicated signal formats. In PMNs, the transmitted signals may also be optimized to enable better implementation of the CACC method. In [14], a mirrored MUSIC algorithm is proposed to handle the image components in the CACC output. Noticing the symmetry of unknown parameters, new signals and basis vectors are constructed by adding the original ones with their sample-reversed versions. This equivalently reduces the unknown parameters by half and improves the estimation performance.

Another method for removing the random phase shifts due to clock asynchronism is using the ratio of CSI measurements across antennas [86, 87]. The advantages of doing this, compared to the CACC method, are that (i) the measurement noise can be largely suppressed [86]; and (ii) the ratio may be better used as more information can be maintained compared to the cross correlation [86, 87]. In [86], a close relationship is established between the ratio of CSI measurements and target movement, which enables the determination of movement direction and distance via the changes of the ratio. This method may be widely used in PMN sensing in the presence of clock asynchronism. For the particular respiration-sensing application in [86], it is shown that the sensing range can be significantly extended with very high accuracy. The work is extended in [87], where respiration sensing for multiple objects is studied. To be able to separate the measurement CSI signals for multiple persons, a blind source separation technique based on the independent component analysis (ICA) is applied. The ICA technique requires that the sources are mixed linearly. The direct ratio of

CSI signals possesses nonlinearity and needs to be modified. In [87], a filter is designed through the genetic algorithm to nullify respiration signals with only static background signals left. Since the timing offset appears as a common phase shift in both the originally received and background signals, the filtered signal can then be used as the denominator in the CSI ratio to suppress the timing offset in the respiration signals. In this way, the modified ratios form linear combinations in the total observations, and ICA can be applied. These techniques may find good applications in PMN, particularly when the main goal is to coherently use CSI measurements for sensing, and/or for extracting the Doppler frequencies.

4.3.7 Pattern Analysis

Using radio signals, high-level application-oriented object, behavior and event recognition and classification can be achieved by combining machine learning and signal processing techniques. They can be realized with or without using the sensing parameter estimation results, which provide location and velocity information.

Although the work on pattern analysis using mobile signals is still in its infancy stage, we have seen some interesting examples, such as [88–90]. We can foresee its booming in the near future, as we have been observing from many successful Wi-Fi sensing applications. Using Wi-Fi signals for object and behavior recognition and classification has been well demonstrated [7, 91, 92]. Mobile signals are more complicated than Wi-Fi signals, and the outdoor propagation environment is also more challenging. However, the PMNs have more advanced infrastructure than Wi-Fi systems, including larger antenna arrays, more powerful signal processing capability, and distributed and cooperative nodes. Using massive MIMO, a PMN BS equivalently possesses a massive number of "pixels" for sensing. It is able to resolve numerous objects at a time and achieve imaging results with better field-of-view and resolution, like optical cameras.

Based on the various approaches developed for Wi-Fi sensing, we can deduce the procedures of applying pattern analysis to mobile signals, as shown in Figure 4.8. They typically involve four steps: signal collection, signal preprocessing, feature extraction, and recognition and classification. In the signal collection step, the signals are collected at the receiver according to the desired rate. In the signal preprocessing step, the collected signals may be stripped, cleaned, and compressed. Signal stripping removes the modulated symbols from the received signal, and hence, the pure CSI is obtained. Multiuser signals may also be decorrelated here. Signal cleaning removes signal distortions associated with, e.g. timing, CFO, and phase noise, and suppresses clutter signals. The purpose is to keep mostly information-carrying signals. Many of the algorithms described before can be applied for this purpose. If signals arrive irregularly in

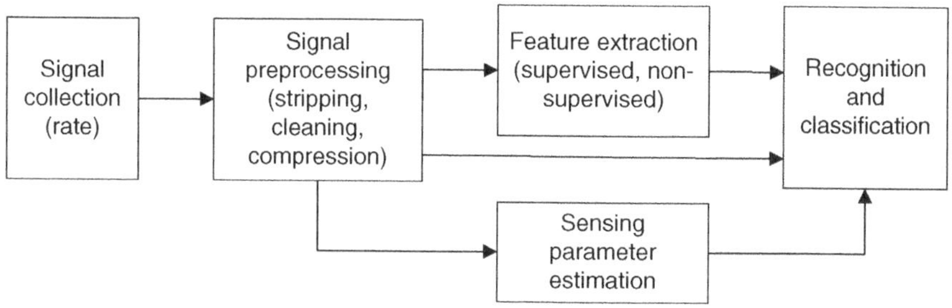

Figure 4.8 Block diagram showing the procedure for pattern recognition.

the first step, the CSI can also be interpolated here if desired. Signal compression makes the signal condense, so that the useful information can be enhanced and the processing complexity in the following steps can be reduced. Common compression techniques include principal component analysis and correlation [93]. Feature signals are then extracted from preprocessed signals, using machine learning techniques such as supervised and nonsupervised deep learning. Finally, recognition and classification are conducted, with inputs from the extracted feature signals, the preprocessed signals, and estimated sensing parameters.

4.3.8 Networked Sensing under Cellular Topology

PMNs provide great opportunities for radio sensing under a cellular structure, which could be well beyond the scale and complexity of distributed radar systems. The main challenge for networked sensing under a cellular topology remains in the way to address competition and cooperation between different nodes for sensing performance characterization and algorithm development. The research in this area is almost blank at the moment. Here, we envision two potential research directions.

4.3.8.1 Fundamental Theories and Performance Bounds for "Cellular Sensing Networks"

This is about investigating the potentials of the cellular structure on improving the spectral efficiency and performance of sensing and developing fundamental theories and performance bounds for such improvement. Similar to communications, a cellular network intuitively also increases frequency reuse factor and hence, the overall "capacity" for sensing. Stochastic geometry model may be an excellent tool for analyzing the dynamics in the sensing network, as have been applied to characterize the aggregated radar interference in an autonomous vehicular network in [94, 95]. New metrics such as the *radar detection coverage probability* [96, 97] can also be introduced to characterize the sensing performance with the application of stochastic geometry models. Both intracell and intercell interference would then be taken into consideration in deriving the MI for networked sensing.

4.3.8.2 Distributed Sensing with Node Grouping and Cooperation

One way of exploiting networked sensing is to develop distributed and cooperative sensing techniques by scheduling and grouping UEs and enable cooperation between RRUs. On the one hand, existing research has shown that distributed radar techniques can improve location resolution and moving target detection by providing large spatial diversity and wide angular observation [2]. Such diversity can be maximized by optimizing both waveform design and the placement of radar nodes. In PMNs, we can group multiple UEs' sensing results to improve uplink sensing. On the other hand, distributed radar can enable high-resolution localization, exploiting coherent phase differences of carrier signals from different distributed nodes [98]. This requires phase synchronization among radar nodes and can only be potentially achieved in downlink sensing by grouping RRUs. For both cases, we may develop distributed sensing techniques, leveraging extensive research works on distributed beamforming and cooperative communications. One example of cooperative detection and localization/sensing is depicted in [4], where a cooperative target detection scheme based on a generalized likelihood ratio test detector is proposed. The impact of network dynamics, including position uncertainty and the number of collaborative nodes, on the detection performance is also evaluated. Another example can be seen from [99], where cooperative passive coherent location is explored in the 5G radio network. The timing-ambiguity problem, as mentioned in Section 4.3.6, may also be efficiently resolved via cooperative sensing, in a way analogous to the triangulation method for the removal of timing asynchronism in current localization systems [100].

4.3.9 Sensing-Assisted Communications

When communications and sensing are integrated, it is important to understand how they can mutually benefit from each other. In the context of PMNs, at least the following techniques can be exploited to improve communications using sensing results: sensing-assisted beamforming and secure communications.

4.3.9.1 Sensing-Assisted Beamforming

Beamforming is an important technique used for concentrating transmission at certain directions to achieve high antenna array gain and is critical for mmWave systems. However, due to the narrow beam width, it is generally time-consuming in mmWave communication systems to find the right beamforming directions and update the pointing directions once the LoS propagation channel is blocked. Techniques exploiting the propagation information of sub-6 GHz signals have been proposed for improving beamforming speed for mmWave communications in mobile networks [101]. These techniques exploit the spatial correlation between channels for the two frequency bands, which, however, is site-specific and needs to be updated in real time because of the environmental dynamics. The translation may also be inaccurate because of the large difference in the signal wavelength between the two bands. Comparatively, using radar operating in a similar mmWave band can potentially provide detailed propagation information, which is ideal for beamforming update and tracking.

In [102, 103], radar-aided beam alignment is investigated for mmWave communications, where a mmWave radar is installed on a BS, colocating with the communication system. A compressive covariance estimation algorithm is proposed to estimate the covariance output at the output of the combiner of a mmWave hybrid receiver, from the covariance matrix of the signals received from radar. Simulation results demonstrate the feasibility of the proposed scheme, while there is a relatively large gap to the upper-bound performance. In [103], it is further shown that using location information obtained via automotive radar can largely reduce the beam training overhead in mmWave communications. Since different frequency bands and two separate systems are used, there are always some limits in the similarity of the channels, and hence the performance is not ideal.

Such limits can be removed by using the mmWave JCAS technology, as both radar sensing and communications are integrated to the same device and use the common signal now. JCAS enables a BS to construct a radio propagation map that limits to one hop in general. This map can provide detailed information for generating initial beamforming and updating it when either the BS or the UE moves, using downlink and uplink sensing [104]. In particular, the multibeam scheme in [23, 41] introduces protocols and algorithms to enable communications and sensing in different directions at the same time even with an analog antenna array, as shown in Figure 4.9. This makes it possible for a JCAS transmitter or a receiver to scan the surrounding environment and update the propagation map while maintaining the communications. The basic idea of the multibeam scheme is to generate beamforming with multiple subbeams, consisting of fixed subbeams primarily for communications and packet-varying subbeams primarily for scanning and sensing. Note that the same data signals are transmitted at the different subbeams here. The basic subbeams can be pregenerated and are combined to readily generate the multibeam in real-time by optimizing the combination coefficients, using the optimization techniques described in Section 4.3.2. The idea can also be extended to hybrid and full-digital arrays.

In addition to achieving environmental awareness, radar sensing in PMNs can also be used for predictive beamforming and tracking. In [105], a predictive beamforming design for a multiuser vehicular-to-infrastructure (V2I) network is proposed. The design introduces a different frame structure, by removing downlink

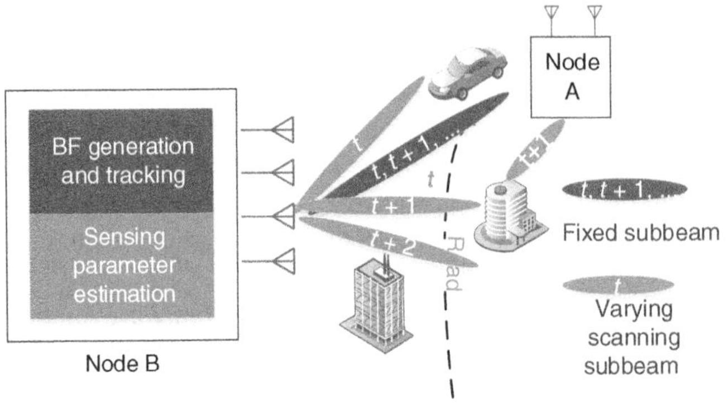

Figure 4.9 Sensing-assisted communications using multibeam where a fixed subbeam points at the communications receiver, and a scanning subbeam varies pointing directions to sense the environment. The multibeam can be realized in all of analog, hybrid, and full-digital arrays. The sensing parameter estimation module in Node B can use the echo signals from both the fixed and the scanning subbeams to establish a radio propagation map.

pilots and uplink feedback parts in conventional training/tracking frames, because the BS (RSU) can actively sense the locations of the vehicles. An extended Kalman filtering (EKF) method is proposed for the tracking and predicting of the state of vehicles based on the sensing results. Thus, instead of conventional beamforming prediction and tracking methods that are purely based on received signals, sensing information on vehicle location and signal propagation paths enables more efficient and direct prediction. To show this, let us consider a mmWave RSU equipped with M_T transmit and M_R receive antennas, which acts as a monostatic radar, and is serving a single-antenna vehicle driving at a nearly constant speed on a straight road. We assume that the RSU communicates with the vehicle over a single LoS path and that all the antenna arrays are adjusted to be parallel to the road. As a consequence, the AoA equals to the AoD in the V2I LoS channel.

At the kth epoch, the RSU transmits a JCAS signal $\mathbf{x}_k(t) = \mathbf{f}_k s_k(t)$ from the RSU to the vehicle, with $\mathbf{f}_k$ being the JCAS beamformer, and $s_k(t)$ being the data stream. The signal is partially received by the vehicle's antenna array and is partially reflected back to the RSU. The received echo signal can be expressed in the form of

$$\mathbf{y}_R(t) = b_k e^{j2\pi f_{D,k} t} \mathbf{a}\left(M_R, \theta_k\right) \mathbf{a}^T\left(M_T, \theta_k\right) \cdot \mathbf{f}_k s_k\left(t - \tau_k\right) + \mathbf{z}_R(t), \tag{4.12}$$

where the beamformer $\mathbf{f}_k$ is designed based on a predicted angle, which is $\mathbf{f}_k = \mathbf{a}^*\left(M_T, \hat{\theta}_{k|k-1}\right)$. Moreover, $\hat{\theta}_{k|k-1}$ is the kth predicted angle based on the $(k-1)$th estimate, and b_k, $f_{D,k}$, θ_k, and τ_k denote the reflection coefficient, the Doppler frequency, the AoA, and the round-trip delay for the vehicle at the kth epoch, respectively. In particular, $f_{D,k}$ and τ_k can be further written as functions of the distance d_k, the velocity v_k, and the AoA θ_k, which are $f_{D,k} = \frac{2v_k \cos\theta_k f_c}{c}$, $\tau_k = \frac{2d_k}{c}$.

By matched-filtering (4.12) with a delayed and Doppler-shifted counterpart of $s_k(t)$, one obtains the estimates of the Doppler and the time-delay, denoted as $\hat{f}_{D,k}$ and $\hat{\tau}_k$. Let us denote the vehicle's state as $\mathbf{q}_k = \left[\theta_k, d_k, v_k, b_k\right]^{\mathrm{T}}$. The sensing measurement and the vehicle state can be connected by a function $\tilde{\mathbf{r}}_k = \mathbf{h}\left(\mathbf{q}_k\right) + \mathbf{z}_k$, which can be expanded as [105]

$$\begin{cases} \tilde{\mathbf{y}}_R = b_k E_k \mathbf{a}\left(M_R, \theta_k\right) \mathbf{a}^T\left(M_T, \theta_k\right) \mathbf{a}^*\left(M_T, \hat{\theta}_{k|k-1}\right) + \mathbf{z}_\theta, \\ \hat{f}_{D,k} = \dfrac{2v_k \cos\theta_k f_c}{c} + z_f, \\ \hat{\tau}_k = \dfrac{2d_k}{c} + z_\tau, \end{cases} \tag{4.13}$$

where E_k is the matched filtering gain, and $\mathbf{z}_k = \left[z_\theta^T, z_f, z_\tau\right]^{\mathrm{T}}$ denotes the measurement noise.

Recalling the assumption that the vehicle is driving on a straight road with a nearly constant speed, the state transition can be modeled as $\mathbf{q}_k = \mathbf{g}\left(\mathbf{q}_{k-1}\right) + \boldsymbol{\omega}_k$, which can be expressed in the form of [105]

$$\begin{cases} \theta_k = \theta_{k-1} + d_{k-1}^{-1} v_{k-1} \Delta T \sin\theta_{k-1} + \omega_{\theta,k}, \\ d_k = d_{k-1} - v_{k-1} \Delta T \cos\theta_{k-1} + \omega_{d,k}, \\ v_k = v_{k-1} + \omega_{v,k}, \\ b_k = b_{k-1}\left(1 + d_{k-1}^{-1} v_{k-1} \Delta T \cos\theta_{k-1}\right) + \omega_{b,k}, \end{cases} \tag{4.14}$$

where $\boldsymbol{\omega}_k = \left[\omega_{\theta,k}, \omega_{d,k}, \omega_{v,k}, \omega_{b,k}\right]^{\mathrm{T}}$ represents the state noise, and Δ_T is the duration of one epoch.

With both models (4.13) and (4.14) above, the RSU can predict and estimate the vehicle's state at each epoch via various approaches, e.g. Kalman filtering and factor graph-based message passing algorithms [105–107]. The predicted angle is then employed to design the JCAS beamformer for the next epoch.

Despite that the works above demonstrate the feasibility and potential of sensing-assisted beamforming in PMN, some major problems are yet to be tackled to make it practical. One problem is how to translate the sensing results to the beamforming design. In particular, the downlink sensing results are associated with the object, while the communication channels are associated with the object's antennas. Given the size of a vehicle, there may be a large offset between them. Another problem is how to deal with multiple reflections. When there are multiple reflections coming from different directions, the phase of signals will also play an important role in beamforming, but they cannot typically be accurately estimated in radar sensing. One potential solution to both problems is combining uplink and downlink sensing in PMN. Downlink sensing can provide quick and coarse information for initial beamforming, while uplink sensing can offer more detailed and accurate information in a complicated propagation environment.

4.3.9.2 Sensing-Assisted Secure Communications

Radio sensing offers informative channel compositions for both active transmitters and passive objects in the surrounding environment. Such detailed channel information can be exploited for secure wireless communications, for which the research is still in its infancy.

One important application of the detailed channel composition information is in physical layer security techniques. Current physical layer security studies are mainly based on CSI [108], using, e.g. artificial-noise-aided, security-oriented beamforming, and physical-layer key generation approaches. This has been attempted in [109], where the artificial noise method is employed to realize secure beamforming in a JCAS system, to combat eavesdroppers. Physical-layer key generation based on CSI has been widely investigated and is shown to be very effective for secure communications. However, the secrecy capacity of CSI is generally limited as it hides the channel propagation details and is the sum of many propagation paths. Comparatively, the sensing results contain more essential information about the environment between a pair of transmitter and receiver. They can motivate more informative secret-key generating methods and agreement in cellular communication networks. As a start, we can characterize the secrecy capacity of PMNs, and develop practical secret-key generating methods for information encryption.

4.4 Conclusions

We have provided a comprehensive review on PMN, which integrates radio sensing into the current communications-only mobile network, using JCAS techniques. Employing 5G NR standard as an example, we have illustrated that uplink and downlink sensing can be realized with different degrees of modifications and enhancement to the current mobile network infrastructure. We have discussed various topics including major challenges, potential solutions, and diverse research opportunities within the context of PMN.

In Chapters 5 and 6, we will provide more technical details on sensing parameter estimations, including sensing techniques for OTFS signals, and for a unified sensing platform applicable to almost all signal formats.

References

1 A. Gupta and R. K. Jha. A survey of 5G network: Architecture and emerging technologies. *IEEE Access*, 3:1206–1232, 2015. doi: 10.1109/ACCESS.2015.2461602.

2 J. Liang and Q. Liang. Design and analysis of distributed radar sensor networks. *IEEE Transactions on Parallel and Distributed Systems*, 22(11):1926–1933, 2011. ISSN 1045-9219. doi: 10.1109/TPDS.2011.45.

3 M. L. Rahman, J. A. Zhang, X. Huang, Y. J. Guo, and R. W. Heath Jr. Framework for a perceptive mobile network using joint communication and radar sensing. *IEEE Transactions on Aerospace and Electronic Systems*, 56(3):1926–1941, 2020.

4 K. Gu, Y. Wang, and Y. Shen. Cooperative detection by multi-agent networks in the presence of position uncertainty. *IEEE Transactions on Signal Processing*, 68:5411–5426, 2020. doi: 10.1109/TSP.2020.3021227.

5 J. A. Zhang, A. Cantoni, X. Huang, Y. J. Guo, and R. W. Heath Jr. Framework for an innovative perceptive mobile network using joint communication and sensing. In *2017 IEEE 85th Vehicular Technology Conference (VTC Spring)*, pages 1–5. IEEE, June 2017. doi: 10.1109/VTCSpring.2017.8108564.

6 J. A. Zhang, X. Huang, Y. J. Guo, and Md. L. Rahman. Signal stripping based sensing parameter estimation in perceptive mobile networks. In *2017 IEEE-APS Topical Conference on Antennas and Propagation in Wireless Communications (APWC)*, pages 67–70. IEEE, 2017.

7 S. Yousefi, H. Narui, S. Dayal, S. Ermon, and S. Valaee. A survey on behavior recognition using WiFi channel state information. *IEEE Communications Magazine*, 55(10):98–104, October 2017. ISSN 0163-6804. doi: 10.1109/MCOM.2017.1700082.

8 M. Braun. *OFDM Radar Algorithms in Mobile Communication Networks.* PhD thesis, Institut fur Nachrichtentechnik des Karlsruher Instituts fur Technologie, Karlsruhe, 2014.

9 F. Liu, L. Zhou, C. Masouros, A. Li, W. Luo, and A. Petropulu. Toward dual-functional radar-communication systems: Optimal waveform design. *IEEE Transactions on Signal Processing*, 66(16):4264–4279, August 2018. ISSN 1053-587X. doi: 10.1109/TSP.2018.2847648.

10 C. B. Barneto, S. D. Liyanaarachchi, M. Heino, T. Riihonen, and M. Valkama. Full duplex radio/radar technology: The enabler for advanced joint communication and sensing. *IEEE Wireless Communications*, 28(1):82–88, 2021. doi: 10.1109/MWC.001.2000220.

11 D. Cataldo, L. Gentile, S. Ghio, E. Giusti, S. Tomei, and M. Martorella. Multibistatic radar for space surveillance and tracking. *IEEE Aerospace and Electronic Systems Magazine*, 35(8):14–30, 2020. doi: 10.1109/MAES.2020.2978955.

12 N. Cao, Y. Chen, X. Gu, and W. Feng. Joint Bi-static radar and communications designs for intelligent transportation. *IEEE Transactions on Vehicular Technology*, 69(11):13060–13071, 2020. doi: 10.1109/TVT.2020.3020218.

13 H. Kuschel, D. Cristallini, and K. E. Olsen. Tutorial: Passive radar tutorial. *IEEE Aerospace and Electronic Systems Magazine*, 34(2):2–19, 2019. doi: 10.1109/MAES.2018.160146.

14 Z. Ni, J. A. Zhang, X. Huang, K. Yang, and J. Yuan. Uplink sensing in perceptive mobile networks with asynchronous transceivers. *IEEE Transactions on Signal Processing*, 69:1287–1300, 2021. doi: 10.1109/TSP.2021.3057499.

15 F. Liu, C. Masouros, A. Li, H. Sun, and L. Hanzo. MU-MIMO communications with MIMO Radar: From co-existence to joint transmission. *IEEE Transactions on Wireless Communications*, 17(4):2755–2770, April 2018. ISSN 1536-1276. doi: 10.1109/TWC.2018.2803045.

16 3GPP TS 38.211. Physical channels and modulation. *V15.2.0*, July 2018.

17 F. Liu, C. Masouros, A. P. Petropulu, H. Griffiths, and L. Hanzo. Joint radar and communication design: Applications, state-of-the-art, and the road ahead. *IEEE Transactions on Communications*, 68(6):3834–3862, 2020. doi: 10.1109/TCOMM.2020.2973976.

18 M. L. Rahman, P. Cui, J. A. Zhang, X. Huang, Y. J. Guo, and Z. Lu. Joint communication and radar sensing in 5G mobile network by compressive sensing. In *2019 19th International Symposium on Communications and Information Technologies (ISCIT)*, pages 599–604, September 2019. doi: 10.1109/ISCIT.2019.8905229.

19 D. E. Hack, L. K. Patton, B. Himed, and M. A. Saville. Detection in passive MIMO radar networks. *IEEE Transactions on Signal Processing*, 62(11):2999–3012, June 2014. ISSN 1053-587X. doi: 10.1109/TSP.2014.2319776.

20 Z. Ni, J. A. Zhang, K. Yang, X. Huang, and T. A. Tsiftsis. Multi-metric waveform optimization for multiple-input single-output joint communication and radar sensing. *IEEE Transactions on Communications*, 70(2):1276–1289, 2022. doi: 10.1109/TCOMM.2021.3132368.

21 Z. Zhang, X. Chai, K. Long, A. V. Vasilakos, and L. Hanzo. Full duplex techniques for 5G networks: Self-interference cancellation, protocol design, and relay selection. *IEEE Communications Magazine*, 53(5):128–137, May 2015. ISSN 0163-6804. doi: 10.1109/MCOM.2015.7105651.

22 C. B. Barneto, T. Riihonen, M. Turunen, L. Anttila, M. Fleischer, K. Stadius, J. Ryynänen, and M. Valkama. Full-duplex OFDM radar with LTE and 5G NR waveforms: Challenges, solutions, and measurements. *IEEE Transactions on Microwave Theory and Techniques*, 67(10):4042–4054, 2019. doi: 10.1109/TMTT.2019.2930510.

23 J. A. Zhang, X. Huang, Y. J. Guo, J. Yuan, and R. W. Heath. Multibeam for joint communication and radar sensing using steerable analog antenna arrays. *IEEE Transactions on Vehicular Technology*, 68(1):671–685, Jan 2019. ISSN 1939-9359. doi: 10.1109/TVT.2018.2883796.

24 C. B. Barneto, S. D. Liyanaarachchi, T. Riihonen, L. Anttila, and M. Valkama. Multibeam design for joint communication and sensing in 5G new radio networks. In *ICC 2020 - 2020 IEEE International Conference on Communications (ICC)*, pages 1–6, 2020. doi: 10.1109/ICC40277.2020.9148935.

25 R. S. Thoma, D. Hampicke, A. Richter, G. Sommerkorn, A. Schneider, U. Trautwein, and W. Wirnitzer. Identification of time-variant directional mobile radio channels. *IEEE Transactions on Instrumentation and Measurement*, 49(2):357–364, 2000. doi: 10.1109/19.843078.

26 M. Landmann, M. Kaske, and R. S. Thoma. Impact of incomplete and inaccurate data models on high resolution parameter estimation in multidimensional channel sounding. *IEEE Transactions on Antennas and Propagation*, 60(2):557–573, 2012. doi: 10.1109/TAP.2011.2173446.

27 A. Liu, Z. Huang, M. Li, Y. Wan, W. Li, T. X. Han, C. Liu, R. Du, D. Tan, K. Pin, J. Lu, Y. Shen, F. Colone, and K. Chetty. A survey on fundamental limits of integrated sensing and communication, 2021. arXiv: 2104.09954.

28 M. Bica, K. Huang, V. Koivunen, and U. Mitra. Mutual information based radar waveform design for joint radar and cellular communication systems. In *2016 IEEE International Conference on Acoustics, Speech and Signal Processing (ICASSP)*, pages 3671–3675, March 2016. doi: 10.1109/ICASSP.2016.7472362.

29 M. R. Bell. Information theory and radar waveform design. *IEEE Transactions on Information Theory*, 39(5):1578–1597, September 1993. ISSN 0018-9448. doi: 10.1109/18.259642.

30 Z. Zhu, S. Kay, and R. S. Raghavan. Information-theoretic optimal radar waveform design. *IEEE Signal Processing Letters*, 24(3):274–278, March 2017. ISSN 1070-9908. doi: 10.1109/LSP.2017.2655879.

31 T. Tian, T. Zhang, L. Kong, G. Cui, and Y. Wang. Mutual information based partial band coexistence for joint radar and communication system. In *2019 IEEE Radar Conference (RadarConf)*, pages 1–5, April 2019. doi: 10.1109/RADAR.2019.8835671.

32 R. Xu, L. Peng, W. Zhao, and Z. Mi. Radar mutual information and communication channel capacity of integrated radar-communication system using MIMO. *ICT Express*, 1(3):102–105, 2015. ISSN 2405-9595. doi: https://doi.org/10.1016/j.icte.2016.01.001. Special Issue on Next Generation (5G/6G) Mobile Communications.

33 A. R. Chiriyath, B. Paul, and D. W. Bliss. Radar-communications convergence: Coexistence, cooperation, and co-design. *IEEE Transactions on Cognitive Communications and Networking*, 3(1):1–12, 2017. ISSN 2332-7731. doi: 10.1109/tccn.2017.2666266.

34 X. Yuan, Z. Feng, J. A. Zhang, W. Ni, R. P. Liu, Z. Wei, and C. Xu. Spatio-temporal power optimization for mimo joint communication and

radio sensing systems with training overhead. *IEEE Transactions on Vehicular Technology*, 1–15, 2020. doi: 10.1109/TVT.2020.3046438.

35 R. Niu, R. S. Blum, P. K. Varshney, and A. L. Drozd. Target localization and tracking in noncoherent multiple-input multiple-output radar systems. *IEEE Transactions on Aerospace and Electronic Systems*, 48(2):1466–1489, 2012.

36 Y. Yang and R. S. Blum. MIMO radar waveform design based on mutual information and minimum mean-square error estimation. *IEEE Transactions on Aerospace and Electronic Systems*, 43(1):330–343, January 2007. ISSN 0018-9251. doi: 10.1109/TAES.2007.357137.

37 B. Tang, J. Tang, and Y. Peng. MIMO radar waveform design in colored noise based on information theory. *IEEE Transactions on Signal Processing*, 58(9):4684–4697, September 2010. ISSN 1053-587X. doi: 10.1109/TSP.2010.2050885.

38 Y. Liu, G. Liao, Z. Yang, and J. Xu. Multiobjective optimal waveform design for OFDM integrated radar and communication systems. *Signal Processing*, 141:331–342, 2017. ISSN 0165-1684. doi: https://doi.org/10.1016/j.sigpro.2017.06.026.

39 Y. Liu, G. Liao, J. Xu, Z. Yang, and Y. Zhang. Adaptive OFDM integrated radar and communications waveform design based on information theory. *IEEE Communications Letters*, 21(10):2174–2177, 2017.

40 Y. Luo, J. A. Zhang, X. Huang, W. Ni, and J. Pan. Optimization and quantization of multibeam beamforming vector for joint communication and radio sensing. *IEEE Transactions on Communications*, 67(9):6468–6482, September 2019. ISSN 1558-0857. doi: 10.1109/TCOMM.2019.2923627.

41 Y. Luo, J. A. Zhang, X. Huang, W. Ni, and J. Pan. Multibeam optimization for joint communication and radio sensing using analog antenna arrays. *IEEE Transactions on Vehicular Technology*, 69(10):11000–11013, 1, 2020.

42 X. Liu, T. Huang, N. Shlezinger, Y. Liu, J. Zhou, and Y. C. Eldar. Joint transmit beamforming for multiuser MIMO communications and MIMO radar. *IEEE Transactions on Signal Processing*, 68:3929–3944, 2020. doi: 10.1109/TSP.2020.3004739.

43 L. Chen, F. Liu, J. Liu, and C. Masouros. Composite signalling for DFRC: Dedicated probing signal or not? 2020. arXiv: 2009.03528.

44 P. Kumari, S. A. Vorobyov, and R. W. Heath. Adaptive virtual waveform design for millimeter-wave joint communication-radar. *IEEE Transactions on Signal Processing*, 68:715–730, 2020. doi: 10.1109/TSP.2019.2956689.

45 G. Hakobyan, M. Ulrich, and B. Yang. OFDM-MIMO radar with optimized nonequidistant subcarrier interleaving. *IEEE Transactions on Aerospace and Electronic Systems*, 56(1):572–584, 2020. doi: 10.1109/TAES.2019.2920044.

46 X. Cui, J. Yang, T. Yan, and R. Yang. Waveform design for integration of MIMO radar and communication based on orthogonal frequency division complex modulation. In *MATEC Web of Conferences*, pages 1–7, 2018.

47 C. Sturm and W. Wiesbeck. Waveform design and signal processing aspects for fusion of wireless communications and radar sensing. *Proceedings of the IEEE*, 99(7):1236–1259, 2011. ISSN 0018-9219 1558-2256. doi: 10.1109/jproc.2011.2131110.

48 R. C. Daniels, E. R. Yeh, and R. W. Heath. Forward collision vehicular radar with IEEE 802.11: Feasibility demonstration through measurements. *IEEE Transactions on Vehicular Technology*, 67(2):1404–1416, February 2018. ISSN 0018-9545. doi: 10.1109/TVT.2017.2758581.

49 Y. Luo, J. A. Zhang, X. Huang, W. Ni, and J. Pan. Multibeam optimization for joint communication and radio sensing using analog antenna arrays. *IEEE Transactions on Vehicular Technology*, 69(10):11000–11013, 2020. doi: 10.1109/TVT.2020.3006481.

50 L. Gaudio, M. Kobayashi, G. Caire, and G. Colavolpe. On the effectiveness of OTFS for joint radar parameter estimation and communication. *IEEE Transactions on Wireless Communications*, 19(9):5951–5965, 2020. doi: 10.1109/TWC.2020.2998583.

51 M. F. Keskin, H. Wymeersch, and A. Alvarado. Radar sensing with OTFS: Embracing ISI and ICI to surpass the ambiguity barrier. In *2021 IEEE International Conference on Communications Workshops (ICC Workshops)*, pages 1–6, 2021. doi: 10.1109/ICCWorkshops50388.2021.9473534.

52 C. Liu, S. Liu, Z. Mao, Y. Huang, and H. Wang. Low-complexity parameter learning for OTFS modulation based automotive radar. In *ICASSP 2021 - 2021 IEEE International Conference on Acoustics, Speech and Signal Processing (ICASSP)*, pages 8208–8212, 2021. doi: 10.1109/ICASSP39728.2021.9414107.

53 A. Hassanien and S. A. Vorobyov. Phased-MIMO radar: A tradeoff between phased-array and MIMO radars. *IEEE Transactions on Signal Processing*, 58(6):3137–3151, June 2010. ISSN 1941-0476. doi: 10.1109/TSP.2010.2043976.

54 C. Qin, J. A. Zhang, X. Huang, and Y. J. Guo. Virtual-subarray-based angle-of-arrival estimation in analog antenna arrays. *IEEE Wireless Communications Letters*, 9(2):194–197, 2020.

55 X. Wang, A. Hassanien, and M. G Amin. Sparse transmit array design for dual-function radar communications by antenna selection. *Digital Signal Processing*, 83:223–234, 2018.

56 X. Wang, A. Hassanien, and M. G. Amin. Dual-function MIMO radar communications system design via sparse array optimization. *IEEE Transactions on Aerospace and Electronic Systems*, 55(3):1213–1226, 2019. doi: 10.1109/TAES.2018.2866038.

57 A. Hassanien, M. G. Amin, E. Aboutanios, and B. Himed. Dual-function radar communication systems: A solution to the spectrum congestion problem. *IEEE Signal Processing Magazine*, 36(5):115–126, September 2019. doi: 10.1109/msp.2019.2900571.

58 K. Wu, J. Guo, X. Huang, and R. W. Heath Jr. Accurate channel estimation for frequency-hopping dual-function radar-communication. In *2020 IEEE International Conference on Communications, CRSS Workshop*, pages 1–6. IEEE.

59 D. Ma, N. Shlezinger, T. Huang, Y. Shavit, M. Namer, Y. Liu, and Y. C. Eldar. Spatial modulation for joint radar-communications systems: Design, analysis, and hardware prototype, 2020.

60 J. A. Zhang, X. Huang, and V. Dyadyuk. Massive hybrid antenna array for millimeter-wave cellular communications. *IEEE Wireless Communications*, 22(1):79–87, 2015.

61 F. Liu and C. Masouros. Hybrid beamforming with sub-arrayed MIMO radar: Enabling joint sensing and communication at mmWave band. In *ICASSP 2019 - 2019 IEEE International Conference on Acoustics, Speech and Signal Processing (ICASSP)*, pages 7770–7774, May 2019. doi: 10.1109/ICASSP.2019.8683591.

62 Y. Liu, J. Zheng, M. Li, Q. Luo, Y. Rui, and Y. J. Guo. Synthesizing beam-scannable thinned massive antenna array utilizing modified iterative FFT for millimeter-wave communication. *IEEE Antennas and Wireless Propagation Letters*, 19(11):1983–1987, 2020. doi: 10.1109/LAWP.2020.3013981.

63 C. Zhou, Y. Gu, X. Fan, Z. Shi, G. Mao, and Y. D. Zhang. Direction-of-arrival estimation for coprime array via virtual array interpolation. *IEEE Transactions on Signal Processing*, 66(22):5956–5971, 2018.

64 H. Chu, L. Zheng, and X. Wang. Super-resolution mmwave channel estimation for generalized spatial modulation systems. *IEEE Journal of Selected Topics in Signal Processing*, 13(6):1336–1347, 2019.

65 M. Di Renzo, A. Zappone, M. Debbah, M.-S. Alouini, C. Yuen, J. de Rosny, and S. Tretyakov. Smart radio environments empowered by reconfigurable intelligent surfaces: How it works, state of research, and the road ahead. *IEEE Journal on Selected Areas in Communications*, 38(11):2450–2525, 2020. doi: 10.1109/JSAC.2020.3007211.

66 J. Hu, H. Zhang, B. Di, L. Li, K. Bian, L. Song, Y. Li, Z. Han, and H. V. Poor. Reconfigurable intelligent surface based RF sensing: Design, optimization, and implementation. *IEEE Journal on Selected Areas in Communications*, 38(11):2700–2716, 2020. doi: 10.1109/JSAC.2020.3007041.

67 S. Buzzi, E. Grossi, M. Lops, and L. Venturino. Foundations of MIMO radar detection aided by reconfigurable intelligent surfaces, 2021. arXiv:2105.09250.

68 S. Gong, X. Lu, D. T. Hoang, D. Niyato, L. Shu, D. I. Kim, and Y.-C. Liang. Toward smart wireless communications via intelligent reflecting surfaces: A contemporary survey. *IEEE Communication Surveys and Tutorials*, 22(4):2283–2314, 2020. doi: 10.1109/COMST.2020.3004197.

69 Z.-M. Jiang, M. Rihan, P. Zhang, L. Huang, Q. Deng, J. Zhang, and E. M. Mohamed. Intelligent reflecting surface aided dual-function radar and communication system. *IEEE Systems Journal*, 1–12, 2021. doi: 10.1109/JSYST.2021.3057400.

70 X. Tong, Z. Zhang, J. Wang, C. Huang, and M. Debbah. Joint multi-user communication and sensing exploiting both signal and environment sparsity. *IEEE Journal of Selected Topics in Signal Processing*, 1, 2021. doi: 10.1109/JSTSP.2021.3111432.

71 X. Wang, Z. Fei, Z. Zheng, and J. Guo. Joint waveform design and passive beamforming for RIS-assisted dual-functional radar-communication system. *IEEE Transactions on Vehicular Technology*, 70(5):5131–5136, 2021. doi: 10.1109/TVT.2021.3075497.

72 X. Wang, Z. Fei, J. Guo, Z. Zheng, and B. Li. RIS-assisted spectrum sharing between MIMO radar and MU-MISO communication systems. *IEEE Wireless Communications Letters*, 10(3):594–598, 2021. doi: 10.1109/LWC.2020.3039369.

73 A. R. Chiriyath and D. W. Bliss. Effect of clutter on joint radar-communications system performance inner bounds. In *2015 49th Asilomar Conference on Signals, Systems and Computers*, pages 1379–1383, November 2015. doi: 10.1109/ACSSC.2015.7421368.

74 A. Sobral and A. Vacavant. A comprehensive review of background subtraction algorithms evaluated with synthetic and real videos. *Computer Vision and Image Understanding*, 122:4–21, 2014. ISSN 1077-3142. doi: https://doi.org/10.1016/j.cviu.2013.12.005.

75 L. Haihan, Y. Li, S. Zhou, and W. Jing. A novel method to obtain CSI based on Gaussian mixture model and expectation maximization. In *2016 8th International Conference on Wireless Communications Signal Processing (WCSP)*, pages 1–5, October 2016. doi: 10.1109/WCSP.2016.7752732.

76 M. L. Rahman, J. A. Zhang, X. Huang, Y. J. Guo, and Z. Lu. Gaussian-mixture-model based clutter suppression in perceptive mobile networks. *IEEE Communications Letters*, 25(1):152–156, 2021. doi: 10.1109/LCOMM.2020.3025880.

77 M. Sørensen and L. De Lathauwer. Multiple invariance ESPRIT for nonuniform linear arrays: A coupled canonical polyadic decomposition approach. *IEEE Transactions on Signal Processing*, 64(14):3693–3704, 2016.

78 Y. Wang, Y. Zhang, Z. Tian, G. Leus, and G. Zhang. Super-resolution channel estimation for arbitrary arrays in hybrid millimeter-wave massive MIMO systems. *IEEE Journal of Selected Topics in Signal Processing*, 13(5):947–960, 2019.

79 M. F. Keskin, H. Wymeersch, and V. Koivunen. MIMO-OFDM joint radar-communications: Is ICI friend or foe? *IEEE Journal of Selected Topics in Signal Processing*, 1, 2021. doi: 10.1109/JSTSP.2021.3109431.

80 Md. L. Rahman, J. A. Zhang, X. Huang, Y. J. Guo, and Z. Lu. Joint communication and radar sensing in 5G mobile network by compressive sensing. *IET Communications*, 14(22):3977–3988, 2020. doi: https://doi.org/10.1049/iet-com.2020.0384.

81 Y. Sun, T. Fei, and N. Pohl. A high-resolution framework for range-Doppler frequency estimation in automotive radar systems. *IEEE Sensors Journal*, 19(23):11346–11358, 2019.

82 Y. Chi, L. L. Scharf, A. Pezeshki, and A. R. Calderbank. Sensitivity to basis mismatch in compressed sensing. *IEEE Transactions on Signal Processing*, 59(5):2182–2195, May 2011. ISSN 1053-587X. doi: 10.1109/TSP.2011.2112650.

83 C. R. Berger, B. Demissie, J. Heckenbach, P. Willett, and S. Zhou. Signal processing for passive radar using OFDM waveforms. *IEEE Journal of Selected Topics in Signal Processing*, 4(1):226–238, February 2010. ISSN 1941-0484. doi: 10.1109/JSTSP.2009.2038977.

84 X. Li, D. Zhang, Q. Lv, J. Xiong, S. Li, Y. Zhang, and H. Mei. IndoTrack: Device-free indoor human tracking with commodity Wi-Fi. *Proceedings of the ACM Interactive, Mobile, Wearable and Ubiquitous Technologies*, 1(3), September 2017. doi: 10.1145/3130940.

85 K. Qian, C. Wu, Y. Zhang, G. Zhang, Z. Yang, and Y. Liu. Widar2.0: Passive human tracking with a single Wi-Fi link. In *Proceedings of the 16th Annual International Conference on Mobile Systems, Applications, and Services*, page 350–361, New York, NY, USA, 2018. Association for Computing Machinery. ISBN 9781450357203. doi: 10.1145/3210240.3210314.

86 Y. Zeng, D. Wu, J. Xiong, E. Yi, R. Gao, and D. Zhang. FarSense: Pushing the range limit of WiFi-based respiration sensing with CSI ratio of two antennas. *Proceedings of the ACM Interactive, Mobile, Wearable and Ubiquitous Technologies*, 3(3), September 2019. doi: 10.1145/3351279.

87 Y. Zeng, D. Wu, J. Xiong, J. Liu, Z. Liu, and D. Zhang. MultiSense: Enabling multi-person respiration sensing with commodity WiFi. *Proceedings of the ACM Interactive, Mobile, Wearable and Ubiquitous Technologies*, 4(3), September 2020. doi: 10.1145/3411816.

88 R. Harris, J.-K. Lam, and E. F. Burroughs. The potential of cell phone radar as a tool for transport applications. Report, Association for European Transport and contributors, 2005.

89 N. David, O. Sendik, H. Messer, and P. Alpert. Cellular network infrastructure: The future of fog monitoring? *Bulletin of the American Meteorological Society*, 96(10):1687–1698, 2015. ISSN 0003-0007 1520-0477. doi: 10.1175/bams-d-13-00292.1.

90 R. Uijlenhoet, A. Overeem, and H. Leijnse. Opportunistic remote sensing of rainfall using microwave links from cellular communication networks. *WIREs Water*, 5(4):e1289, 2018. doi: 10.1002/wat2.1289.

91 W. Wang, A. X. Liu, M. Shahzad, K. Ling, and S. Lu. Device-free human activity recognition using commercial WiFi devices. *IEEE Journal on Selected Areas in Communications*, 35(5):1118–1131, May 2017. ISSN 0733-8716. doi: 10.1109/JSAC.2017.2679658.

92 G. Wang, Y. Zou, Z. Zhou, K. Wu, and L. M. Ni. We can hear you with Wi-Fi! *IEEE Transactions on Mobile Computing*, 15(11):2907–2920, November 2016. ISSN 1536-1233. doi: 10.1109/TMC.2016.2517630.

93 Z. Shi, J. A. Zhang, R. Xu, and Q. Cheng. Deep learning networks for human activity recognition with CSI correlation feature extraction. In *ICC 2019 - 2019 IEEE International Conference on Communications (ICC)*, pages 1–6, May 2019. doi: 10.1109/ICC.2019.8761445.

94 A. Al-Hourani, R. J. Evans, S. Kandeepan, B. Moran, and H. Eltom. Stochastic geometry methods for modeling automotive radar interference. *IEEE Transactions on Intelligent Transportation Systems*, 19(2):333–344, 2018.

95 P. Chu, A. Zhang, X. Wang, Z. Fei, G. Fang, and D. Wang. Interference characterization and power optimization for automotive radar with directional antenna. *IEEE Transactions on Vehicular Technology*, 69(4):3703–3716, 2020.

96 S. S. Ram, G. Singh, and G. Ghatak. Estimating radar detection coverage probability of targets in a cluttered environment using stochastic geometry. In *2020 IEEE International Radar Conference (RADAR)*, pages 665–670, 2020. doi: 10.1109/RADAR42522.2020.9114637.

97 S. S. Ram, G. Singh, and G. Ghatak. Optimization of radar parameters for maximum detection probability under generalized discrete clutter conditions using stochastic geometry, 2021. arXiv: 2101.12429.

98 A. M. Haimovich, R. S. Blum, and L. J. Cimini. MIMO radar with widely separated antennas. *IEEE Signal Processing Magazine*, 25(1):116–129, 2008. ISSN 1053-5888. doi: 10.1109/MSP.2008.4408448.

99 R. S. Thoma, C. Andrich, G. D. Galdo, M. Dobereiner, M. A. Hein, M. Kaske, G. Schafer, S. Schieler, C. Schneider, A. Schwind, and P. Wendland. Cooperative passive coherent location: A promising 5G service to support road safety. *IEEE Communications Magazine*, 57(9):86–92, 2019. doi: 10.1109/MCOM.001.1800242.

100 Y. Wang, K. Gu, Y. Wu, W. Dai, and Y. Shen. NLOS effect mitigation via spatial geometry exploitation in cooperative localization. *IEEE Transactions on Wireless Communications*, 19(9):6037–6049, 2020. doi: 10.1109/TWC.2020.2999667.

101 N. Gonzalez-Prelcic, A. Ali, V. Va, and R. W. Heath. Millimeter-wave communication with out-of-band information. *IEEE Communications Magazine*, 55(12):140–146, 2017. doi: 10.1109/MCOM.2017.1700207.

102 N. González-Prelcic, R. Méndez-Rial, and R. W. Heath. Radar aided beam alignment in mmWave V2I communications supporting antenna diversity.

In *2016 Information Theory and Applications Workshop (ITA)*, pages 1–7, January 2016. doi: 10.1109/ITA.2016.7888145.

103 A. Ali, N. Gonzalez-Prelcic, R. W. Heath, and A. Ghosh. Leveraging sensing at the infrastructure for mmWave communication. *IEEE Communications Magazine*, 58(7):84–89, 2020. doi: 10.1109/MCOM.001.1900700.

104 C. Jiao, Z. Zhang, C. Zhong, and Z. Feng. An indoor mmwave joint radar and communication system with active channel perception. In *2018 IEEE International Conference on Communications (ICC)*, pages 1–6, May 2018. doi: 10.1109/ICC.2018.8422132.

105 F. Liu, W. Yuan, C. Masouros, and J. Yuan. Radar-assisted predictive beamforming for vehicular links: Communication served by sensing. *IEEE Transactions on Wireless Communications*, pages 1, 2020. doi: 10.1109/TWC.2020.3015735.

106 W. Yuan, F. Liu, C. Masouros, J. Yuan, D. W. K. Ng, and N. González-Prelcic. Bayesian predictive beamforming for vehicular networks: A low-overhead joint radar-communication approach. *IEEE Transactions on Wireless Communications*, pages 1, 2020. doi: 10.1109/TWC.2020.3033776.

107 F. Liu and C. Masouros. A tutorial on joint radar and communication transmission for vehicular networks - part III: Predictive beamforming without state models (invited paper). *IEEE Communications Letters*, pages 1, 2020. doi: 10.1109/LCOMM.2020.3025331.

108 Y. Zou, J. Zhu, X. Wang, and L. Hanzo. A survey on wireless security: Technical challenges, recent advances, and future trends. *Proceedings of the IEEE*, 104(9):1727–1765, 2016. doi: 10.1109/JPROC.2016.2558521.

109 N. Su, F. Liu, and C. Masouros. Secure radar-communication systems with malicious targets: Integrating radar, communications and jamming functionalities. *IEEE Transactions on Wireless Communications*, pages 1, 2020. doi: 10.1109/TWC.2020.3023164.

5

Integrating Low-Complexity and Flexible Sensing into Communication Systems: A Unified Sensing Framework

As illustrated in Chapter 4, joint communications and sensing (JCAS) can substantially improve the energy consumption, cost, and spectral efficiency of systems that need both communication and sensing functions. Integrating sensing into standardized communication systems can potentially benefit many consumer applications requiring radio frequency functions. However, such integration may not achieve the expected gains in cost and energy efficiency without an effective sensing method. In this chapter, a flexible sensing framework is introduced, which has a low complexity that is only dominated by the Fourier transform and also provides the flexibility in adapting to different sensing needs. In the framework, a whole block of the echo signal is segmented evenly into subblocks; adjacent ones are allowed to overlap. A virtual cyclic prefix (VCP) is introduced to facilitate the use of two common ways for removing the impact of communication data symbols on sensing and generating two types of range-Doppler maps (RDMs). A comprehensive analysis of the signal components in the RDMs is also performed, showing that the interference-plus-noise (IN) terms of the RDMs are approximately Gaussian distributed. The statistical properties of the distributions are derived, which leads to analytical comparisons between the two RDMs and between the prior and the sensing methods presented. Moreover, the impact of the lengths of subblock, VCP and overlapping signal on sensing performance is analyzed. Some guidance for designing these lengths to achieve good sensing performance is also provided.

5.1 Problem Statement and Signal Model

This section shall first establish the signal model of the considered JCAS scenario based on the orthogonal time-frequency space (OTFS) modulation. Then, a classical OFDM sensing (COS) scheme [1] is briefly reviewed. This elicits several

Joint Communications and Sensing: From Fundamentals to Advanced Techniques, First Edition.
Kai Wu, J. Andrew Zhang, and Y. Jay Guo.

important issues in terms of sensing performance. Solutions to the issues will be developed in Sections 5.2 and 5.3.

5.1.1 Signal Model

We consider the case that a communication-only node is turned into a JCAS node by incorporating a sensing receiver. While the communication signals are transmitted, the receiver collects target echoes for sensing. As is common in communication-centric (CC) JCAS works [1–4], we can ignore the self-interference, i.e. the signal leakage directly from the transmitter to the receiver. As for the communication waveform, the OTFS modulation is employed here, not only because it is a potential waveform candidate for future mobile communications [5] but also due to its capability in representing other common multicarrier waveforms, e.g. OFDM and DFT-S-OFDM.

Let d_i $(i = 0, 1, \dots, I-1)$ denote the data symbols to be transmitted, where the data symbols are independently drawn from the same constellation, e.g. 64-QAM, and I denotes the total number of data symbols. In OTFS modulation, the I data symbols are first placed in a two-dimensional delay-Doppler plane. Let the delay and Doppler dimensions be discretized into M and N grids, respectively. Denoting the time duration of M data symbols as T, the sampling frequency along the Doppler dimension is then $\frac{1}{T}$, which leads to a Doppler resolution of $\frac{1}{NT}$. Since there are M grids along the delay dimension, the corresponding resolution is $\frac{T}{M}$. In the OTFS modulation, the data symbols can be mapped from the delay-Doppler domain into the frequency-time domain via the following transform [6],

$$S[m,n] = \frac{1}{\sqrt{MN}} \sum_{k=0}^{N-1} \sum_{l=0}^{M-1} d_{kM+l} e^{j2\pi(\frac{nk}{N} - \frac{mT}{M} l\Delta_f)}, \tag{5.1}$$

where j denotes the imaginary unit and Δ_f is the subcarrier frequency interval. At a fixed l, d_{kM+l} is a signal sequence of k. Then the above k-summation acts like the discrete Fourier transform of d_{kM+l} over k, and the variable in the dual domain is denoted by n. Moreover, by fixing k, a similar observation can be obtained over the l-summation.

The frequency-time-domain signal $S[m,n]$ is then transformed into the time domain by performing the inverse discrete Fourier transform (DFT) (IDFT) w.r.t. m for each n. This leads to

$$s[l,n] = \sum_{m=0}^{M-1} S[m,n] \mathcal{Z}_M^{-ml}, \ \forall n, \tag{5.2}$$

where $\mathcal{Z}_M^{-ml}$ denotes the DFT basis, as given by

$$\mathcal{Z}_a^{bc} = e^{-j\frac{2\pi bc}{a}} / \sqrt{a}. \tag{5.3}$$

When the critical sampling is employed, i.e. $T\Delta_f = 1$ in (5.1), the IDFT performed in (5.2) will cancel the l-related transform in (5.1). Treating l as the row index and n the column index, the signal $s[l,n]$ will be transmitted column-by-column and, in each column, the entries $l = 0, 1, \dots, M-1$ are transmitted sequentially. Before going through the digital-to-analog converter, some extra processing on $s[l,n]$, such as frequency up-conversion and pulse-shaping, would be necessary. To prevent intersymbol interference (ISI), the cyclic prefix is generally used in multicarrier transmissions. There are two types of CP in the OTFS literature.

In the first type, every M data symbols have a CP added [7], which is referred to as CP-OTFS hereafter. Let Q denote the number of samples in a CP. Based on $s[l,n]$ given in (5.2), the signal to be transmitted can be given by

$$\tilde{s}_{\mathrm{CP}}[i] = s\left[\left\langle \langle i\rangle_{M+Q} - Q\right\rangle_M, \lfloor i/(M+Q)\rfloor\right],\ i = 0, 1, \dots, N(M+Q)-1, \quad (5.4)$$

where $\langle x\rangle_y$ takes x modulo y and $\lfloor x\rfloor$ rounds toward negative infinity. The indexes on the RHS of (5.4) indicate that every $(M+Q)$ samples of $\tilde{s}_{\mathrm{CP}}[i]$ are obtained by copying the last Q samples from $s[0,n], s[1,n], \dots, s[M-1,n]$ at some n and pasting to the beginning.

In the second-type CP, the whole block of MN data symbols have a single CP added, which is known as the reduced CP-OTFS (RCP-OTFS) [8]. The signal to be transmitted in RCP-OTFS can be given by

$$\tilde{s}_{\mathrm{RCP}}[i] = s\left[\left\langle \tilde{i}\right\rangle_M, \lfloor \tilde{i}/M\rfloor\right],\ \text{s.t. } \tilde{i} = \langle i - Q\rangle_{MN},\ i = 0, 1, \dots, MN+Q-1, \quad (5.5)$$

where the indexes on the RHS indicate that the last Q samples from $s[0, N-1]$, $s[1, N-1], \dots, s[M-1, N-1]$ are copied and pasted to the beginning of $s[l, 0]$. The CP-added signal will go through a digital-to-analog conversion (DAC) and other analog-domain processing, e.g. frequency up-conversion and power amplification, before being transmitted. A pulse-shaping filter is generally performed to reduce the out-of-band (OOB) emission. For illustration convenience, the filter is not included in the signal model. However, practical pulse-shaping filters will be used in the simulations.

As mentioned earlier, the sensing receiver is colocated with the communication transmitter. Therefore, it is reasonable to assume prefect synchronization and zero frequency offset for sensing. Consider P targets. The scattering coefficient, time delay, and Doppler frequency of the p-th target are denoted by α_p, τ_p and ν_p, respectively. Let $\tilde{s}[i]$ be either $\tilde{s}_{\mathrm{CP}}[i]$ or $\tilde{s}_{\mathrm{RCP}}[i]$ (the set of i varies accordingly). The target echo, as a sum of the scaled and delayed versions of $\tilde{s}[i]$, can be modeled as

$$x[i] = \sum_{p=0}^{P-1} \tilde{\alpha}_p \tilde{s}\left[i - \mathfrak{l}_p\right] e^{\mathrm{j}2\pi i \tilde{\mathfrak{k}}_p} + w[i],\ \text{s.t. } \tilde{\alpha}_p = \alpha_p e^{-\mathrm{j}2\pi\nu_p\tau_p};\ \mathfrak{l}_p = \tau_p/T_{\mathrm{s}};\ \tilde{\mathfrak{k}}_p = \nu_p T_{\mathrm{s}}, \quad (5.6)$$

where T_s is the sampling interval, and $w[i] \sim \mathcal{CN}\left(0, \sigma_w^2\right)$ is the additive noise conforming to a circularly symmetric complex centered Gaussian distribution.

It is worthwhile to note some features of the above signal model. *First*, the signal $\tilde{s}_{\text{CP}}[i]$ can represent DFT-S-OFDM and OFDM with slight changes made on $S[m, n]$. In particular, DFT-S-OFDM can be obtained when the Fourier transform with regards to k is suppressed in (5.1), while OFDM is obtained when both Fourier transforms in (5.1) are suppressed. *Second*, $s[l, n]$ obtained in (5.2) approximately conforms to a complex centered Gaussian distributions, as denoted by $s[l, n] \sim \mathcal{CN}(0, \sigma_d^2)$, where σ_d^2 is the power of d_i; see (5.1). The above result can be attained using (5.1) and (5.2) in combination with another two facts: the complex envelope of an uncoded OFDM system converges in distribution to a complex Gaussian random process [9]; the unitary DFT does not change the statistical properties and the whiteness of a Gaussian process [10]. *Third*, the above two facts can be used to validate that $s[l, n] \sim \mathcal{CN}(0, \sigma_d^2)$ also holds for DFT-S-OFDM and OFDM.

5.1.2 Classical OFDM Sensing (COS)

COS was developed about a decade ago and has been widely adopted in the sensing literature; see [11, 12] and their references. Below, we first briefly review COS and then highlight some long-standing issues. As COS is originally developed for CP-OFDM, we assume that $\tilde{s}_{\text{CP}}[i]$ is transmitted and used $x[i]$ to describe the method.

In COS, the $(M + Q)N$ numbers of echo samples $x[i]$ are divided into N consecutive symbols, each having $(M + Q)$ samples. Removing the first Q samples in each symbol and taking the M-point unitary DFT of the remaining samples in the symbol, we obtain a common echo signal model

$$X_n[m] \approx \sum_{p=0}^{P-1} \tilde{\alpha}_p S[m, n] e^{-j\frac{2\pi m l_p}{M}} e^{j2\pi n(M+Q)\tilde{k}_p} + W_n[m], \tag{5.7}$$

where $W_n[m]$ denotes the DFT of the background noise. Note that the intrasymbol Doppler effect is suppressed in (5.7) due to its negligible value in general.

Dividing $X_n[m]$ by $S[m, n]$ in a pointwise manner, we can remove the communication data symbols. Then, a two-dimensional Fourier transform can be performed over m and n, leading to the following range-Doppler map (RDM),

$$\begin{aligned} U_k^{\mathrm{r}}[l] &= \sum_{n=0}^{N-1} \sum_{m=0}^{M-1} X_n[m] \,/\, S[m, n] \mathcal{Z}_M^{-ml} \mathcal{Z}_N^{nk}, \\ &= \sum_{p=0}^{P-1} \tilde{\alpha}_p \mathcal{S}_M \left(l - l_p\right) \mathcal{S}_N \left((M + Q)N\tilde{k}_p - k\right) + W_k^{\mathrm{r}}[l], \end{aligned} \tag{5.8}$$

where l indexes the range bin, k indicates the Doppler channel, the superscript $\{\cdot\}^{\mathrm{r}}$ stands for 'ratio' to differentiate with another way of removing $S[m,n]$, as to be illustrated in Section 5.1.3.3; $\mathcal{Z}_a^{bc}$ is the unitary DFT basis, as defined in (5.3); $W_k[l]$ is the two-dimensional Fourier transform of $W_n^{\mathrm{r}}[m]/S[m,n]$; and $\mathcal{S}_x(y)$ is introduced to denote the DFT results of the two exponential signals in (5.7). The general form of $\mathcal{S}_x(y)$ is given by

$$\mathcal{S}_x(y) = \frac{1}{\sqrt{x}} \frac{\sin\left(\frac{x}{2}\frac{2\pi y}{x}\right)}{\sin\left(\frac{1}{2}\frac{2\pi y}{x}\right)} e^{\mathrm{j}\frac{x-1}{2}\frac{2\pi y}{x}}. \tag{5.9}$$

The function $\mathcal{S}_x(y)$ is localized around $y = 0$, and hence, $|U_k^{\mathrm{r}}[l]|$ can present P dominant peaks in the range-Doppler domain, if $\tilde{\alpha}_p\ \forall p$ is sufficiently large. Thus, a threshold detector based on, e.g. likelihood ratio test (LRT), can be developed for target detection, from which coarse estimations of target parameters can be attained.

5.1.3 Problem Statement

COS has been widely applied due to its low complexity. However, COS and many of its variants can have limited sensing performance due to the strict compliance with the underlying communication systems. Some of those issues are illustrated below.

5.1.3.1 CP-limited Sensing Distance

CP plays a nontrivial role in COS. Specifically, CP makes each received symbol consist of a cyclically shifted version of the transmitted symbol. This then enables us to attain the convenient echo model given in (5.7) and further facilitates the removal of $S[m,n]$ to generate the RDM given in (5.8). However, CP also puts a constraint on sensing. Namely, the round-trip delay of the maximum sensing distance should be smaller than the time duration of the CP. Such limitation stands even when we have a sufficient link budget for sensing a longer distance. Moreover, for the communication waveform with a reduced CP, as modeled in (5.5), COS is not directly applicable.

5.1.3.2 Communication-limited Velocity Measurement

While the sensing distance is limited by CP, the velocity measurement performance can be constrained by the values of M and N. Substituting $\tilde{k}_p = \nu_p T_s$ into (5.7), we see that the Doppler frequency ν_p becomes the frequency of the exponential signal of n and $(M+Q)T_s$ is the sampling interval. Thus, the maximum (unambiguous) measurable value of the Doppler frequency, as denoted by $\nu_{\max}$, and its resolution, as denoted by Δ_ν, can be given by

$$\nu_{\max} = 1/\left(2(M+Q)T_s\right);\ \Delta_\nu = 1/\left(N(M+Q)T_s\right).$$

While a small M can give us a large unambiguous region for Doppler measurement, a large N is necessary to keep a small Δ_ν. However, assigning the values of M and N in a sensing-favorable way may degrade the performance of the underlying communication system, e.g. 5G [13], that generally has stringent requirements on the two parameters.

5.1.3.3 COS Adapted for DFT-S-OFDM

As shown in (5.8), communication data symbols are removed via pointwise divisions in COS. For CP-OFDM, there is no problem in doing so, as $S[m,n]$, directly drawn from a constellation, does not take zero in general. However, for DFT-S-OFDM and OTFS, $S[m,n]$ conforms to a complex centered Gaussian distribution, as illustrated in Section5.1.1. This means a certain portion of $S[m,n]$ is centered around the origin and the direct division can lead to severe noise enhancement. To address the issue, a time-domain CCC is proposed in [2] to replace the frequency-domain division. The RDM under CCC can be written based on (5.8), leading to

$$U_k^{\mathrm{c}}[l] = \sum_{n=0}^{N-1} \sum_{m=0}^{M-1} X_n[m] S^*[m,n] \mathcal{Z}_M^{-ml} \mathcal{Z}_N^{nk}, \tag{5.10}$$

where a closed-form result, as in the second line of (5.8), is not available, due to the randomness of $S^*[m,n]$. Note that $S[m,n]$ here is not the same as in (5.8). As stated in Section5.1.1, for DFT-S-OFDM, $S[m,n]$ can be obtained by suppressing the k-related Fourier transform in (5.1). *Now that we have two ways of generating RDMs, a question follows naturally: which one gives the better sensing performance?*

To address the issues highlighted in Sections 5.1.3.1 and 5.1.3.2, a more advanced flexible sensing framework is introduced in Section 5.2. Performance analysis for the sensing framework will be conducted in Section 5.3, which helps answer the question asked in Section 5.1.3.3.

5.2 A Low-Complexity Sensing Framework

The new sensing framework is illustrated in Figure 5.1. A communication transmitter (C-Tx) transmits $\tilde{s}[i]$ which can be either $\tilde{s}_{\mathrm{CP}}[i]$ in (5.4) with regular CPs (e.g. OFDM and DFT-S-OFDM) or $\tilde{s}_{\mathrm{RCP}}[i]$ in (5.5) with a RCP (e.g. RCP-OTFS). A copy of $\tilde{s}[i]$ is given to the sensing receiver (S-Rx), which, as mentioned earlier, colocates with C-Tx and shares the same clock. While $\tilde{s}[i]$ is transmitted, S-Rx receives the target echo, i.e. $x[i]$ given in (5.6). The sensing framework solely relies on $x[i]$ and $\tilde{s}[i]$ without requiring any cooperation or changes from C-Tx.[1]

1 Since the design does not affect the underlying communication system at all, we shall only focus on the sensing aspects in this chapter.

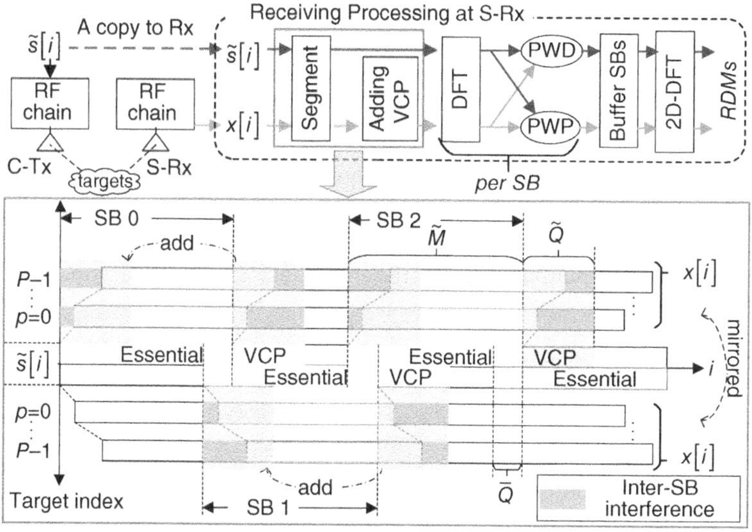

Figure 5.1 The schematic diagram of a sensing framework, where a sensing receiver (S-Rx) is colocated with a communication transmitter (C-Tx) and receives the echo signal $x[i]$ for wireless sensing. Substantially differentiating the design from the previous ones, e.g. COS, is the highlighted block segmentation and virtual cyclic prefix (VCP). These operations turn the block of echo signal $x[i]$, as well as the copy of transmitted signal $\tilde{s}[i]$, into $\tilde{N}$ subblocks (SBs) each having $\tilde{M}$ samples. Afterward, two common ways of generating RDMs, using pointwise division (PWD) [1] and product (PWP) [2], can be performed.

The sensing receiver starts with segmenting $x[i]$ and $\tilde{s}[i]$ into multiple consecutive subblocks, then removes or exploits the communication data symbols for generating an RDM. Substantially differentiating the sensing framework from previous sensing methods, e.g. COS, is the way a block of samples is segmented, as detailed next.

Instead of fully complying with the underlying communication systems, we segment $x[i]$ into $\tilde{N}$ subblocks, each having $\tilde{M}$ samples, where $\tilde{N} = N$ is not required in the presented design. Moreover, we allow the consecutive subblocks to overlap by $\overline{Q}$ samples, where $\overline{Q}$ is either zero or a positive integer. As also shown in Figure 5.1, we segment the communication-transmitted signal $\tilde{s}[i]$, *a copy at the sensing receiver*, in the same way as described above and call each segment the essential signal of the subblock. Due to the propagation delay of a target, part of the essential signal is not within the received subblock but right after it. To preserve the essential signal in each subblock, we add the $\tilde{Q}$ samples right after a subblock onto the first $\tilde{Q}$ samples within the subblock, creating a VCP. Note that VCPs are samples from the block of received echo signals with fixed and known positions. Also, note that VCP is independent of the original CP of the underlying communication system. It is only introduced for the sensing receiver and will not incur any change to the communication transmitter.

As seen from Figure 5.1, adding VCP can make each received subblock comprised of cyclically shifted versions of its essential signal part, as long as $\tilde{Q}$, *not* Q *anymore*, is greater than the maximum target delay. Since the value of $\tilde{Q}$ is not limited to the original CP length Q, we can design the maximum sensing distance flexibly subject to a sufficient link budget. Next, the above description is further elaborated on using the signal model provided in Section 5.1.

Let $s_n[l]$ denote the essential signal of the n-th subblock. Based on the above illustration, we can write $s_n[l]$ as follows:

$$s_n[l] = \tilde{s}[n(\tilde{M} - \overline{Q}) + l],\ l = 0, 1, \dots, \tilde{M} - 1,\ n = 0, 1, \dots, \tilde{N} - 1,\ \tilde{N} = \left\lfloor \frac{(I - \tilde{Q} - \overline{Q})}{(\tilde{M} - \overline{Q})} \right\rfloor, \tag{5.11}$$

where $\tilde{s}[\cdot]$ on the right-hand side can be either $\tilde{s}_{\mathrm{CP}}[i]$ in (5.4) or $\tilde{s}_{\mathrm{RCP}}[i]$ in (5.5) and $\tilde{N}$ is the total number of subblocks. Take the three subblocks in Figure 5.1 for an illustration. By excluding the last $\overline{Q}$ samples of subblock two and its $\tilde{Q}$-sample VCP, we see that each of the first three subblocks has $(\tilde{M} - \overline{Q})$ unique samples. This can be generalized into the expression of $\tilde{N}$ given in (5.11).

With reference to Figure 5.1, after adding VCP, the received signal in subblock n becomes

$$x_n[l] \approx \sum_{p=0}^{P-1} \tilde{\alpha}_p s_n\left[\left\langle l - \mathfrak{l}_p \right\rangle_{\tilde{M}}\right] e^{\mathrm{j}2\pi n(\tilde{M}-\overline{Q})\tilde{\mathfrak{k}}_p} + w_n[l] + z_n^{(p)}[l] g_{\tilde{Q}}[l],\ l = 0, 1, \dots, \tilde{M} - 1, \tag{5.12}$$

where $\tilde{\alpha}_p$, $\mathfrak{l}_p$ and $\tilde{\mathfrak{k}}_p$ are given in (5.6). Similar to (5.7), the approximation here is also due to the suppression of the intrasubblock Doppler impact. We emphasize that, due to the $\overline{Q}$-sample overlapping of consecutive subblocks, the Doppler phase is $2\pi(\tilde{M} - \overline{Q})\tilde{\mathfrak{k}}_p$ *not* $2\pi\tilde{M}\tilde{\mathfrak{k}}_p$. In (5.12), $z_n^{(p)}[l] g_{\tilde{Q}}[l]$ denotes the interference term and $g_{\tilde{Q}}[l]$ is a rectangular window function which takes one at $l = 0, 1, \dots, \tilde{Q} - 1$, and zero elsewhere. Moreover, the noise term $w_n[l]$ in (5.12) is obtained by first segmenting $w[i]$ given in (5.6) as done in (5.11) and then adding VCP. Since the addition of two i.i.d. Gaussian variables are still Gaussian with the variance doubled, we have

$$w_n[l] \sim \begin{cases} \mathcal{CN}(0, 2\sigma_w^2) & \text{for } l = 0, 1, \dots, \tilde{Q} - 1 \\ \mathcal{CN}(0, \sigma_w^2) & \text{for } l = \tilde{Q}, \dots, \tilde{M} - 1 \end{cases} \tag{5.13}$$

Note that the interference term in (5.12) is the price paid for having VCP (equivalently, sensing flexibility). Though the interference will make the white background noise become colored, we will show in Section 5.3 that the overall IN background in the RDM still approximates a white Gaussian distribution under certain conditions.

As in OFDM, the cyclic shift of the essential signal preserves the subcarrier orthogonality. Therefore, taking the $\tilde{M}$-point DFT of $x_n[l]$ w.r.t. l leads to

$$X_n[m] = \sum_{p=0}^{P-1} \tilde{\alpha}_p S_n[m] e^{-j\frac{2\pi m l_p}{\tilde{M}}} e^{j\frac{2\pi n k_p}{\tilde{N}}} + W_n[m] + Z_n[m],$$

$$\text{s.t. } \mathrm{k}_p = \tilde{N}(\tilde{M} - \overline{Q})\tilde{\mathrm{k}}_p,\ F_n[m] = \sum_{l=0}^{\tilde{M}-1} f_n[l] \mathcal{Z}_{\tilde{M}}^{lm},\ (F, f) \in \{(S, s), (W, w)\},$$

$$Z_n[m] = \sum_{p=0}^{P-1} \tilde{\alpha}_p \sum_{l=0}^{\tilde{M}-1} z_n^{(p)}[l] g_{\tilde{Q}}[l] \mathcal{Z}_{\tilde{M}}^{lm}, \tag{5.14}$$

where $S_n[m]$, $W_n[m]$, and $Z_n[m]$ are the DFTs of the respective terms in (5.12). Note again that the unitary DFT basis, as defined in (5.3), is used. Since $s_n[l]$ is known, $S_n[m]$ can be readily calculated. Corresponding to (5.8), we can divide both sides of $X_n[m]$ by $S_n[m]$ and take the two-dimensional DFT w.r.t. n and m, attaining the following ratio-based RDM:

$$V_k^{\mathrm{r}}[l] = \sum_{p=0}^{P-1} \tilde{\alpha}_p S_{\overline{M}}\left(l - \mathrm{l}_p\right) S_{\tilde{N}}\left(\mathrm{k}_p - k\right) + W_k^{\mathrm{r}}[l] + Z_k^{\mathrm{r}}[l],$$

$$\text{s.t. } \mathrm{X}_k^{\mathrm{r}}[l] = \sum_{n=0}^{N-1} \sum_{m=0}^{M-1} \frac{\mathrm{X}_n[m]}{S_n[m]} \mathcal{Z}_M^{-ml} \mathcal{Z}_N^{nk},\ \mathrm{X} \in \{W, Z\} \tag{5.15}$$

where $S_x(y)$ is defined in (5.9). Corresponding to (5.10), we can multiply both sides of $X_n[m]$ by the conjugate of $S_n[m]$ and take the same DFT w.r.t. n and m as above, obtaining the CCC-based RDM:

$$V_k^{\mathrm{c}}[l] = \sum_{p=0}^{P-1} \tilde{\alpha}_p S_k^{\mathrm{c}}[l] + W_k^{\mathrm{c}}[l] + Z_k^{\mathrm{c}}[l],$$

$$\text{s.t. } S_k^{\mathrm{c}}[l] = \sum_{n=0}^{\tilde{N}-1} \sum_{m=0}^{\tilde{M}-1} |S_n[m]|^2 e^{-j\frac{2\pi m l_p}{\tilde{M}}} e^{j\frac{2\pi n k_p}{\tilde{N}}} \mathcal{Z}_{\tilde{M}}^{-ml} \mathcal{Z}_{\tilde{N}}^{nk};$$

$$\mathrm{X}_k^{\mathrm{r}}[l] = \sum_{n=0}^{\tilde{N}-1} \sum_{m=0}^{\tilde{M}-1} \mathrm{X}_n[m] S_n^*[m] \mathcal{Z}_{\tilde{M}}^{-ml} \mathcal{Z}_{\tilde{N}}^{nk},\ \mathrm{X} \in \{W, Z\}. \tag{5.16}$$

It should be noted that the CCC-based RDM is different from the ratio-based one, particularly when $S_n[m]$ has nonconstant amplitudes over m. As pointed out in [2], $S_k^{\mathrm{c}}[l]$ is the CCC between $s_n[l]$ given in (5.11) and $x_n[l]$ given in (5.12). This can be readily shown by writing $|S_n[m]|^2 = S_n[m]S_n^*[m]$, replacing $S_n[m]$ with its

DFT expression $\sum_{l'=0}^{M-1} s_n[l']\mathcal{Z}_{\tilde{M}}^{l'm}$ and rewriting the remaining m-related summation. CCC is essentially equivalent to matched filtering in conventional radar signal processing [14]. Thus, CCC will generate a peak at a target delay, facilitating target detection and estimation.

Based on the RDMs obtained in (5.15) and (5.16), target detection and parameter estimation can be performed for various sensing applications. Moreover, as will be proved in Section 5.3, Propositions 5.1 and 5.2 in specific, the IN signals in both RDMs, i.e. $W_k^{\text{r}}[l] + Z_k^{\text{r}}[l]$ and $W_k^{\text{c}}[l] + Z_k^{\text{c}}[l]$, over range-Doppler grids, i.e. k and l, approximately conform to i.i.d. Gaussian distributions. This enables many existing target detectors and parameter estimators to be directly applicable under the sensing framework. To validate the new design and analysis, the cell-averaging constant false-alarm rate detector (CA-CFAR) [14, Chapter 16] will be performed in simulation analysis. The detector is also described in Section 2.4.4.

Next, we summarize the sensing framework in Algorithm 5.1, where CA-CFAR is also briefly described. From the input of Algorithm 5.1, we see some extra parameters, e.g. $\tilde{M}$, $\tilde{Q}$, and $\overline{Q}$ that are not owned by COS. These parameters endow the sensing framework with better flexibility and adaptability than COS. Their design criteria will be illustrated in Section 5.3. In Algorithm 5.1, Steps (1) and (2) perform the block segmentation and VCP. Step (3) transforms the time-domain signal into the frequency domain. Steps (4) and (5) show two different ways of

---

Algorithm 5.1 A Flexible Sensing Framework

Input: $\tilde{M}$, $\tilde{Q}$, $\bar{Q}$, $S_n[m]$ and $\tilde{N}$ given in (5.11), and $x[i]$ given in (5.6):

1. Segment $x[i]$ into $\tilde{N}$ subblocks (SBs): the n-th SB starts from the $n(\tilde{M} - \bar{Q})$ $(n = 0, 1, \dots, \tilde{N} - 1)$ and has $\tilde{M}$ samples.
2. Add the $\tilde{Q}$ samples after each SB onto the first $\tilde{Q}$ within the SB.
3. Take the $\tilde{M}$-point DFT of each SB, attaining $X_n[m]$ given in (5.14).
4. If the ratio-based RDM is preferred, divide $X_n[m]$ by $S_n[m]$ pointwise and take a two-dimensional DFT w.r.t. n and m, leading to (5.15).
5. If the CCC-based RDM is chosen, multiply $X_n[m]$ with $S_n^*[m]$ pointwise and take a two-dimensional DFT, yielding (5.16).
6. Provided P_{F}, N_{g}^k, N_{g}^l, N_{r}^k and N_{r}^l, enumerate each range-Doppler grid by performing the following steps:
   a) Estimate the power of the local IN background according to (5.17).
   b) Calculate the detecting threshold T based on (5.18).
   c) If a power of the grid under test is greater than T, a target exists; otherwise, no target.If a target exists, the coarse estimates of its parameters can be obtained; see (5.19).

---

removing communication data symbols and accordingly generating RDMs. Step (6) and its substeps implement the CA-CFAR.

In Step (6), P_{F} is the expected false-alarm rate; N_{g}^{k} and N_{g}^{l} denote the number of gap samples on each side of the grid under test (GUT) along the k- and l-dimensions; likewise, N_{r}^{k} and N_{r}^{l} denote the number of reference samples. The gap samples will be excluded while the reference samples will be used, when estimating the power of local IN background. Given a Gaussian IN background, the maximum likelihood estimate of the power is the mean of the signal power of the selected reference grids, i.e.

$$\hat{\sigma}_{k^*,l^*}^2 = \frac{1}{|\Omega_{k^*,l^*}^{\mathrm{r}}|} \sum_{(k,l)\in\Omega_{k^*,l^*}^{\mathrm{r}}} |V_k^{\mathsf{X}}[l]|^2, \ \mathsf{X} \in \{\mathrm{r},\mathrm{c}\},$$
$$\Omega_{k^*,l^*}^{\mathrm{r}} = \left\{ (k,l) \middle| \begin{array}{l} k = k^* - N_{\mathrm{r}}^{k} - N_{\mathrm{g}}^{k}, \dots, k^* + N_{\mathrm{r}}^{k} + N_{\mathrm{g}}^{k}; \\ l = l^* - N_{\mathrm{r}}^{l} - N_{\mathrm{g}}^{l}, \dots, l^* + N_{\mathrm{r}}^{l} + N_{\mathrm{g}}^{l} \end{array} \right\} \setminus$$
$$\left\{ \middle| (k,l) \begin{array}{l} k = k^* - N_{\mathrm{g}}^{k}, \dots, k^* + N_{\mathrm{g}}^{k}; \\ l = l^* - N_{\mathrm{g}}^{l}, \dots, l^* + N_{\mathrm{g}}^{l} \end{array} \right\}, \tag{5.17}$$

where (k^*, l^*) denotes the index of GUT, $\Omega_{k^*,l^*}^{\mathrm{r}}$ denotes the index set of reference grids, $\{\}\setminus\{\}$ gives the set difference, and $|\Omega|$ denotes the number of entries in the set Ω. The above equation looks complicated, but is simple in essence; refer to Figure 2.3 for an intuitive interpretation.

Using $\hat{\sigma}_{k^*,l^*}^2$, we can set the CA-CFAR threshold as [14, (16.23)],

$$\mathsf{T} = \beta \hat{\sigma}_{k^*,l^*}^2, \ \beta = |\Omega_{k^*,l^*}^{\mathrm{r}}| \left(\mathsf{P}_{\mathrm{F}}^{-1/|\Omega_{k^*,l^*}^{\mathrm{r}}|} - 1 \right). \tag{5.18}$$

If $|V_{k^*}^{\mathsf{X}}[l^*]|^2 \geq \mathsf{T}$, we report the presence of a target at (k^*, l^*). The coarse estimates of the delay and Doppler frequency of the target, say the p-th, can be obtained as follows:

$$\hat{\tau}_p = l^* T_{\mathrm{s}}; \ \hat{\nu}_p = k^* / \left((\tilde{M} - \overline{Q}) \tilde{N} T_{\mathrm{s}} \right), \tag{5.19}$$

where the relationship among relevant variables, as given in (5.6) and (5.14), is used for the above result. For applications requiring high-accuracy estimations of target location and velocity, various methods for parameter refinement are available in the literature, such as the conventional multiple signal clarification (MUSIC) [15] and the DFT-interpolation-based estimators introduced in Chapter 3.

Illustrated next is the computational complexity of the sensing framework. From Algorithm 5.1, we see that the explicit complexity is dominated by the computations performed in Steps (3)–(6). An implicit complexity, however, lies in the computation of $S_n[m]$ which is an extra compared with previous designs, e.g. COS [1] and C-COS [2]. Since $S_n[m]$ is the result of $\tilde{N}$ numbers of $\tilde{M}$-point DFTs, computing it incurs the complexity of $\mathcal{O}\{\tilde{N}\tilde{M}\log\tilde{M}\}$, where $\mathcal{O}\{\tilde{M}\log\tilde{M}\}$ is complexity of the $\tilde{M}$-point DFT (under the fast implementation [16]). Step 3) shares the same complexity of $\mathcal{O}\{\tilde{N}\tilde{M}\log\tilde{M}\}$. The complexity of generating an RDM, performing either Step (4) or Step (5), is dominated by the two-dimensional DFT and can be given by $\mathcal{O}\{\tilde{N}\tilde{M}\log\tilde{M}+\tilde{M}\tilde{N}\log\tilde{N}\}$. Step (6) essentially processes the RDM by a two-dimensional filter, and hence, can be performed through the same two-dimensional Fourier transform as in Steps (4) and (5) [17]. Consequently, we can say that the overall computational complexity of the sensing framework is $\mathcal{O}\{\tilde{N}\tilde{M}\log\tilde{M}+\tilde{M}\tilde{N}\log\tilde{N}\}$.

5.3 Performance Analysis

In what follows, we shall analyze the interference and noise background in the two RDMs obtained in (5.15) and (5.16). The signal-to-interference-plus-noise ratio (SINR) of the RDMs are also derived, analyzed, and compared, through which the question asked in Section 5.1.3.3 will be answered. Moreover, some insights into the parameter design for the sensing framework will also be provided.

5.3.1 Preliminary Results

From (5.15), we see that both RDMs are obtained based on $X_n[m]$ given in (5.14). Thus, we analyze first its three signal components, i.e. $S_n[m]$, $Z_n[m]$ and $W_n[m]$. Their useful features are provided here. In particular, their distributions are provided in Lemma 5.1 with the proof given in Appendix A.1. The independence of $S_n[m]$, $Z_n[m]$ and $W_n[m]$ over n is given in Lemma 5.2; see Appendix A.2 for the proof. In addition, the independence of the signals over m is illustrated in Lemma 5.3; see Appendix A.3 for the proof.

Lemma 5.1 *The useful signal and the noise in (5.14) satisfy*

$$\begin{aligned} S_n[m] &\sim \mathcal{CN}(0,\sigma_{\mathrm{d}}^2);\\ W_n[m] &\sim \mathcal{CN}\left(0,\sigma_W^2\right),\ \sigma_W^2=\left(1+\tilde{Q}/\tilde{M}\right)\sigma_w^2, \end{aligned} \tag{5.20}$$

where σ_{d}^2 is the power of communication data symbols, i.e. d_{i} given in (5.1), and σ_w^2 is power of the receiver noise, i.e. $w[i]$ given in (5.6). Moreover, provided that $\alpha_0,\alpha_1,\dots,\alpha_{P-1}$ are uncorrelated, the interference term in (5.14) conforms to

$$Z_n[m]\sim\mathcal{CN}\left(0,\sigma_Z^2\right),\ \sigma_Z^2=\frac{\tilde{Q}\sigma_{\mathrm{d}}^2\sigma_P^2}{\tilde{M}},\ \sigma_P^2=\sum_{p=0}^{P-1}\sigma_p^2, \tag{5.21}$$

where σ_{p}^2 is the power of the p-th scattering coefficient, i.e. α_{p} given in (5.6).

Lemma 5.2 *Given $\tilde{M} > (\tilde{Q} + \overline{Q})$ and at any m, $Z_n[m]$ is i.i.d. over n, whereas $S_n[m]$ and $W_n[m]$ are each independent over either the set of odd n's or that of even n's. In addition, we have, at any m,*

$$\begin{aligned}\mathbb{C}\left(S_n[m], S_{n+1}[m]\right) &= \overline{Q}/\tilde{M};\\ \mathbb{C}\left(W_n[m], W_{n+1}[m]\right) &= (\tilde{Q} + \overline{Q})/\tilde{M}.\end{aligned} \tag{5.22}$$

where $n = 0, 1, \ldots, \tilde{N} - 2$ and $\mathbb{C}(x, y) = \frac{|\mathbb{E}\{xy^*\}|}{\sqrt{\mathbb{E}\{|x|^2\}\mathbb{E}\{|y|^2\}}}$ is the absolute correlation coefficient between x and y.

Lemma 5.3 *For any n, $S_n[m]$, and $W_n[m]$ are independent over m, while $Z_n[m]$ is not and satisfies*

$$\mathbb{C}\left(Z_n[m_1], Z_n[m_2]\right) = \frac{\sin\left(\frac{2\pi}{\tilde{M}}\frac{\tilde{Q}(m_1-m_2)}{2}\right)}{\tilde{Q}\sin\left(\frac{2\pi}{\tilde{M}}\frac{(m_1-m_2)}{2}\right)}. \tag{5.23}$$

We notice that the SINR improvement is maximized when the IN background is independent over the range-Doppler grids, also known as "white." However, we see from Lemmas 5.2 and 5.3, the interference and noise signals are somewhat dependent over range-Doppler grids. More interestingly, there is a trade off in this regard caused by $\frac{\tilde{Q}}{\tilde{M}}$. To reduce the correlation of $S_n[m]$ and $W_n[m]$ along n, it is preferred that $\tilde{M} \gg (\tilde{Q} + \overline{Q})$, which also means $\tilde{M} \gg \tilde{Q}$. However, according to (5.23), reducing $\frac{\tilde{Q}}{\tilde{M}}$ will heavily increase the correlation of $Z_n[m]$ over m. In an extreme case, consider $\tilde{Q}$ takes one, the smallest value. We then have $\mathbb{C}\left(Z_n[m_1], Z_n[m_2]\right) = 1$ $(\forall m_1, m_2)$. As will be shown shortly, the dependence of $S_n[m]$, $W_n[m]$, and $Z_n[m]$ over n and m makes it difficult to analyze the distribution of the IN background in the RDMs. This, nevertheless, can be conquered.

5.3.2 Analyzing Signal Components in Two RDMs

Let us start with analyzing the distribution of the IN background, i.e., $W_k^{\mathrm{r}}[l] + Z_k^{\mathrm{r}}[l]$, in the ratio-based RDM. From (5.15), $W_k^{\mathrm{r}}[l] + Z_k^{\mathrm{r}}[l]$ can be rewritten as follows:

$$Z_k^{\mathrm{r}}[l] + W_k^{\mathrm{r}}[l] = \sum_{n=0}^{\tilde{N}-1}\sum_{m=0}^{\tilde{M}-1} \overbrace{\frac{\left(Z_n[m] + W_n[m]\right) z_{\tilde{M}}^{-ml} z_{\tilde{N}}^{nk}}{S_n[m]}}^{D_{n,m}^{k,l}}. \tag{5.24}$$

Since $Z_n[m]$ and $W_n[m]$ are independent Gaussian variables, their sum is also Gaussian distributed. Moreover, given $\forall k, m, n, l$, $z_{\overline{M}}^{-lm}$ and $z_{\tilde{N}}^{nk}$ have deterministic values, as defined in (5.3). Accordingly, applying Lemma 5.1, we obtain

$$D_{n,m}^{k,l} \sim \mathcal{CN}\left(0, (\sigma_Z^2 + \sigma_W^2) / \tilde{M}\tilde{N}\right), \tag{5.25}$$

where the coefficient of the variance is from the two DFT bases; see (5.3). Then, the summand in (5.24) becomes the ratio of two uncorrelated complex Gaussian variables. Such a ratio conforms to a Cauchy distribution [18]. Now that $Z_k^{\mathrm{r}}[l] + W_k^{\mathrm{r}}[l]$ becomes the sum of Cauchy variables, one would think of using the central limit theorem (CLT) to approximate the summation as a Gaussian distribution. *Unfortunately, CLT does not apply to Cauchy variables, as they have infinite variances* [18]. The following remedy can be applied because the CLT applies to the truncated Cauchy distributions [19].

Instead of using $S_n[m]$ as divisor directly, we can use $\alpha S_n[m]$, where α a real positive coefficient. Since $\alpha S_n[m] \sim \mathcal{CN}\left(0, \alpha^2\sigma_{\mathrm{d}}^2\right)$ according to Lemma 5.1, we can take a sufficiently large α such that the probability of the event $|\alpha S_n[m]| < 1$ can be reduced to a small value, say ϵ. Moreover, if $\epsilon I < 1$, then out of I samples of $\mathcal{CN}\left(0, \alpha^2\sigma_{\mathrm{d}}^2\right)$, the event $|\alpha S_n[m]| < 1$ may not happen at all. According to Lemma 6.3 to be illustrated in Chapter 6, the critical value of α, leading to $\epsilon = 1/I$, can be given by

$$\alpha_{\mathrm{c}} = 1 \Bigg/ \left(\sigma_d \sqrt{\ln \frac{I-1}{I}}\right). \tag{5.26}$$

Based on the above illustration, we can revise the ratio-based RDM as follows:

$$\begin{aligned}
\tilde{V}_k^{\mathrm{r}}[l] &= \sum_{n=0}^{N-1}\sum_{m=0}^{M-1} \mathbb{I}_{\mathcal{E}}\left\{\frac{X_n[m]}{\alpha S_n[m]}\right\} Z_M^{-ml} Z_N^{nk} \\
&\approx \sum_{p=0}^{P-1} \frac{\tilde{\alpha}_p}{\alpha} S_{\overline{M}}\left(l - l_p\right) S_{\tilde{N}}\left(k_p - k\right) + \tilde{W}_k^{\mathrm{r}}[l] + \tilde{Z}_k^{\mathrm{r}}[l], \\
\text{s.t. } \tilde{\mathrm{X}}_k^{\mathrm{r}}[l] &= \sum_{n=0}^{N-1}\sum_{m=0}^{M-1} \mathbb{I}_{\mathcal{E}}\left\{\frac{X_n[m]}{\alpha S_n[m]}\right\} Z_M^{-ml} Z_N^{nk}, \mathrm{X} \in \{W, Z\} \\
&\mathbb{I}_{\mathcal{E}}\{\cdot\} = 1 \text{ if event } \mathcal{E} \text{ happens; otherwise } \mathbb{I}_{\mathcal{E}}\{\cdot\} = 0,\ \mathcal{E} \triangleq \{|\alpha S_n[m]| \geq 1\},
\end{aligned} \tag{5.27}$$

where $X_n[m]$ is given in (5.14) and $\alpha S_n[m]$ is used as the divisor compared with using $S_n[m]$. Note that the approximation is based on that $\mathbb{I}_{\mathcal{E}} = 0$ can barely happen with a sufficiently large α. For the same reason, we will drop the operator $\mathbb{I}_{\mathcal{E}}\{\cdot\}$ below for notation simplicity. But bear in mind that $\mathbb{I}_{\mathcal{E}}\{\cdot\}$ has indeed been applied.

It looks like we can invoke the CLT based on (5.27). However, there is one more trap – the summands under the CLT need to be i.i.d., while, as indicated by Lemmas 5.2 and 5.3, the i.i.d. condition is not satisfied here. To this end, we resort to the case $\tilde{M} \gg (\tilde{Q} + \overline{Q})$, under which the correlation of $S_n[m]$ and $W_n[m]$ over

n can be negligibly weak. As illustrated at the end of Section 5.3.1, $\tilde{M} \gg (\tilde{Q} + \overline{Q})$ can severely increase the correlation of $Z_n[m]$ over m. Nevertheless, we discover that applying the CLT along the n-dimension first can approximately remove the correlation of the resulted Gaussian distributions over m. More details are given in Appendix A.4, while the results are summarized in the following proposition.

Proposition 5.1 *Provided that $\tilde{N}$ is large and $\tilde{M} \gg (\tilde{Q} + \overline{Q})$, the IN background of the ratio-based RDM obtained in (5.27) approaches a complex centered Gaussian distribution which satisfies*

$$\tilde{W}_k^r[l] + \tilde{Z}_k^r[l] \sim \mathcal{CN}\left(0, \frac{(\sigma_Z^2 + \sigma_W^2)\beta(\epsilon)}{\alpha^2\sigma_d^2}\right), \text{ s.t.}$$

$$\beta(\epsilon) = 2\ln\left(2(1-\epsilon)/\left(\epsilon\sqrt{\epsilon(2-\epsilon)}\right)\right), \tag{5.28}$$

where σ_Z^2 and σ_W^2 are given in Lemma 5.1, ϵ is a sufficiently small number and e denotes the base of the natural logarithm.

Despite the complex expression of the variance in (5.28), it actually has a clear structure. Specifically, the fraction $\frac{(\sigma_Z^2+\sigma_W^2)}{\alpha^2\sigma_d^2}$ is the ratio between the variance of $D_{n,m}^{k,l}$ and that of $\alpha S_n[m]$; in parallel with that $\tilde{W}_k^r[l] + \tilde{Z}_k^r[l]$ is the ratio between the two random variables. Such a ratio is known to have a heavy-tail PDF [18] and hence, the coefficient $\beta(\epsilon)$, greater than one in general, acts as a penalty factor to account for the heavy tail. Next, we elaborate more on $\beta(\epsilon)$. According to [20, Appendix D], ϵ is the probability that $\left|\Re\{\tilde{W}_k^r[l] + \tilde{Z}_k^r[l]\}\right|$ is larger than a threshold, where $\Re\{\}$ takes the real part of a complex number. Regardless of the specific expression of the threshold, we hope ϵ is such a small probability that out of $\tilde{M}\tilde{N}$ samples of $\tilde{W}_k^r[l] + \tilde{Z}_k^r[l]$, less than one sample can have the magnitude of its real part exceed the threshold. So a critical value of ϵ is $1/\left(\tilde{M}\tilde{N}\right)$. Substituting the value into (5.28) leads to

$$\beta_c = \beta\left(1/\left(\tilde{M}\tilde{N}\right)\right) = 2\left(\ln\left(\frac{2(\tilde{M}\tilde{N}-1)}{\sqrt{2\tilde{M}\tilde{N}-1}}\right) - 1\right). \tag{5.29}$$

Note that $\tilde{M}\tilde{N} \approx I$, the number of samples in the whole block can be tens of thousands and even greater.

With reference to the analysis yielding Proposition 5.1, we can similarly analyze the distributions of the signal components in the CCC-based RDM obtained in (5.16). This time, the CLT is directly applicable to the summands in (5.16), as each is a product of two Gaussian variables and has a limited variance [21]. However, the useful signal in the CCC-based RDM is substantially different from that in the ratio-based RDM; see (5.16) and (5.27). Thus, we provide some more analysis in

Appendix A.5, focusing on the useful signal in the CCC-based RDM. The following proposition summarizes the analytical results:

Proposition 5.2 *Provided large $\tilde{M}$ and $\tilde{N}$ as well as $\tilde{M} \gg (\tilde{Q} + \overline{Q})$, the signal components of the RDM obtained in (5.16) approach Gaussian distributions that approximately satisfy*

$$S_k^{\mathrm{c}}[l] \sim \begin{cases} \mathcal{N}\left(\sigma_{\mathrm{d}}^2\sqrt{\tilde{M}\tilde{N}}, \sigma_{\mathrm{d}}^4\right) & l = \mathrm{l}_p \text{ and } k = \mathrm{k}_p \\ \mathcal{N}\left(0, \sigma_{\mathrm{d}}^4\right) & l \neq \mathrm{l}_p \text{ or } k \neq \mathrm{k}_p \end{cases}; \tag{5.30}$$

$$W_k^{\mathrm{c}}[l] + Z_k^{\mathrm{c}}[l] \sim \mathcal{CN}\left(0, \sigma_{\mathrm{d}}^2(\sigma_Z^2 + \sigma_W^2)\right). \tag{5.31}$$

Detecting in Gaussian IN background has been widely studied for decades. Thus, Propositions 5.1 and 5.2 allow us to employ many existing detectors to detect targets from the RDMs obtained under the sensing framework. The CA-CFAR has been briefly reviewed in Section 5.2. The two propositions also allow us to analyze and compare the SINRs in the two RDMs and draw insights into sensing parameter design. This is carried out next.

5.3.3 Comparison and Insights

The SINRs in the two RDMs are first derived based on Propositions 5.1 and 5.2. Based on (5.27), the power of the useful signal in the ratio-based RDM can be given by $\sigma_P^2\tilde{M}\tilde{N}/\alpha^2$, where $\sigma_P^2 = \sum_{p=0}^{P-1}\sigma_p^2$. Then, combining the power of the IN terms derived in Proposition 5.1, we obtain the SINR of the ratio-based RDM, as given by

$$\gamma_V^{\mathrm{r}} = \frac{\tilde{M}\tilde{N}\sigma_P^2\sigma_{\mathrm{d}}^2}{(\sigma_Z^2 + \sigma_W^2)\beta(\epsilon)} = \frac{\tilde{M}\tilde{N}}{\left(\frac{\tilde{Q}}{\tilde{M}} + \left(1 + \frac{\tilde{Q}}{\tilde{M}}\right)\frac{1}{\gamma_0\sigma_P^2}\right)\beta(\epsilon)}, \quad \text{s.t. } \gamma_0 = \sigma_{\mathrm{d}}^2 / \sigma_w^2,\ \sigma_P^2 = \sum_{p=0}^{P-1}\sigma_p^2, \tag{5.32}$$

where the expressions of σ_Z^2 and σ_W^2 given in Lemma 5.1 are used to get the result. The above SINR can be simplified under certain asymptotic conditions. In particular, we have

$$\gamma_V^{\mathrm{r}} \begin{cases} \overset{\gamma_0 \ll \frac{1}{\sigma_P^2}}{\approx} \dfrac{\tilde{M}\left(\frac{(I-\tilde{Q}-\overline{Q})}{(\tilde{M}-\overline{Q})}\right)\gamma_0\sigma_P^2}{\left(1+\frac{\tilde{Q}}{\tilde{M}}\right)\beta(\epsilon)} \overset{(a)}{\approx} \dfrac{I\gamma_0\sigma_P^2}{\left(1-\frac{\overline{Q}}{\tilde{M}}\right)\left(1+\frac{\tilde{Q}}{\tilde{M}}\right)\beta(\epsilon)}; \\ \overset{\gamma_0 \gg \frac{1}{\sigma_P^2}}{\approx} I \Big/ \left(\left(1 - \frac{\overline{Q}}{\tilde{M}}\right)\frac{\tilde{Q}}{\tilde{M}}\beta(\epsilon)\right) \end{cases} \tag{5.33}$$

where $\overset{\gamma_0 \ll \frac{1}{\sigma_P^2}}{\approx}$ is obtained by (I) suppressing $\frac{\tilde{Q}}{\tilde{M}}$ from the denominator of (5.32) as $\gamma_0 \ll \frac{1}{\sigma_P^2}$ leads to $\frac{\tilde{Q}}{\tilde{M}} \ll \left(1 + \frac{\tilde{Q}}{\tilde{M}}\right)\frac{1}{\gamma_0\sigma_P^2}$; and (II) replacing $\tilde{N}$ with its expression given in (5.11) while suppressing the flooring operator. Moreover, $\overset{(a)}{\approx}$ is due to $I = (M+Q)N \gg (\tilde{Q} + \overline{Q})$. The second line in (5.33) can be obtained similarly.

For the CCC-based RDM, its SINR can be obtained by applying Proposition 5.2 in (5.16). In particular, we have

$$\gamma_V^{\mathrm{c}} = \frac{(\tilde{M}\tilde{N}+1)\sigma_P^2\sigma_d^4}{(\sigma_Z^2+\sigma_W^2)\sigma_d^2+\sigma_P^2\sigma_d^4} = \frac{(\tilde{M}\tilde{N}+1)}{\frac{\tilde{Q}}{\tilde{M}}+\left(1+\frac{\tilde{Q}}{\tilde{M}}\right)\frac{1}{\gamma_0\sigma_P^2}+1}, \tag{5.34}$$

where $\sigma_P^2\sigma_d^4$ in the denominator of the middle result is the interference caused by $S_k^{\mathrm{c}}[l]$ at $l \neq \mathsf{l}_p$ or $k \neq \mathsf{k}_p$. With reference to the way (5.33) is obtained, we can also attain the asymptotic γ_V^{c}, as given by

$$\gamma_V^{\mathrm{c}} \begin{cases} \overset{\gamma_0 \ll \frac{1}{\sigma_P^2}}{\approx} I\gamma_0\sigma_P^2 \Big/ \left(\left(1-\frac{\overline{Q}}{\tilde{M}}\right)\left(1+\frac{\tilde{Q}}{\tilde{M}}\right)\right) \\ \overset{\gamma_0 \gg \frac{1}{\sigma_P^2}}{\approx} I \Big/ \left(\left(1-\frac{\overline{Q}}{\tilde{M}}\right)\left(1+\frac{\tilde{Q}}{\tilde{M}}\right)\right) \end{cases} \tag{5.35}$$

The SINRs derived in (5.32) and (5.34) can be adapted for the RDMs obtained in the framework of COS, i.e. (5.8) and (5.10). As reviewed in Section 5.1.2, COS complies with the underlying communication system. Thus, we can take $\tilde{M} = M$ and $\tilde{N} = N$ in (5.32) and (5.34). Moreover, since COS uses the original communication CP, σ_Z^2 in (5.32) and (5.34), which is the power of the interference caused by VCP, can be suppressed, and σ_W^2 can be replaced by σ_w^2. Under the above changes, the SINRs of the RDMs in (5.8) and (5.10) can be, respectively, given by

$$\gamma_U^{\mathrm{r}} = \frac{MN\gamma_0\sigma_P^2}{\beta(\epsilon)} \overset{(a)}{=} \frac{I\gamma_0\sigma_P^2}{\left(1+\frac{Q}{M}\right)\beta(\epsilon)} \ (\forall\gamma_0);\ \gamma_U^{\mathrm{c}} = \frac{(MN+1)\gamma_0\sigma_P^2}{1+\gamma_0\sigma_P^2} \begin{cases} \overset{\gamma_0 \ll \frac{1}{\sigma_P^2}}{\approx} I\gamma_0\sigma_P^2 \Big/ \left(1+\frac{Q}{M}\right) \\ \overset{\gamma_0 \gg \frac{1}{\sigma_P^2}}{\approx} I \Big/ \left(1+\frac{Q}{M}\right) \end{cases} \tag{5.36}$$

where $\overset{(a)}{=}$ is obtained by replacing N with $\frac{I}{(M+Q)}$, the same replacement is also performed for γ_U^{c}, and the approximations are similarly attained, as done in (5.33). Now, we are ready to make some comparisons using the SINR expressions.

Remark 5.1 For the ratio-based RDM, the following comparisons can be made between COS and the sensing framework illustrated in this chapter:

5.1a) In low-SNR regions, such that $\gamma_0 = \frac{\sigma_d^2}{\sigma_w^2} \ll \frac{1}{\sigma_P^2}$, the sensing framework has a greater SINR than COS with a gain no less than $\frac{1}{1-\frac{\overline{Q}}{\tilde{M}}}$, provided $\frac{\tilde{Q}}{\tilde{M}} \leq \frac{Q}{M}$;

5.1b) Provided the maximum round-trip delay of a target is smaller than the communication CP duration, i.e. $\max_{\forall p}\{\tau_p\} \leq QT_s$, COS can have a greater SINR than the sensing framework for $\gamma_0 > \frac{1+\frac{Q}{M}}{\left(1-\frac{\tilde{Q}}{\tilde{M}}\right)\frac{\tilde{Q}}{\tilde{M}}\sigma_P^2}$;

5.1c) Provided $\max_{\forall p}\{\tau_p\} > QT_s$, the result in Remark 5.1b) may not hold any more; moreover, the sensing framework can have a greater SINR than COS.

The first two results can be readily attained based on (5.33) and (5.36). It is noteworthy that the condition $\max_{\forall p}\{\tau_p\} \leq QT_s$ is implicitly required by COS to remove communication data symbols for generating RDMs; see the review in Section 5.1.2. If the condition is unsatisfied, the SINR of COS, as given in (5.36), becomes invalid. However, because COS cannot effectively remove communication data symbols while the sensing framework can, we attain the result in Remark 5.1c). As will be validated by Figure 5.3 in Section 5.4, the SINRs of the two RDMs under COS degrade severely, as $\max_{\forall p}\{\tau_p\}$ exceeds QT_s.

Remark 5.2 For the CCC-based RDM, provided $\frac{\tilde{Q}}{\tilde{M}} \leq \frac{Q}{M}$, the sensing framework always has a greater SINR than COS regardless of γ_0 and the SINR gain is no less than $\frac{1}{1-\frac{Q}{M}}$. The results can be easily validated using (5.35) and (5.36). An intriguing question is why the relationship between COS and the sensing framework presented in this chapter is substantially different under the two RDMs. This is caused by the different ways the communication data symbols are removed for generating the two RDMs. For the ratio-based RDM, the pointwise division, as shown in (5.28), magnifies the IN background by introducing the multiplicative coefficient $\beta(\epsilon)$. In contrast, the pointwise product for the CCC-based RDM, see (5.16), introduces an additive interference $\sigma_P^2\sigma_d^4$; see (5.34). While the VCP-incurred interference does not bother COS, the CCC-incurred interference exists in COS and the sensing framework presented in this chapter.

Remark 5.3 Some comparisons between the ratio- and CCC-based RDMs are made here. Based on (5.33) and (5.35), we can attain the following results for the sensing framework presented in this chapter:

5.3a) In low-SNR regions where $\gamma_0 \ll 1/\sigma_P^2$, the CCC-based RDM has an SINR that is $\beta(\epsilon)$ times the SINR in the ratio-based RDM, where $\beta(\epsilon) > 1$ in general; see (5.29).

5.3b) In high-SNR regions where $\gamma_0 \gg 1/\sigma_P^2$, the ratio-based RDM can have a greater SINR than the CCC-based RDM, provided $\beta(\epsilon) \leq \frac{\tilde{M}}{\tilde{Q}}$.

5.3c) Regardless of γ_0, the CCC-based RDM always has a greater SINR than the ratio-based RDM, if $\beta(\epsilon) > \frac{\tilde{M}}{\tilde{Q}} + 1$.

Based on (5.36), similar results as above can be given for COS:

5.3d) The result in 5.3a) directly applies to COS;
5.3e) In high-SNR regions where $\gamma_0 \gg 1/\sigma_P^2$, the CCC-based RDM has a greater SINR than the ratio-based RDM, if $\beta(\epsilon) > \gamma_0\sigma_P^2$, while if $\beta(\epsilon) < \gamma_0\sigma_P^2$ the ratio-based RDM has a greater SINR.

5.3.4 Criteria for Setting Key Sensing Parameters

Unlike COS that follows the underlying communication system, the sensing framework presented in this chapter has the flexibility to cater different sensing needs via adjusting several key parameters: $\tilde{M}$, $\tilde{Q}$ and $\overline{Q}$ ($\tilde{N}$ is determined given the former three). Below, we illustrate the criteria for setting these parameters to optimize sensing performance.

First, we can set $\tilde{Q}$ based on the required maximum sensing distance, as denoted by $r_{\max}$. From Section 5.2, the sensing distance of the design here is given by $\frac{C\tilde{Q}T_s}{2}$, which, equating with $r_{\max}$, yields

$$\tilde{Q} = 2r_{\max}/CT_s. \tag{5.37}$$

It is worth noting that the issue of CP-limited sensing, as described in Section 5.1.3.1, is addressed by introducing $\tilde{Q}$. Unlike in COS and its variants where $r_{\max}$ is determined by Q, we now can set $\tilde{Q}$ to satisfy $r_{\max}$ (provided a sufficient link budget).

Second, we determine $\tilde{M}$ given the requirements on velocity measurement. Applying the analysis in Section 5.1.3.2, the maximum measurable value and the resolution of Doppler frequency of the sensing framework presented in this chapter are given by

$$\nu_{\max} = 1/\left(2(\tilde{M} - \overline{Q})T_s\right); \; \Delta_\nu \approx 1/\left(IT_s\right). \tag{5.38}$$

Thus, to cater the expected $\nu_{\max}^*$ we need to keep $\tilde{M} \leq 1/\left(2\nu_{\max}^* T_s\right) + \overline{Q}$. Moreover, we prefer to have a relatively large[2] $\tilde{M}$ which can lead to a small $\frac{\overline{Q}}{\tilde{M}}$ and hence a high SINR in both RDMs; see (5.33) and (5.35). It is noteworthy that the issue of limited velocity measurement, as illustrated in Section 5.1.3.2, is now addressed by introducing $\tilde{M}$ and $\overline{Q}$. Instead of having an M-limited $\nu_{\max}$, we now have the flexibility of configuring $\tilde{M}$ to satisfy the requirement on $\nu_{\max}$.

Third, given $\tilde{M}$, we can then set $\overline{Q}$. To increase the SINR in both RDMs, we expect to have $\overline{Q}$ as large as possible; see (5.33) and (5.35). However, the larger $\overline{Q}$ the more

2 Note that, if $\overline{Q}/\tilde{M}$ is fixed, the larger $\tilde{M}$ the smaller $\tilde{N}(\approx I/(\tilde{M}(1 - \overline{Q}/\tilde{M})))$ will become. This may affect the precision of results given in Propositions 5.1 and 5.2 which expect both $\tilde{M}$ and $\tilde{N}$.

correlated the signals between adjacent subblocks can be; see Lemmas 5.2 and 5.3. As seen from Propositions 5.1 and 5.2, the correlation can make the results less precise. The detailed impact, however, is difficult to analyze. As will be shown through the simulations in Figures 5.4 and 5.5, the derivations and analysis in Sections 5.3.2 and 5.3.3 are consistently precise when $\overline{Q}$ takes from a small value to the one as large as $\tilde{M}/2 - \tilde{Q}$.

5.4 Simulation Results

Simulations are performed in this section to validate the design in this chapter. The simulation parameters are set with reference to [2] and are summarized in Table 5.1. The root raised cosine (RRC) filter with the roll-off coefficient of 0.2 is used at both the communication transmitter and the sensing receiver. In generating target echo signals, a four-times upsampling is performed by the transmitter RRC filter; the target delay and Doppler frequency are added at the high sampling rate; a four-times decimating is performed at the receiver RRC filter. This generates off-grid range and Doppler values, making the simulations comply with practical scenarios. Further, the Swerling 0 target model [14, Table 7-3] is employed in the simulation. Namely, $\alpha_p = \sqrt{\sigma_p^2}$ is taken over independent trials. However, to have independent scattering coefficients, the phase of α_p ($\forall p$) is uniformly drawn from $[0, 2\pi]$, yet independently over targets and trials.

Table 5.1 Simulation parameters

Var.	Description	IEEE 802.11ad
f_c	Carrier frequency	60.48 GHz
B	Bandwidth	1.825 GHz
M	No. of subcarriers per symbol	512
Q	CP length	128
N	No. of symbols	143 (0.05 ms packet)
I	Total No. of samples; see (5.4)	$N(M+Q) = 91520$
σ_d^2	Power of data symbol d_i; see (5.1)	0 dB
σ_p^2	Power of α_p; see (5.6)	$[0, -10, -20]$ dB
r_p	Target range	$\mathcal{U}_{[0,10]}$ m †
v_p	Target velocity	$\mathcal{U}_{[-139,139]}$ m/s
σ_w^2	Variance of AWGN $w[i]$; see (5.6)	-20 dB

a) $\mathcal{U}_{[x,y]}$ denotes a uniform distribution in the region given by the subscript.

The benchmark sensing framework is COS [1], as reviewed in Section 5.1.2. The original COS, as developed for OFDM [1], uses the ratio-based RDM given in (5.8), while the variant of COS, as developed for DFT-S-OFDM [2], employs the CCC-based RDM given in (5.10). As illustrated in Section5.1.1, for both OFDM and DFT-S-OFDM, the time-domain transmitted communication signals conform to Gaussian distributions, which is similar to OTFS. Therefore, we unitedly use OTFS modulation for all methods to be simulated for a fair comparison. In essence, it is the way a block of the echo signal is segmented, rather than the communication waveforms, that differentiates COS and the sensing framework presented in this chapter. Substantially differentiating the sensing framework here from the existing ones of the same kind, e.g. [1] and [2], is mainly the way how the RDMs are generated. Therefore, following the most related previous works [1] and [2], the SINRs of the RDMs and the target detection performance that is directly related to SINRs, are employed as the main performance metrics in the following simulations.

In the legends of the simulation results, we use "r" to indicate the "ratio-based RDM," "c" the "CCC-based RDM," "sim" the simulated result and "pp" the "presented design."

5.4.1 Illustrating SINRs in RDMs

Figure 5.2 plots the SINRs of the ratio- and CCC-based RDMs versus γ_0, under different values of $\tilde{M}$. In this simulation, $\tilde{Q} = Q$, $\overline{Q} = 150$ and other parameters are given in Table 5.1. Overall, the derived SINRs can precisely describe the actual (simulated) SINRs. This validates the analysis in Section 5.3.2. More specifically, we see from Figure 5.2a that the design given in this chapter achieves higher SINRs in the ratio-based RDM than COS in the case of $\gamma_0 \ll 1$, which validates Remark 5.1a). We also see Figure 5.2a that as $\tilde{M}$ increases, the gap between the design given in this chapter and COS becomes smaller for $\gamma \ll 1$. This is consistent with the SINR expression derived in (5.32). We further see that, when $\gamma_0 \gg 1$, COS can outperform the design given in this chapter, which complies with Remark 5.1b). We see from Figure 5.2b that the SINR achieved in the CCC-based RDM first increases with γ_0 and then converges for large γ_0's. This is consistent with (5.34). Moreover, we see that, for the CCC-based RDM, the design given in this chapter achieves a higher SINR across the whole region of γ_0 compared with COS. This aligns with Remark 5.2. The comparison between the ratio- and CCC-based RDMs validates the analysis given in Remark 5.3.

It is noteworthy that the value of $\tilde{M}$ determines the maximum measurable Doppler frequency of the sensing framework presented in this chapter, as specified in (5.38). Approximately, the maximum Doppler frequency under $\tilde{M} = 600$ is the triple of that under $\tilde{M} = 1800$. Figure 5.2 shows that the design given in this

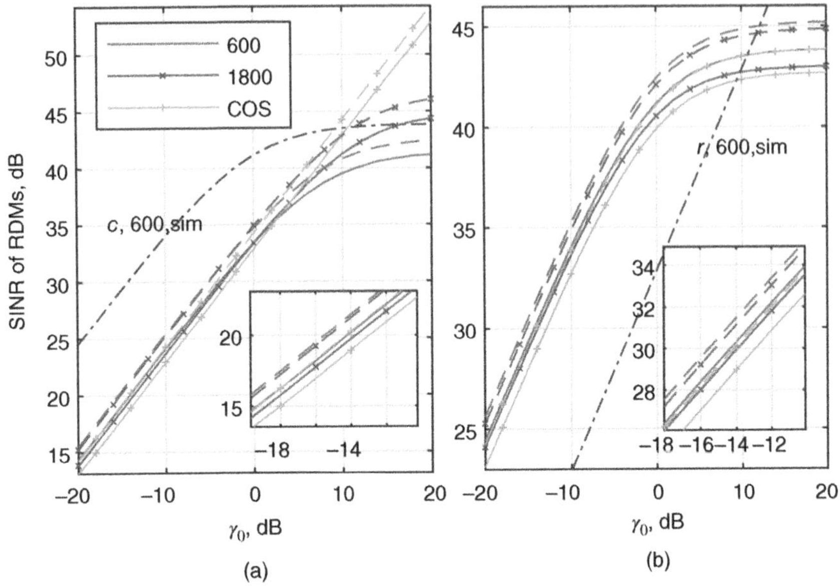

Figure 5.2 SINRs in the RDMs versus γ_0 defined in (5.32), where three targets are set as specified in Table 5.1. The ratio-based RDMs are shown in Figure 5.2(a), while the CCC-based RDMs in Figure 5.2(b). The two subfigures share the same legend, where the numbers are the values of $\tilde{M}$ used for the design given in this chapter. Corresponding to the solid curves, the dash ones are the theoretical SINRs derived in (5.32), (5.34) and (5.36). For comparison convenience, the curve "c, 600, sim" in Figure 5.2a is copied from Figure 5.2b and the curve "r, 600, sim" in Figure 5.2b is from Figure 5.2a.

chapter achieves consistent SINRs under the two substantially different cases. This manifests the flexibility of the sensing framework presented in this chapter in adapting to different requirements on Doppler measurement. Such flexibility is not owned by any existing designs of a similar kind, e.g. COS [1] and C-COS [2].

Figure 5.3 illustrates the SINRs achieved by the sensing framework presented in this chapter in the two types of RDMs under different values of $\tilde{Q}$. The figure is dedicated to demonstrating the ability of the sensing framework presented in this chapter to extend the maximum sensing distance (a function of $\tilde{Q}$ given in (5.37)). Since the ability is independent of the number of targets, we only consider a single target in this simulation. As said in the caption of the figure, we keep the ratios $\frac{\tilde{Q}}{\bar{M}}$ and $\frac{\bar{Q}}{\tilde{M}}$ fixed under different $\tilde{Q}$'s. Then, according to (5.32) and (5.34), we know that the SINRs achieved by the design given in this chapter should be the same over $\tilde{Q}$, which is clearly validated by Figure 5.3. This illustrates the great flexibility of the proposed design in extending or reducing sensing distance as per practical sensing

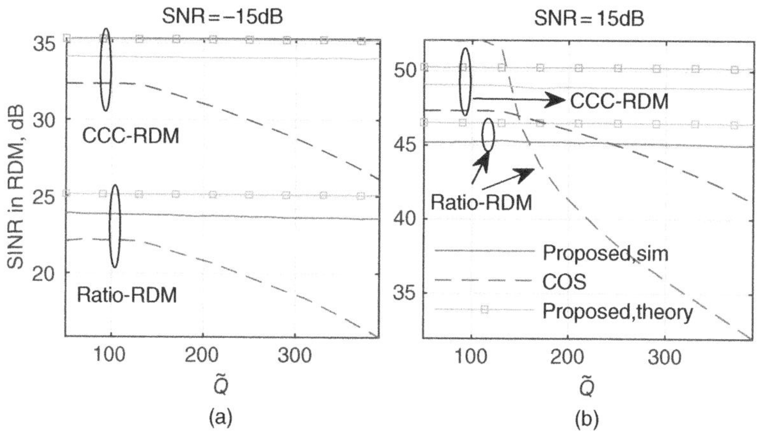

Figure 5.3 SINRs in RDMs versus $\tilde{Q}$ (the length of VCP), where the values of $\tilde{M}$ and $\overline{Q}$ are changed with $\tilde{Q}$ to keep $\frac{\tilde{Q}}{\tilde{M}} = \frac{1}{4}$ and $\frac{\overline{Q}}{\tilde{M}} = \frac{1}{3}$. A single target is simulated with its distance set as $\frac{(\tilde{Q}-1)C}{2B}$ m, while other parameters are set as in Table 5.1.

needs. In contrast, COS degrades severely when $\tilde{Q}$ exceeds $Q = 128$. This is because COS strictly follows the underlying communication system and cannot sense well when the echo delay is larger than the CP length; see (5.7).

In Figure 5.3, the gap between the theoretical and simulated results is less than 2 dB and is caused by the approximation used in deriving the theoretical SINR. In particular, the theoretical gain of the signal power is taken as $\tilde{M}\tilde{N}$ in Section 5.3.2, which, according to (5.14) and (5.27), is only accurate when $\mathfrak{l}_p$ ($\forall p$) and $\mathfrak{k}_p$ ($\forall p$) are integers. As mentioned at the beginning of Section 5.4, we employ an upsampling and downsampling method to generate off-grid range and Doppler values, resulting in the gap seen in Figure 5.3.

Figures 5.4 and 5.5 demonstrate another great flexibility of the sensing framework presented in this chapter by showing the SINRs achieved in the two types of RDMs under different values of $\overline{Q}$. Note that three targets are set as detailed in Table 5.1. Overall, we see from the two figures that SINRs increase with $\overline{Q}$, but the slopes decrease as $\tilde{M}$ becomes larger. This can be seen from the analytical SINR expressions derived in (5.32) and (5.34). Moreover, we see from Figure 5.4 that the SINR performance under low and high SNRs are different, while such a phenomenon is not seen in Figure 5.5. The rationale for this result can be seen from Remark 5.2. It is worth highlighting that, as seen from the figure, the analytical results match the simulated ones in the whole region of $\overline{Q}$. This manifests the high flexibility of the design given in this chapter.

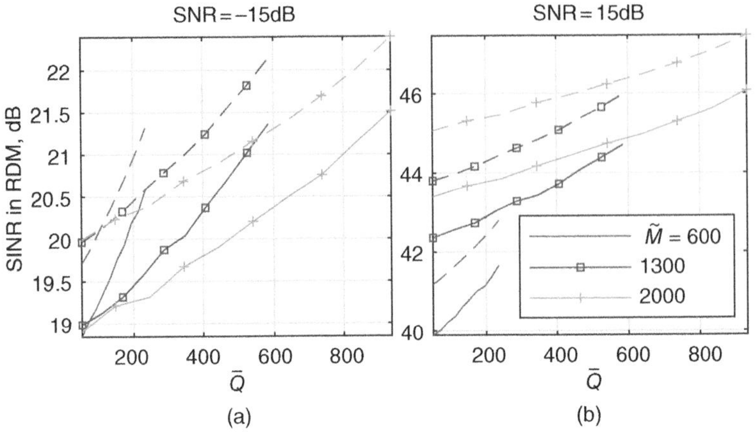

Figure 5.4 SINRs in the ratio-based RDM versus $\bar{Q}$ (the number of overlapping samples between adjacent subblocks), where $\tilde{Q} = Q$. The solid curves are simulated results, while the dash ones are theoretical.

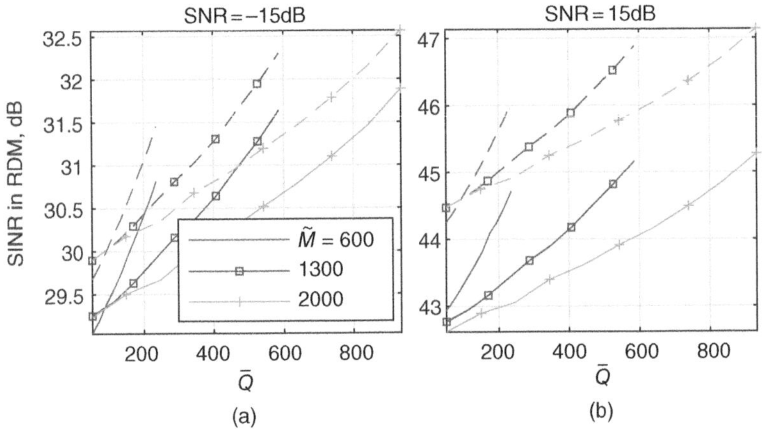

Figure 5.5 SINRs in the CCC-based RDM versus $\bar{Q}$ (the number of overlapping samples between adjacent subblocks), where $\tilde{Q} = Q$. The solid curves are simulated results while the dash ones are theoretical.

5.4.2 Illustration of Target Detection

Next, we translate the SINR results obtained above into the actual detecting performance of the sensing framework presented in this chapter. To do so, we perform the CA-CFAR according to the steps given in Algorithm 5.1. In the following simulations, the parameters in Step (6), Algorithm 5.1 are set as follows: $N_g^k = N_g^l = 3$, $N_r^k = 2$ and $N_r^l = 5$. For a better time efficiency of simulating and calculating the detecting probability, we make two changes in the target scenario. *First*, we increase the number of targets to $P = 10$ and have their ranges linearly spaced in [0, 10] m. The velocities of the targets are still uniformly distributed, as illustrated in Table 5.1. *Second*, the powers of the targets are set as follows: $\sigma_0^2 = 0$ dB, $\sigma_p^2 = -20$ dB for $p = 1, \dots, 4$ and $\sigma_p^2 = -30$ dB for $p = 5, \dots, 9$. The setting of 10 targets applies to all simulation results in this section, i.e. Figures 5.6–5.11.

Figure 5.6 shows the detecting probability of COS and the sensing framework presented in this chapter under different $\tilde{M}$. Figure 5.7 shows $\mathrm{P_F}$ versus γ_0 corresponding to each curve in Figure 5.6. Jointly observing the two figures, we see that the sensing framework presented in this chapter achieves better detecting performance than COS for all cases of the ratio-based RDM and most cases of the CCC-based RDM. Moreover, the improvement of the detecting probability is

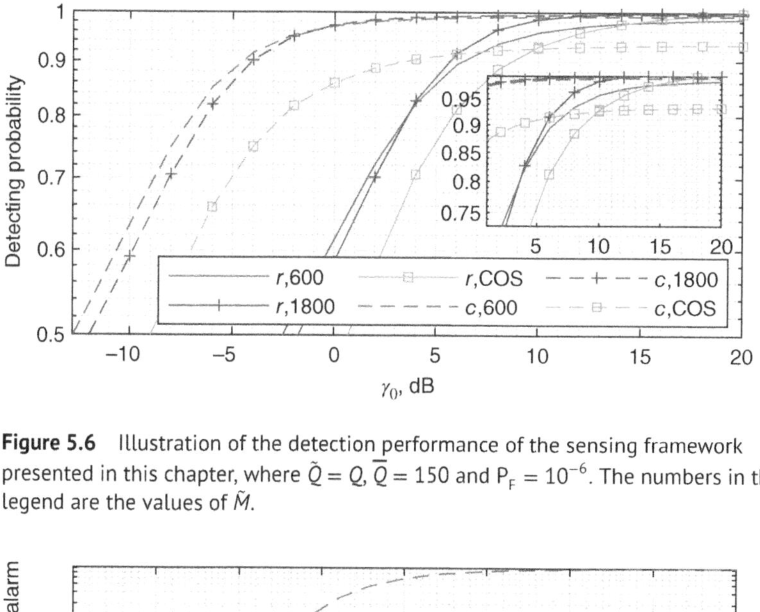

Figure 5.6 Illustration of the detection performance of the sensing framework presented in this chapter, where $\tilde{Q} = Q$, $\overline{Q} = 150$ and $\mathrm{P_F} = 10^{-6}$. The numbers in the legend are the values of $\tilde{M}$.

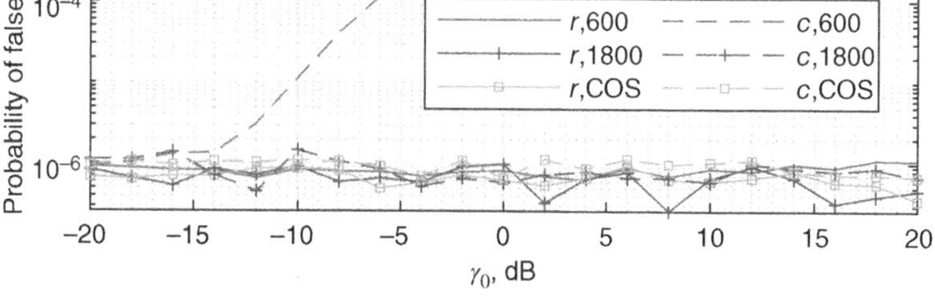

Figure 5.7 Illustrating $\mathrm{P_F}$ corresponding to the curves in Figure 5.6.

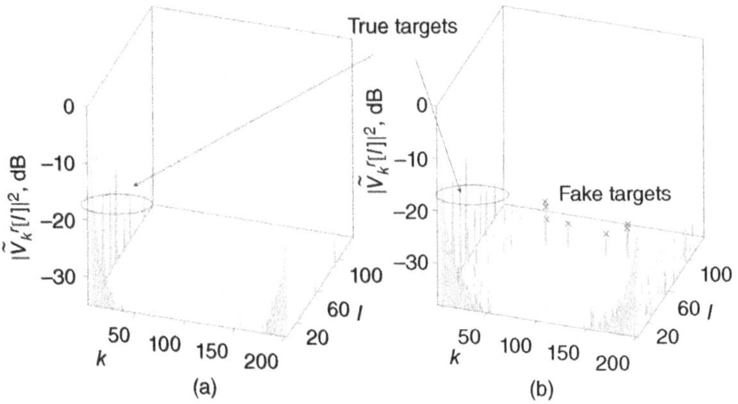

Figure 5.8 Comparing the ratio- and CCC-based RDMs, as given in Figures 5.8(a) and 5.8(b), respectively. For illustration clarity, both RDMs are noise-less, and the *z*-axis is limited to avoid heavy interference background.

precisely predicted by the SINR results observed in Figure 5.2. This validates the analysis and derivations provided in Section 5.3. From Figure 5.7, we see that $\mathrm{P_F} = 10^{-6}$ is not achieved for the CCC-based RDM under $\tilde{M} = 600$. In such a case, $\mathrm{P_F}$ increases with γ_0. The reason is that the overlapping of consecutive subblocks makes the essential signal of a block partially correlated with the interference from its previous subblock; see Figure 5.1. This is validated in Figure 5.8, where we can see the fake targets in the CCC-based RDM.

Figures 5.9 and 5.10 illustrate the receiver operating characteristic (ROC) of the sensing framework presented in this chapter in comparison with that of COS. The cases of $\tilde{Q} = 100$ and 400 in Figure 5.3 are considered here, corresponding to the maximum ranges of 8 m and 33 m, respectively. From Figures 5.9 and 5.10, we see that the design given in this chapter is robust under different maximum ranges, while COS, as predicted in Figure 5.3, degrades severely when the maximum range exceeds that specified by underlying communication systems, i.e. 10 m. This demonstrates the superior flexibility of the design given in this chapter in handling different sensing requirements.

Figure 5.11 shows another flexibility of the design given in this chapter by introducing $\overline{Q}$. We see that, in overall, the detecting performance of the design given in this chapter becomes better as $\overline{Q}$ increases. This is consistent with the SINR results observed in Figures 5.4 and 5.5. We also see that for the low SNR shown in Figure 5.11a, the impact of $\overline{Q}$ is more prominent compared with that in the high SNR case shown in Figure 5.11b. This is reasonable as $\overline{Q}$ is introduced to increase the number of subblocks, and hence, the SINR in the RDM. However, such an improvement is limited as the larger $\overline{Q}$, the higher correlation of the IN background over subblocks, as illustrated in Section 5.3.1.

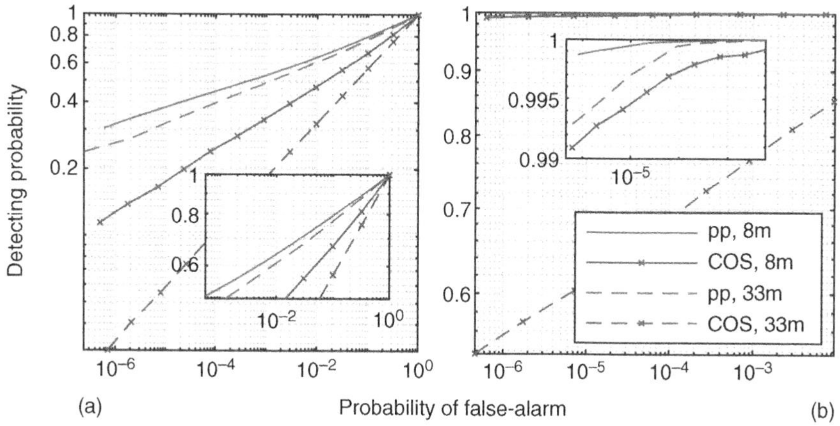

Figure 5.9 Comparing the receiver operating characteristic (ROC) of COS and the presented (pp) sensing framework using the ratio-based RDM, where $\gamma_0 = -15$ dB is set for Figure 5.9(a) and $\gamma_0 = 15$ dB for Figure 5.9(b). In the legend, 8 m and 33 m are the maximum ranges of targets.

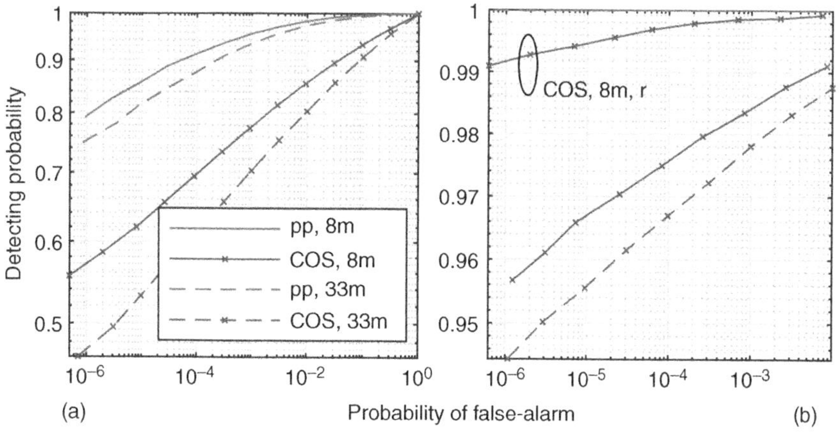

Figure 5.10 ROC curves under the CCC-based RDMs, corresponding to Figure 5.9. The one labeled "COS, 8m, r" is copied from Figure 5.9(b) for comparison.

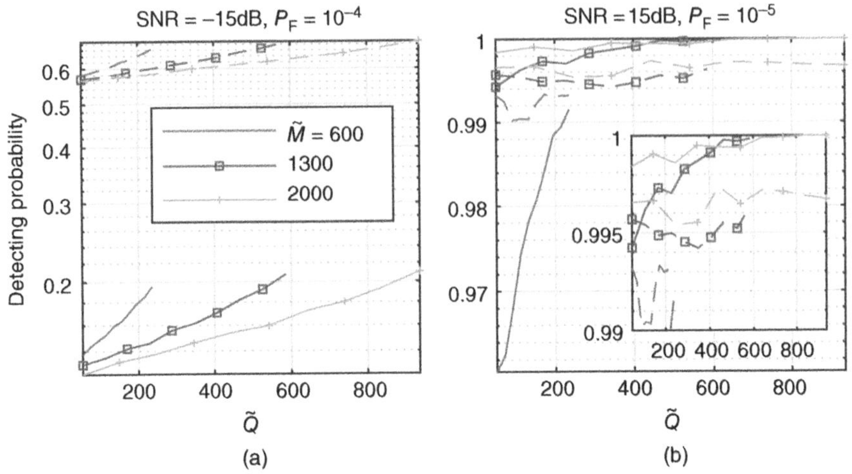

Figure 5.11 Illustration of the detecting probability versus $\overline{Q}$ corresponding to the results presented in Figures 5.4 and 5.5. The solid curves are for the ratio-based RDMs while the dash ones are for the CCC-based RDMs.

5.5 Conclusions

In this chapter, a unified sensing framework is presented employing communication waveforms. It applies to not only cyclic prefixed waveforms, such as OFDM and DFT-S-OFDM, but also those with reduced CP, like RCP-OTFS. Unlike COS and its variants that have been widely applied so far, the presented sensing framework is much less restricted by the underlying communication system and has the flexibility of adapting to different sensing needs. This is achieved by a new block segmentation design that segments a whole block of echo signal evenly into multiple subblocks that can overlap between adjacent ones. This is also accomplished by a newly introduced VCP that allows us to attain the ratio- and CCC-based RDMs under any block segmentation. Comprehensive analyses show that the IN terms in both RDMs approximately conform to centered Gaussian distributions whose variances are also derived. Further analyses comparing COS and the presented sensing framework are also provided. Extensive simulations validate the flexibility of the sensing framework presented in this chapter and its superiority over COS and C-COS.

For illustration clarity and simplicity, we have ignored the potential impact of practical transceivers on the performance of the sensing framework presented in this chapter. In particular, as seen from Figure 5.1, a copy of $\tilde{s}[i]$, the time-domain transmitted communication signal, is shared with the sensing receiver before being processed by the RF chain. However, for the sensing echo signals, $\tilde{s}[i]$ would go through the transmitter RF chain and the receiver one. No two RF chains can be identical for sure. Thus, directly using $\tilde{s}[i]$ for sensing can cause performance loss. (We would mention that this seems to be a common problem in current JCAS works of a similar kind.) It is interesting and important to further explore how these practical hardware factors affect sensing performance.

References

1 C. Sturm and W. Wiesbeck. Waveform design and signal processing aspects for fusion of wireless communications and radar sensing. *Proceedings of the IEEE*, 99(7):1236–1259, 2011.

2 Y. Zeng, Y. Ma, and S. Sun. Joint radar-communication with cyclic prefixed single carrier waveforms. *IEEE Transactions on Vehicular Technology*, 69(4):4069–4079, 2020. doi: 10.1109/TVT.2020.2975243.

3 P. Kumari, J. Choi, N. González-Prelcic, and R. W. Heath. IEEE 802.11ad-based radar: An approach to joint vehicular communication-radar system. *IEEE Transactions on Vehicular Technology*, 67(4):3012–3027, 2018. doi: 10.1109/TVT.2017.2774762.

4 E. Grossi, M. Lops, L. Venturino, and A. Zappone. Opportunistic radar in IEEE 802.11ad networks. *IEEE Transactions on Signal Processing*, 66(9):2441–2454, May 2018. ISSN 1941-0476. doi: 10.1109/TSP.2018.2813300.

5 Z. Wei, W. Yuan, S. Li, J. Yuan, G. Bharatula, R. Hadani, and L. Hanzo. Orthogonal time-frequency space modulation: A full-diversity next generation waveform. *arXiv preprint arXiv:2010.03344*, 2020.

6 Z. Wei, W. Yuan, S. Li, J. Yuan, and D. W. K. Ng. Transmitter and receiver window designs for orthogonal time-frequency space modulation. *IEEE Transactions on Communications*, 69(4):2207–2223, 2021. doi: 10.1109/TCOMM.2021.3051386.

7 S. S. Das, V. Rangamgari, S. Tiwari, and S. C. Mondal. Time domain channel estimation and equalization of CP-OTFS under multiple fractional Dopplers and residual synchronization errors. *IEEE Access*, 9:10561–10576, 2021. doi: 10.1109/ACCESS.2020.3046487.

8 P. Raviteja, Y. Hong, E. Viterbo, and E. Biglieri. Practical pulse-shaping waveforms for reduced-cyclic-prefix OTFS. *IEEE Transactions on Vehicular Technology*, 68(1):957–961, 2019. doi: 10.1109/TVT.2018.2878891.

9 S. Wei, D. L. Goeckel, and P. A. Kelly. Convergence of the complex envelope of bandlimited OFDM signals. *IEEE Transactions on Information Theory*, 56(10):4893–4904, 2010. doi: 10.1109/TIT.2010.2059550.

10 A. V. Oppenheim. *Discrete-Time Signal Processing*. Pearson Education India, 1999.

11 F. Roos, J. Bechter, C. Knill, B. Schweizer, and C. Waldschmidt. Radar sensors for autonomous driving: Modulation schemes and interference mitigation. *IEEE Microwave Magazine*, 20(9):58–72, 2019. doi: 10.1109/MMM.2019.2922120.

12 G. Hakobyan and B. Yang. High-performance automotive radar: A review of signal processing algorithms and modulation schemes. *IEEE Signal Processing Magazine*, 36(5):32–44, 2019. doi: 10.1109/MSP.2019.2911722.

13 S. Ahmadi. *5G NR: Architecture, Technology, Implementation, and Operation of 3GPP New Radio Standards*. Academic Press, 2019.

14 M. A. Richards, J. Scheer, W. A. Holm, and W. L. Melvin. *Principles of Modern Radar*. CiteSeer, 2010.

15 H. L. Van Trees. *Optimum Array Processing: Part IV of Detection, Estimation, and Modulation Theory*. John Wiley & Sons, 2004.

16 K. Wu, J. A. Zhang, X. Huang, and Y. J. Guo. A low-complexity method for FFT-based OFDM sensing. *arXiv preprint arXiv:2105.13596*, 2021.

17 M. Kronauge and H. Rohling. Fast two-dimensional CFAR procedure. *IEEE Transactions on Aerospace and Electronic Systems*, 49(3):1817–1823, 2013.

18 R. J. Baxley, B. T. Walkenhorst, and G. Acosta-Marum. Complex Gaussian Ratio distribution with applications for error rate calculation in fading channels with imperfect CSI. In *2010 IEEE Global Telecommunications Conference GLOBECOM 2010*, pages 1–5, 2010. doi: 10.1109/GLOCOM.2010.5683407.

19 F. Hampel and E. Zurich. Is statistics too difficult? *Canadian Journal of Statistics*, 26(3):497–513, 1998.

20 K. Wu, J. A. Zhang, X. Huang, and Y. J. Guo. OTFS-based joint communication and sensing for future industrial IoT. *IEEE Internet of Things Journal*, 1, early access, 2021. doi: 10.1109/JIOT.2021.3139683.

21 N. O'Donoughue and J. M. F. Moura. On the product of independent complex Gaussians. *IEEE Transactions on Signal Processing*, 60(3):1050–1063, 2012. doi: 10.1109/TSP.2011.2177264.

6

Sensing Framework Optimization

This chapter optimizes the parameters of the sensing framework introduced in Chapter 5 to further improve its sensing performance. Chapter 5 presents a unified sensing framework that performs low-complexity and flexible sensing based on a block of consecutive communication symbols. With the framework established, the signal and interference-plus-noise signals have been analyzed with their distributions unveiled and statistics derived. However, we have ignored the impact of intercarrier interference (ICI) hitherto. Such interference is inevitable in some circumstances, and its impact can be nontrivial in multicarrier waveform-based joint communications and sensing (JCAS) systems. Thus, a more in-depth analysis is provided in this chapter to uncover the impact of ICI on the sensing framework introduced in Chapter 5. We also show how to optimize the parameters of the sensing framework introduced in Chapter 5 to further improve its sensing performance.

6.1 Echo Preprocessing

Here, we continue using the signal model established in Section 5.1. This means the sensing receiver is now having the time-domain echo signal $x[i]$ given in (5.6) and knows the communication-transmitted signal $\tilde{s}[i]$. As shown in (5.6), $x[i]$ is underlain by $\tilde{s}[i]$. As also stated in the text above (5.6), $\tilde{s}[i]$ can be either $\tilde{s}_{\mathrm{CP}}[i]$ given in (5.4) or $\tilde{s}_{\mathrm{RCP}}[i]$ in (5.5). According to the sensing framework built in Section 5.2, we shall first segment $x[i]$ in a sensing-favorable manner. Note from Figure 5.1 that $\bar{Q}$ is an auxiliary variable in the sensing framework; it decides the number of overlapping samples between any adjacent segmentations. Previously, we only consider the nonnegative values of $\bar{Q}$. Here, we show a different case of $\bar{Q} < 0$ which can also be of interest in practice, particularly when less interference can be tolerated.

Joint Communications and Sensing: From Fundamentals to Advanced Techniques, First Edition.
Kai Wu, J. Andrew Zhang, and Y. Jay Guo.

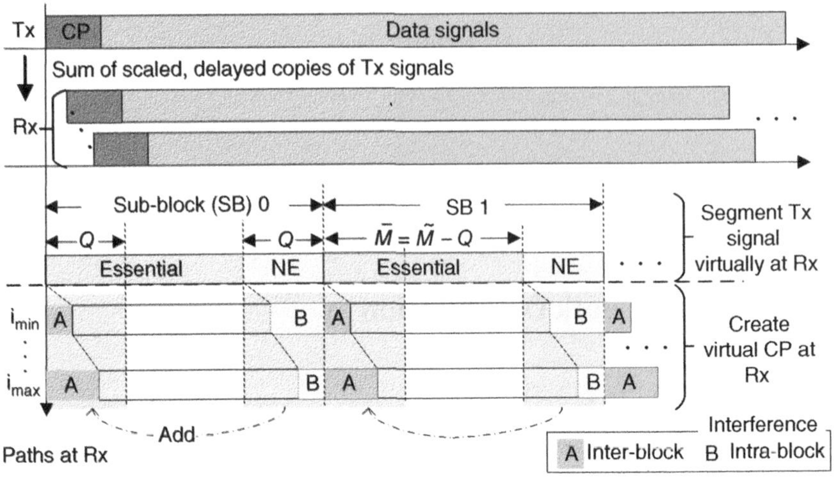

Figure 6.1 Illustrating virtual cyclic prefix (VCP), where i_{min} denotes the minimum sample delay, i_{max} the maximum, and "NE" stands for "non-essential." We emphasize that segmenting and adding VCP are performed on the sensing receiver side, without requiring any changes made on the OTFS transmitter.

6.1.1 Reshaping

With reference to Figure 5.1, the case of $\bar{Q} < 0$ is replotted in Figure 6.1. Moreover, we let $\bar{Q} = \tilde{Q}$, where $\tilde{Q}$ is the length of virtual cyclic prefix (VCP) shown in Figure 5.1. The relation is not a must, but it can help simplify notations. In particular, it allows us to drop the accents of Q in $\tilde{Q}$ and $\bar{Q}$ and simply use Q. However, one should keep in mind that the Q samples shown in Figure 6.1 entails the two functions of the $\tilde{Q}$ and $\bar{Q}$ samples shown in Figure 5.1. Next, we briefly apply the steps illustrated in Section 5.2 to generate the ratio-based range-doppler map (RDM) under the case. Let $\tilde{N}$ denote the number of subblocks and $\tilde{M}$ the number of samples in a subblock. Note that it is not required to have $\tilde{M} = M$ or $\tilde{N} = N$, but $\tilde{M}\tilde{N} \leq MN$ is necessary. Here, M and N are the numbers of communication-transmitted symbols and subcarriers per symbol, respectively.

The variables to be employed below have been defined in Chapter 5. For convenience, we recall that ν_p denotes the Doppler frequency of the pth target, T_s the sampling interval, $\tilde{\alpha}_p$ the target response, l_p target delay, $w_n[m]$ the additive Gaussian noise.

Provided that the following condition holds,

$$\nu_p \tilde{M} T_s \approx 0, \quad \forall p, \tag{6.1}$$

the target echo in the nth subblock can be written into

$$x_n[m] = \sum_{p=0}^{P-1} \tilde{\alpha}_p s\left[\left\langle n\tilde{M} + m - l_p \right\rangle_{MN}\right] e^{j2\pi\nu_p(n\tilde{M}+m)T_s} + w_n[m],$$

$$\forall n \in [0, \tilde{N}-1],\ m \in [0, \tilde{M}-1], \tag{6.2a}$$

$$\approx \sum_{p=0}^{P-1} \tilde{\alpha}_p s[n\tilde{M} + m - l_p] e^{j2\pi\nu_p n\tilde{M}T_s} + w_n[m], \tag{6.2b}$$

where $\tilde{\alpha}_p = \alpha_p e^{-j2\pi\nu_p l_p T_s}$ and $w_n[m] = w[n\tilde{M} + m]$. The approximation is obtained by plugging (6.1) into (6.2a). *For now, we employ the approximated result to go through the steps in the sensing framework introduced in Chapter 5. Later in this chapter, we will revisit the accurate echo signal form given above to investigate the impact of ICI on sensing.*

6.1.2 Virtual Cyclic Prefix (VCP)

As illustrated in Figure 6.1, each subblock is divided into two parts: the first part, consisting of $(\tilde{M} - Q)$ samples of each transmitted block, is the essential part, while the second part, consisting of the remaining Q samples, is the nonessential part. At the sensing receiver, a subblock is delayed to a different extent after being reflected by different targets. From Figure 6.1, we see that as long as Q is no smaller than the maximum sample delay, we can add the sampled signals in the nonessential segment onto the first Q samples of the essential part and then obtain the circularly shifted versions of the signals in the essential part. Therefore, the nonessential part of each subblock acts as a VCP. Note that Q does not have to take the same value as the CP length of the underlying communication system.

Note that the practical implementation of the above sensing design is similar to the conventional continuous-waveform radar. Following Section 5.1, we still consider the synchronized transceiver here. Thus, the receiver receives echo signals while the communication transmitter transmits orthogonal time-frequency space (OTFS) signals. In practice, the segmentation parameters, i.e. $\tilde{M}$ and Q, can be predetermined based on the analysis to be performed in Section 6.3. So, within the synchronized timing frame, the receiver can segment both the originally transmitted signal (as shared by the transmitter) and the echo signal and can add VCPs for subblocks, as described in Figure 6.1. After signal preprocessing, the receiver can then estimate target parameters using the method to be developed in Section 6.2.

Let $\tilde{s}_n[m]$ denote the essential signal part of the nth transmitted subblock. Based on Figure 6.1, we have

$$\tilde{s}_n[m] = s\left[n\tilde{M} + m\right],\ m \in [0, \bar{M}-1]\ \ (\bar{M} = \tilde{M} - Q), \tag{6.3}$$

where $\bar{M}$ is introduced to simplify notation. After adding VCP for the nth subblock, the target echo can be rewritten based on (6.2), as given by

$$\tilde{x}_n[m] = \sum_{p=0}^{P-1} \tilde{\alpha}_p \tilde{s}_n \left[\langle m - l_p \rangle_{\bar{M}} \right] e^{j2\pi v_p n \tilde{M} T_s} + z_n^{\mathrm{A}}[m] + z_n^{\mathrm{B}}[m] + w_n[m],$$
$$m = 0,1,\ldots,\bar{M}-1, \tag{6.4}$$

where $z_n^{\mathrm{A}}[m]$ denotes the interblock interference for the nth subblock and $z_n^{\mathrm{B}}[m]$ the intrablock interference. Based on the illustration in Figure 6.1, $z_n^{\mathrm{A}}[m]$ can be expressed as follows:

$$z_n^{\mathrm{A}}[m] = \sum_{p=0}^{P-1} \tilde{\alpha}_p \tilde{s}_{n-1} \left[m + \tilde{M} - l_p \right] g_{l_p}[m] e^{j2\pi v_p (n-1)\tilde{M} T_s}, \quad m = 0,1,\ldots, \mathrm{i}_{\max} - 1, \tag{6.5}$$

and $z_n^{\mathrm{B}}[m]$ can be given by

$$z_n^{\mathrm{B}}[m] = \sum_{p=0}^{P-1} \tilde{\alpha}_p \tilde{s}_n \left[m + \bar{M} \right] (1 - g_{l_p}[m]) \times e^{j2\pi v_p n \tilde{M} T_s}, \quad m = \mathrm{i}_{\min}, \ldots, Q-1, \tag{6.6}$$

where $g_x[u]$ $(x \le Q-1)$ is defined as follows:

$$g_x[u] = 1, \quad \text{for } u \in [0, x-1]; \quad g_x[u] = 0 \text{ for } u \in [x, Q-1].$$

Note that the two interference terms, $z_n^{\mathrm{A}}[m]$ and $z_n^{\mathrm{B}}[m]$, are the price paid for introducing VCP. Nevertheless, we notice that $z_n^{\mathrm{A}}[m]$ and $z_n^{\mathrm{B}}[m]$ can be treated as Gaussian noise uncorrelated with the essential signal part of the nth subblock. Lemma 6.1 provides some useful features of the different signal components in (6.4). In addition, we point out that the impact of the two interference terms can be minimized by properly configuring $\tilde{M}$, as will be seen in Section 6.3.

Lemma 6.1 *The signal components in (6.4), i.e.* $\tilde{s}_n[m]$, $z_n^A[m]$, $z_n^B[m]$, *and* $w_n[m]$, *are approximately independent over* $m = 0,1,\ldots,\bar{M}-1$, *and are mutually uncorrelated complex centered Gaussian variables satisfying*

$$\tilde{s}_n[m] \sim \mathcal{CN}(0,\sigma_{\mathrm{d}}^2), \quad z_n^A[m] \sim \mathcal{CN}(0,\sigma_{z^A}^2[m]),$$
$$z_n^B[m] \sim \mathcal{CN}(0,\sigma_{z^B}^2[m]), \quad \text{and} \quad w_n[m] \sim \mathcal{CN}(0,\sigma_w^2), \tag{6.7}$$

where σ_{d}^2 *is the average power of communication data* d_{i} $(\forall \mathrm{i})$, $\sigma_{z^A}^2[m]$ *takes the discrete values summarized in Table 6.1, and the possible values of* $\sigma_{z^B}^2[m]$ *are given in Table 6.2.*[1]

1 Note that the impact of Q on the interference variances is implicitly shown in the number of different variance values, i.e. the columns of the tables.

Table 6.1 Values of $\sigma^2_{z^{\mathrm{A}}}[m]$.

m	$(i_{P-2}, i_{P-1}]$	$(i_{P-3}, i_{P-2}]$	$\cdots$	$(0, i_0]$
$\sigma^2_{z^{\mathrm{A}}}[m]$	$\sigma^2_d \sigma^2_{P-1}$	$\sigma^2_d \sum_{p=P-2}^{P-1} \sigma^2_p$	$\cdots$	$\sigma^2_d \sum_{p=0}^{P-1} \sigma^2_p$

Table 6.2 Values of $\sigma^2_{z^{\mathrm{B}}}[m]$.

m	$[i_0, i_1)$	$[i_1, i_2)$	$\cdots$	$[i_{P-1}, Q-1)$
$\sigma^2_{z^{\mathrm{B}}}[m]$	$\sigma^2_d \sigma^2_0$	$\sigma^2_d \sum_{p=0}^{1} \sigma^2_p$	$\cdots$	$\sigma^2_d \sum_{p=0}^{P-1} \sigma^2_p$

Proof: As illustrated in Section 5.1.1, $S[m,n] \sim \mathcal{CN}(0, \sigma^2_d)$ is satisfied and independent over $\forall m, n$. Given the relation between $S[m,n]$ and $s_n[m]$ depicted in Section 5.1, we know that the latter preserves the white Gaussian features of the former. Since $\tilde{s}_n[m]$ is the sampled version of $s_n[m]$; see (6.3), we have $\tilde{s}_n[m] \sim \mathcal{CN}(0, \sigma^2_d)$ and is independent over $\forall m, n$. Reflecting $\tilde{s}_n[m] \sim \mathcal{CN}(0, \sigma^2_d)$ $(\forall n, m)$ in (6.5) and (6.6), we see that $z^{\mathrm{A}}_n[m]$ and $z^{\mathrm{B}}_n[m]$ are also centered complex Gaussian variables.

Based on (6.5), the variance of $z^{\mathrm{A}}_n[m]$ can be calculated as follows:

$$
\begin{aligned}
\sigma^2_{z^{\mathrm{A}}}[m] &= \mathbb{E}\left\{\left|z^{\mathrm{A}}_n[m]\right|^2\right\} \\
&= \mathbb{E}\left\{\left(\sum_{p_1=0}^{P-1} \tilde{\alpha}^*_{p_1} \tilde{s}^*_{n-1}\left[m+\tilde{M}-i_{p_1}\right] g_{i_{p_1}}[m] e^{-\mathrm{j}2\pi\nu_{p_1}(n-1)\tilde{M}T_s}\right)\right. \\
&\quad \left.\times\left(\sum_{p_2=0}^{P-1} \tilde{\alpha}_{p_2} \tilde{s}_{n-1}\left[m+\tilde{M}-i_{p_2}\right] g_{i_{p_2}}[m] e^{\mathrm{j}2\pi\nu_{p_2}(n-1)\tilde{M}T_s}\right)\right\} \\
&\overset{(a)}{=} \mathbb{E}\left\{\left(\sum_{p=0}^{P-1} \left|\tilde{\alpha}_p\right|^2 \left|\tilde{s}_{n-1}\left[m+\tilde{M}-i_p\right]\right|^2 g_{i_p}[m]\right)\right\} = \sum_{p=0}^{P-1} \sigma^2_p \sigma^2_d g_{i_p}[m],
\end{aligned}
\tag{6.8}
$$

where $\overset{(a)}{=}$ is obtained by suppressing the cross-terms at $p_1 \neq p_2$. The suppression is enabled by the uncorrelated target scattering coefficients. We see from (6.8) that the variance of $z^{\mathrm{A}}_n[m]$ can change with m due to the various support of the rectangular function associated with each path, i.e. $g_{i_p}[m]$. As shown in Figure 6.1, the number of paths involved in $z^{\mathrm{A}}_n[m]$ decreases, as m becomes larger. This leads to the discrete values of $\sigma^2_{z^A}[m]$, as summarized in Table 6.1. Similar to the above analysis for $z_n^A[m]$, the variance of $z_n^B[m]$ can be obtained, as summarized in Table 6.2. On the other hand, we notice from Figure 6.1 that $z_n^A[m]$, $z_n^B[m]$, and $\tilde{s}_n[m]$ have nonoverlapping support. Therefore, the three components are mutually uncorrelated.

The whiteness of $z_n^A[0], z_n^A[1], \dots, z_n^A[\bar{M}-1]$ can be validated based on (6.5). In particular, we can show that

$$\mathbb{E}\left\{z_n^A[m_1](z_n^A[m_2])^*\right\} = \sum_{p=0}^{P-1}\sigma_p^2 \times \mathbb{E}\left\{\tilde{s}_{n-1}\left[m_1+\tilde{M}-l_p\right]\tilde{s}^*_{n-1}\left[m_2+\tilde{M}-l_p\right]\right\},$$
$$= 0, \quad \forall m_1 \neq m_2, \tag{6.9}$$

where the rectangular functions in $z_n^A[m_1]$ and $z_n^A[m_2]$ are dropped for brevity, and the cross-terms are suppressed directly due to $\mathbb{E}\left\{\alpha_{p_1}\alpha^*_{p_2}\right\} = 0$. Similarly, we can validate the whiteness of $z_n^A[m]$ along the n-dimension given $\forall m$, and moreover, the whiteness of $z_n^B[m]$ along n- and m-dimensions.

6.1.3 Removing Communication Information

Taking the $\bar{M}$-point DFT of $\tilde{x}_n[m]$ obtained in (6.4) w.r.t. m, we obtain

$$X_n[l] = \sum_{p=0}^{P-1}\tilde{\alpha}_p\tilde{S}_n[l]e^{-j\frac{2\pi l l_p}{\bar{M}}}e^{j2\pi\nu_p n\tilde{M}T_s} + Z_n^A[l] + Z_n^B[l] + W_n[l], \quad l = 0,1,\dots,\bar{M}-1, \tag{6.10}$$

where $\tilde{S}_n[l]$, $Z_n^A[l]$, $Z_n^B[l]$, and $W_n[l]$ are the $\bar{M}$-point DFTs (w.r.t. m) of $\tilde{s}_n[m]$, $z_n^A[m]$, $z_n^B[m]$, and $w_n[m]$, respectively. Since $\tilde{s}_n[m]$ is known to the sensing receiver, $\tilde{S}_n[l]$ can be calculated as follows:

$$\tilde{S}_n[l] = \sum_{m=0}^{\bar{M}-1}\tilde{s}_n[m]\mathcal{Z}_{\bar{M}}^{ml}, \quad \text{s.t. } \mathcal{Z}_x^{ji} = e^{-j\frac{2\pi ij}{x}}/\sqrt{x}, \tag{6.11}$$

where $\mathcal{Z}_x^{ji}$ is the (j,i)th entry of an x-point unitary DFT matrix. As will be frequently used, the following features of the signal components in (6.17) are worth highlighting.

Lemma 6.2 *Given $\forall n, l$, the signal components in (6.10) satisfy*

$$\tilde{S}_n[l] \sim \mathcal{CN}(0,\sigma_d^2), \quad W_n[l] \sim \mathcal{CN}(0,\sigma_w^2),$$
$$Z_n^A[l] \sim \mathcal{CN}(0,\sigma^2_{Z^A}), \text{ and } Z_n^B[l] \sim \mathcal{CN}(0,\sigma^2_{Z^B}), \tag{6.12}$$

where each component is independent over $\forall l$. *Assuming that* l_p $(\forall p)$ *is uniformly distributed in* $[0, Q-1]$, *we then have*

$$\sigma^2_{Z^{\mathrm{A}}} = \frac{\mathrm{i}_{\max}\sigma_d^2\sum_{p=0}^{P-1}(p+1)\sigma_p^2}{P\bar{M}}, \quad \sigma^2_{Z^{\mathrm{B}}} = \frac{(Q-\mathrm{i}_{\min})\sigma_d^2\sum_{p=0}^{P-1}(P-p)\sigma_p^2}{P\bar{M}}, \tag{6.13}$$

where $\mathrm{i}_{\max}$ *and* $\mathrm{i}_{\min}$ *are the maximum and minimum sample delay, respectively.*

Proof: Each component in (6.12) is a unitary DFT of the corresponding component given in (6.7). Thus, with reference to the proof of Lemma 6.1, we can readily show the white Gaussian nature of the four components in (6.12). The details are suppressed for brevity. Next, we show the derivation of the variance of $\sigma^2_{Z^{\mathrm{A}}}$.

Similar to (6.11), $Z_n^{\mathrm{A}}[l]$ can be written as $Z_n^{\mathrm{A}}[l] = \sum_{m=0}^{\mathrm{i}_{\max}-1} z_n^{\mathrm{A}}[m]\mathcal{Z}_{\bar{M}}^{ml}$, where $\mathcal{Z}_{\bar{M}}^{ml}$ is the DFT basis, as given in (6.11). The variance of $Z_n^{\mathrm{A}}[l]$ can be calculated as follows:

$$\sigma^2_{Z^{\mathrm{A}}} = \sum_{m=0}^{\mathrm{i}_{\max}-1} \frac{1}{\bar{M}}\mathbb{E}\left\{\sigma^2_{z^{\mathrm{A}}}[m]\right\}, \tag{6.14}$$

where the cross-terms are suppressed due to the whiteness of $z_n^{\mathrm{A}}[m]$, as illustrated in Lemma 6.1. As also shown in Lemma 6.1, $\sigma^2_{z^{\mathrm{A}}}[m]$ is a discrete function of m. Based on the uniform distribution of l_p as stated in the condition of Lemma 6.2, $\sigma^2_{z^{\mathrm{A}}}[m]$ takes each value in Table 6.1 with the same probability $\frac{1}{P}$. This further leads to

$$\mathbb{E}\left\{\sigma^2_{z^{\mathrm{A}}}[m]\right\} = \frac{\sigma_d^2}{P}\left(\sigma^2_{P-1} + \sum_{p=P-2}^{P-1}\sigma_p^2 + \cdots + \sum_{p=0}^{P-1}\sigma_p^2\right) = \frac{\sigma_d^2}{P}\sum_{p=0}^{P-1}(p+1)\sigma_p^2. \tag{6.15}$$

Substituting (6.15) into (6.14), we obtain the variance of $Z_n^{\mathrm{A}}[l]$, as given in (6.13). Similarly, $\sigma^2_{Z^{\mathrm{B}}}$ can be derived. The details are suppressed here.

From (6.10), we see that communication information can be suppressed by dividing both sides of the equation by $\tilde{S}_n[l]$. This, however, can yield significant bursts in the resulted signal due to the Gaussian randomness of $\tilde{S}_n[l]$. To reduce the bursts, we can use $\mathsf{k}\tilde{S}_n[l]$ as the divisor. The rationale of introducing k is *multiplying* $\mathsf{k}(>1)$ *can increase* $\mathbb{V}\left\{\mathsf{k}\tilde{S}_n[l]\right\}$ *and then decrease* $\mathbb{C}\left\{|\mathsf{k}\tilde{S}_n[l]| \leq 1,\ \forall l\right\}$. Here, $\mathbb{V}\{x\}$ denotes the variance of a random variable x and $\mathbb{C}\{\cdot\}$ the event count. The following lemma helps configure k.

Lemma 6.3 *For* $\mathbb{P}\left\{|\mathsf{k}\tilde{S}_n[l]| \leq 1\right\} = \epsilon$ $(\forall n, l)$, *we can set*

$$\mathsf{k} = 1\Big/\left(\sigma_d\sqrt{\ln\frac{1}{1-\epsilon}}\right), \tag{6.16}$$

where ϵ *denotes a sufficiently small probability.*

Proof: The probability $\mathbb{P}\left\{|\mathsf{k}\tilde{S}_n[l]| \leq 1\right\}$ is equivalent to $\mathbb{P}\left\{|\mathsf{k}\tilde{S}_n[l]|^2 \leq 1\right\}$ which can be calculated as follows:

$$
\begin{aligned}
\frac{1}{\tilde{M}} &= \mathbb{P}\left\{|\mathsf{k}\tilde{S}_n[l]| \leq 1\right\} = \mathbb{P}\left\{|\mathsf{k}\tilde{S}_n[l]|^2 \leq 1\right\} \\
&= \mathbb{P}\left\{ \underbrace{\mathsf{k}^2\frac{\sigma_d^2}{2}}_{1/z} \underbrace{\left(\frac{\Re\{\tilde{S}_n[l]\}^2}{\sigma_d^2/2} + \frac{\Im\{\tilde{S}_n[l]\}^2}{\sigma_d^2/2}\right)}_{x} \leq 1 \right\} = \mathbb{P}\{x \leq z\} \\
&= 1 - e^{-\frac{1}{\mathsf{k}^2\sigma_d^2}},
\end{aligned}
$$

where $\tilde{S}_n[l] \sim \mathcal{CN}(0, \sigma_d^2)$ is illustrated in Lemma 6.2, x then conforms to a chi-squared distribution with 2 degrees of freedom, and the last result is based on the fact that $\mathbb{P}\{x \leq z\} = 1 - e^{-z/2}$ [1]. Solving the above equation results in (6.16).

Dividing both sides of (6.10) by $\mathsf{k}\tilde{S}_n[l]$, we obtain

$$
\begin{aligned}
\tilde{X}_n[l] &= \mathbb{1}_n[l]\left(\sum_{p=0}^{P-1}\frac{\tilde{\alpha}_p}{\mathsf{k}} e^{-j\frac{2\pi l l_p}{\tilde{M}}} e^{j2\pi\nu_p n\tilde{M}T_s} + \tilde{Z}_n^{\mathrm{A}}[l] + \tilde{Z}_n^{\mathrm{B}}[l] + \tilde{W}_n[l]\right), \\
\mathbb{1}_n[l] &= \begin{cases} 0, & \text{if } |\mathsf{k}\tilde{S}_n[l]| \leq 1, \\ 1, & \text{otherwise}, \end{cases}
\end{aligned}
\tag{6.17}
$$

where the intermediate variables are given by

$$
\tilde{Z}_n^{\mathrm{A}}[l] = Z_n^{\mathrm{A}}[l]/(\mathsf{k}\tilde{S}_n[l]), \quad \tilde{Z}_n^{\mathrm{B}}[l] = Z_n^{\mathrm{B}}[l]/(\mathsf{k}\tilde{S}_n[l]), \quad \tilde{W}_n[l] = W_n[l]/(\mathsf{k}\tilde{S}_n[l]).
\tag{6.18}
$$

In case $|\mathsf{k}\tilde{S}_n[l]| < 1$ (although unlikely given a properly selected k), $\mathbb{1}_n[l]$ can help prevent noise enhancement.

We remark that the way we remove communication information is similar to the widely used orthogonal frequency-division multiplexing (OFDM) sensing [2]. However, we emphasize that it is the designs presented in Sections 6.1.1 and 6.1.2 that enable OTFS sensing to be performed in such a simple manner. On the other hand, when dividing communication data symbols in OFDM sensing, the noise enhancement is not generally an issue, as phase shift keying (PSK) (with constant modulus) is assumed in most OFDM sensing works. In contrast, we are dealing with OTFS modulation that obtains the signals in the time-frequency domain through a symplectic Fourier transform; see (5.1), making the noise enhancement issue rather severe. But thanks to the design and analysis in this section, the issue can be greatly relieved.

6.2 Target Parameter Estimation

In Chapter 5, we validate the performance of the sensing framework through target detection. Here, as a complement, we illustrate a target parameter estimation scheme employing the frequency estimators introduced in Chapter 3. Enabled by the preprocessing on the target echo in Section 6.1, the resulted signal $\tilde{X}_n[l]$ presents a clear structure. In particular, we see from (6.17) that $e^{-j\frac{2\pi l l_p}{\tilde{M}}}$ and $e^{j2\pi \nu_p n\tilde{M}T_s}$ are two single-tone signals whose frequencies are related to target range and velocity, respectively. Therefore, parameter estimation in OTFS sensing is turned into estimating the center frequencies of the multitone signal $\tilde{X}_n[l]$. We present a low-complexity and high-accuracy method to fulfill the task, combining the commonly used estimate-and-subtract strategy [3, 4] with the frequency estimator for a single-tone signal, as introduced in Section 3.3. Below, we first illustrate the parameter estimation method and then analyze the computational complexity (CC) of the whole OTFS-sensing scheme.

6.2.1 Parameter Estimation Method

The overall estimation procedure is summarized in Table 6.3.[2] In each iteration, we always estimate the parameters of the presently strongest target, as done in Steps (3)–(6) of the table; reconstruct the echo signal of the target and subtract the reconstructed echo signal, as done in Step (2); and re-perform the above three steps for the next strongest target. For ease of illustration, we assume

Table 6.3 Overall parameter estimation.

Input: $\tilde{X}_n[l]$, as given in (6.17), and P (the number of targets)
(1) Initialize $p = 0$;
(2) Update $\tilde{X}_n^{(p)}[l]$ as done in (6.19);
(3) Estimate $(\tilde{l}_p, \tilde{n}_p)$ as done in (6.22);
(4) Estimate δ_p and ϵ_p by running Table 6.4;
(5) Obtain the estimates of l_p and ν_p as given in (6.30);
(6) Estimate α_p as done in (6.31);
(7) End if $p = P - 1$; otherwise, $p = p + 1$ and go to Step (2).

2 As commonly assumed in the work centered on parameter estimation [3, 4], the total number of targets, i.e. P, is taken as a known input. In practice, P can be estimated through well-developed techniques like the Akaike information criterion (AIC) and the minimum description length (MDL) [5]. However, we remark that a low-complexity detection method that is suitable for IoT devices is worth investigating.

that the pth ($\forall p$) target has stronger echo than the p'th ($\forall p' > p$) target, namely $|\tilde{\alpha}_0| > |\tilde{\alpha}_1| > \cdots > |\tilde{\alpha}_{P-1}|$. Thus, the signal used for estimating the pth target can be given by

$$\tilde{X}_n^{(p)}[l] = \tilde{X}_n^{(p-1)}[l] - \frac{\hat{\alpha}_{p-1}}{\mathrm{k}} e^{-\mathrm{j}\frac{2\pi l \hat{l}_{p-1}}{\tilde{M}}} e^{\mathrm{j}2\pi \hat{\nu}_{p-1} n \tilde{M} T_s},$$
$$\text{s.t. } p \geq 1, \quad \tilde{X}_n^{(0)}[l] = \tilde{X}_n[l] \text{ given in (6.17)}, \tag{6.19}$$

where $\tilde{X}_n^{(p-1)}[l]$ denotes the signal used for estimating the $(p-1)$th target, and $\hat{\alpha}_{p-1}, \hat{l}_{p-1}$, and $\hat{\nu}_{p-1}$ denote the estimated parameters of the target.

As mentioned earlier, $e^{-\mathrm{j}\frac{2\pi l l_p}{\tilde{M}}}$ and $e^{\mathrm{j}2\pi \nu_p n \tilde{M} T_s}$ in (6.17) are exponential signals along l and n, respectively. Therefore, we can identify their center frequencies through a two-dimensional DFT:

$$\check{X}_{\tilde{n}}^{(p)}[\tilde{l}] = \sum_{n=0}^{\tilde{N}-1} \sum_{l=0}^{\tilde{M}-1} \tilde{X}_n^{(p)}[l] Z_{\tilde{M}}^{-l\tilde{l}} Z_{\tilde{N}}^{n\tilde{n}} = \frac{\tilde{\alpha}_p}{\mathrm{k}} \mathcal{S}_{\tilde{M}}\left(\tilde{l} - l_p\right) \mathcal{S}_{\tilde{N}}\left(\underbrace{\tilde{N}\tilde{M}T_s \nu_p}_{n_p} - \tilde{n}\right) + \Xi_{\tilde{n}}^{(p)}[\tilde{l}], \tag{6.20}$$

where $Z_{\tilde{M}}^{-l\tilde{l}}$ and $Z_{\tilde{N}}^{n\tilde{n}}$ are the DFT bases defined in (6.11). Note that all irrelevant terms, including interference plus noise terms given in (6.17) and the echoes of weaker targets p' $(= p+1, \ldots, P-1)$, are absorbed in $\Xi_{\tilde{n}}^{(p)}[\tilde{l}]$ for brevity. Moreover, the function $\mathcal{S}_x(y)$ in (6.20) is defined as follows:

$$\mathcal{S}_x(y) = \frac{1}{\sqrt{x}} \frac{\sin\left(\frac{x}{2}\frac{2\pi y}{x}\right)}{\sin\left(\frac{1}{2}\frac{2\pi y}{x}\right)} e^{\mathrm{j}\frac{x-1}{2}\frac{2\pi y}{x}}. \tag{6.21}$$

Note that $\mathcal{S}_x(y)$ is a discrete sinc function which is maximized at $y = 0$. Therefore, the integer parts of (l_p, n_p) can be estimated by identifying the maximum of $\left|\check{X}_{\tilde{n}}^{(p)}[\tilde{l}]\right|^2$, i.e.

$$(\tilde{l}_p, \tilde{n}_p) : \max_{\forall \tilde{l}, \tilde{n}} \left|\check{X}_{\tilde{n}}^{(p)}[\tilde{l}]\right|^2. \tag{6.22}$$

To obtain high-accuracy estimations of target parameters, the fractional parts of (l_p, n_p) $\forall p$, as denoted by (δ_p, ϵ_p), also need to be estimated. With reference to Section 3.3, we develop below the methods for estimating δ_p and ϵ_p.

We start with δ_p. As summarized in Table 6.4, its estimation is performed iteratively. At iteration $i(\geq 1)$, we have the estimate of δ_p from the iteration $(i-1)$, as denoted by $\hat{\delta}_p^{(i-1)}$. Let $\xi_p^{(i)} = \delta_p - \hat{\delta}_p^{(i-1)}$ denote the estimation error which is

Table 6.4 Estimating δ_p (or ϵ_p [a]).

(1) Input: $\tilde{X}_n^{(p)}[l]$ given in (6.19), $(\tilde{l}_p, \tilde{n}_p)$ in (6.22) and N_{iter} (the maximum number of iterations);
(2) Initialize: $\hat{\delta}_p^{(0)} = 0.25 d_p$ and $i = 1$, where d_p is given in (6.26);
(3) Interpolate the DFT coefficients at $\tilde{l} = \tilde{l}_p^{\pm} \stackrel{\Delta}{=} \tilde{l}_p + \hat{\delta}_p^{(i-1)} \pm 0.25$, leading to $\check{X}_{\tilde{n}_p}^{(p)}[\tilde{l}_p^{\pm}]$ given in (6.23);
(4) Construct the ratio of $\rho_p^{(i)} = \frac{|x_+|^2 - |x_-|^2}{|x_+|^2 + |x_-|^2}$, s.t. $x_{\pm} \stackrel{\Delta}{=} \check{X}_{\tilde{n}_p}^{(p)}[\tilde{l}_p^{\pm}]$;
(5) Calculate r_i $(i = 0,1,2)$ as done in (6.25);
(6) Estimate $\xi_p^{(i)}$ as $\hat{\xi}_p^{(i)} = r_{i^*}$, s.t. $i^* = \text{argmin}_{\,i=0,1,2}\ |r_i|$;
(7) Update $\hat{\delta}_p^{(i)} = \hat{\delta}_p^{(i-1)} + \xi_p^{(i)}$;
(8) Set $i = i + 1$ and go back to Step (3), if $i < N_{\text{iter}}$;
(9) Output: the final estimate $\hat{\delta}_p = \hat{\delta}_p^{(i-1)} + \hat{\xi}_p^{(i)}$.

a) When estimating ϵ_p, δ_p above is replaced by ϵ_p, $\xi_p^{(i)}$ by $\eta_p^{(i)}$, $\hat{\delta}_p^{(i)}$ by $\hat{\epsilon}_p^{(i)}$; see (6.27). Moreover, $\check{X}_{\tilde{n}_p}^{(p)}[\tilde{l}_p^{\pm}]$ in Steps (3) and (4) becomes $\check{X}_{\tilde{n}_p^{\pm}}^{(p)}[\tilde{l}_p]$ given in (6.28). The remaining steps can be run without changes. However, note that c_1, c_3, and c_5 required in (6.25) now become the Taylor coefficients of the function given in (6.29).

estimated in Steps (3)–(6). In Step (3), the interpolated DFT coefficients are given by

$$\check{X}_{\tilde{n}_p}^{(p)}[\tilde{l}_p^{\pm}] = \sum_{n=0}^{\tilde{N}-1} \sum_{l=0}^{\tilde{M}-1} \tilde{X}_n^{(p)}[l] \mathcal{Z}_{\tilde{M}}^{-l\tilde{l}_p^{\pm}} \mathcal{Z}_{\tilde{N}}^{n\tilde{n}_p} = \Xi_{\tilde{n}_p}^{(p)}[\tilde{l}_p^{\pm}] \frac{\tilde{\alpha}_p}{\mathrm{k}} S_{\tilde{M}}\left(\tilde{l}_p^{\pm} - l_p\right) S_{\tilde{N}}\left(n_p - \tilde{n}_p\right), \quad (6.23)$$

where $(\tilde{l}_p, \tilde{n}_p)$ is obtained in (6.22) and $\tilde{l}_p^{\pm} \stackrel{\Delta}{=} \tilde{l}_p + \hat{\delta}_p^{(i-1)} \pm 0.25$. In Step (4), the ratio $\rho_p^{(i)}$ is constructed such that it can be regarded as a noisy value of the function $f(\xi_p^{(i)})$, as given by

$$f(\xi_p^{(i)}) = \frac{\left|\frac{\tilde{\alpha}_p}{\mathrm{k}} S_{\tilde{M}}\left(\tilde{l}_p^{+} - l_p\right) S_{\tilde{N}}\left(n_p - \tilde{n}_p\right)\right|^2 - \left|\frac{\tilde{\alpha}_p}{\mathrm{k}} S_{\tilde{M}}\left(\tilde{l}_p^{-} - l_p\right) S_{\tilde{N}}\left(n_p - \tilde{n}_p\right)\right|^2}{\left|\frac{\tilde{\alpha}_p}{\mathrm{k}} S_{\tilde{M}}\left(\tilde{l}_p^{+} - l_p\right) S_{\tilde{N}}\left(n_p - \tilde{n}_p\right)\right|^2 + \left|\frac{\tilde{\alpha}_p}{\mathrm{k}} S_{\tilde{M}}\left(\tilde{l}_p^{-} - l_p\right) S_{\tilde{N}}\left(n_p - \tilde{n}_p\right)\right|^2}$$

$$\stackrel{(a)}{=} \frac{\left|S_{\tilde{M}}\left(\xi_p^{(i)} - 0.25\right)\right|^2 - \left|S_{\tilde{M}}\left(\xi_p^{(i)} + 0.25\right)\right|^2}{\left|S_{\tilde{M}}\left(\xi_p^{(i)} - 0.25\right)\right|^2 + \left|S_{\tilde{M}}\left(\xi_p^{(i)} + 0.25\right)\right|^2}. \quad (6.24)$$

Note that the result "$\stackrel{(a)}{=}$" in (6.24) is obtained by suppressing the common terms of the numerator and denominator and by replacing $\tilde{l}_p^{\pm} - l_p$ with

$$\tilde{l}_p^{\pm} - l_p = \tilde{l}_p + \hat{\delta}_p^{(i-1)} \pm 0.25 - (\tilde{l}_p + \delta_p) = -\xi_p^{(i)} \pm 0.25.$$

Equating $f(\xi_p^{(i)})$ and $\rho_p^{(i)}$, $\xi_p^{(i)}$ can be estimated by solving the equation. As derived in Section 3.3, there are three roots,

$$\begin{aligned}
&r_1 = -k_2/3 + 2B, \quad r_2 = -k_2/3 - B + D, \quad r_3 = -k_2/3 - B - D, \\
&\text{s.t. } B = (S+T)/2, \quad D = \sqrt{3}(S-T)\mathrm{j}/2, \\
&S = \sqrt[3]{R + \sqrt{D}}, \quad T = \sqrt[3]{R - \sqrt{D}}, \quad D = Q^3 + R^2, \\
&R = (9k_1k_2 - 27k_0 - 2k_2^3)/54, \quad Q = (3k_1 - k_2^2)/9, \\
&k_2 = -\rho b_2/a_3, \quad k_1 = a_1/a_3, \quad k_0 = -\rho/a_3, \\
&a_1 = c_1, \quad a_3 = c_3 - c_1c_5\,/c_3, \quad b_2 = c_5\,/c_3,
\end{aligned} \tag{6.25}$$

where c_1, c_3, and c_5 are the coefficients of the first, third, and fifth power terms in the Taylor series of $f(\xi_p^{(i)})$ at $\xi_p^{(i)} = 0$. Only one root provides the estimate of $\xi_p^{(i)}$, as determined in Step (6). Next, we update the estimate of δ_p in Step (7), check the stop criterion in Step (8) and, if unsatisfied, run another iteration. Above is the general iteration procedure. As for the initialization of the algorithm, d_p is the result of the following sign test:

$$d_p = \text{sign}\left\{ \left[\check{X}_{\tilde{n}_p}^{(p)}[\tilde{l}_p - 1] - \check{X}_{\tilde{n}_p}^{(p)}[\tilde{l}_p + 1] \right] \left(\check{X}_{\tilde{n}_p}^{(p)}[\tilde{l}_p] \right)^* \right\}, \tag{6.26}$$

where $\check{X}_{\tilde{n}_p}^{(p)}[\tilde{l}_p]$ is obtained by plugging $\tilde{n} = \tilde{n}_p$ and $\tilde{l} = \tilde{l}_p$ into (6.20) and likewise $\check{X}_{\tilde{n}_p}^{(p)}[\tilde{l}_p \pm 1]$ is obtained. Above, the estimation steps are given with the rationales suppressed for brevity. Interested readers may refer to Section 3.3 for more details.

We notice that ϵ_p can also be estimated as done in Table 6.4, with necessary changes pointed out in the table. For ϵ_p, the estimation error in the iteration $i(\geq 1)$ is denoted by

$$\eta_p^{(i)} = \epsilon_p - \hat{\epsilon}_p^{(i-1)}, \tag{6.27}$$

where $\hat{\epsilon}_p^{(i-1)}$ is the estimate from the previous iteration. The DFT interpolation in Step (3) now happens at $\tilde{n} = \tilde{n}_p^{\pm} \overset{\Delta}{=} \tilde{n}_p + \hat{\epsilon}_p^{(i-1)} \pm 0.25$ and the interpolated coefficients are given by

$$\check{X}_{\tilde{n}_p^{\pm}}^{(p)}[\tilde{l}_p] = \sum_{n=0}^{\tilde{N}-1}\sum_{l=0}^{\tilde{M}-1} \tilde{X}_n^{(p)}[l] z_{\tilde{M}}^{-l\tilde{l}_p} z_{\tilde{N}}^{n\tilde{n}_p^{\pm}} = \Xi_{\tilde{n}_p^{\pm}}^{(p)}[\tilde{l}_p] \frac{\tilde{\alpha}_p}{\mathsf{k}} S_{\tilde{M}}\left(\tilde{l}_p - l_p\right) S_{\tilde{N}}\left(n_p - \tilde{n}_p^{\pm}\right), \tag{6.28}$$

which can be likewise interpreted as $\check{X}^{(p)}_{\tilde{n}_p}[\tilde{l}^{\pm}_p]$ given in (6.23). Moreover, corresponding to $f(\xi^{(i)}_p)$ given in (6.24), we now have

$$f(\eta^{(i)}_p) = \frac{\left|S_{\tilde{N}}\left(\eta^{(i)}_p - 0.25\right)\right|^2 - \left|S_{\tilde{N}}\left(\eta^{(i)}_p + 0.25\right)\right|^2}{\left|S_{\tilde{N}}\left(\eta^{(i)}_p - 0.25\right)\right|^2 + \left|S_{\tilde{N}}\left(\eta^{(i)}_p + 0.25\right)\right|^2}, \tag{6.29}$$

which is used for computing the three roots according to (6.25).

After running Table 6.4, we obtain $\hat{\delta}_p$ and $\hat{\epsilon}_p$. Adding them with $\tilde{l}_p$ and $\tilde{n}_p$ given in (6.22), the estimates of l_p and ν_p can be written as follows:

$$\hat{l}_p = \tilde{l}_p + \hat{\delta}_p,\ \hat{\nu}_p = (\tilde{n}_p + \hat{\epsilon}_p)/(\tilde{N}\tilde{M}T_s), \tag{6.30}$$

where the relation between n_p and ν_p is shown in (6.20). Substituting $\hat{l}_p$ and $\hat{\nu}_p$ into (6.20), the complex coefficient $\tilde{\alpha}_p$ can be estimated as follows:

$$\hat{\alpha}_p = \mathrm{k}\check{X}^{(p)}_{\tilde{n}_p}[\tilde{l}_p] / \left(S_{\bar{M}}\left(\tilde{l}_p - \hat{l}_p\right) S_{\tilde{N}}\left(\hat{n}_p - \tilde{n}_p\right)\right), \tag{6.31}$$

where the values of the two discrete sinc functions can be readily calculated based on (6.21).

6.2.2 Computational Complexity

Next, we analyze the CC of the OTFS sensing presented above and compare it with the existing methods. For preprocessing the echo, as illustrated in Section 6.1, the only computation-intensive computations are given in (6.10) and (6.17) which perform $\tilde{N}$ numbers of $\bar{M}$-dimensional DFT and an $\tilde{N}\bar{M}$-size pointwise division, respectively. Thus, the CC of echo preprocessing is in the order of $\mathcal{O}\{\tilde{N}\bar{M}\log\bar{M}\}$. For the parameter estimation method, as summarized in Table 6.3, its CC is analyzed below.

- Step (3) involves a computation of a two-dimensional DFT, as done in (6.20), incurring the CC of $\mathcal{O}\{\tilde{N}\bar{M}\log(\bar{M}\tilde{N})\}$. Step (3) also requires P times of searching for the peak in an $\tilde{N}\times\bar{M}$-dimensional range-Doppler profile, yielding the CC of $\mathcal{O}\{P\tilde{N}\log\bar{M}\}$. Since $\tilde{N}\bar{M}\log(\bar{M}\tilde{N}) \gg P\tilde{N}\log\bar{M}$, the overall CC of Step (3) in Table 6.3 is $\mathcal{O}\{\tilde{N}\bar{M}\log(\bar{M}\tilde{N})\}$;
- Step (4) runs Table 6.4 for P times, one for each targets. In each time, the algorithm is performed twice, one for range estimation and another for Doppler. Moreover, the CC of Table 6.4 is dominated by the interpolation in Step (3) [6] and is given by $\mathcal{O}\{(2N_{\text{iter}}+1)\bar{M}\}$ and $\mathcal{O}\{(2N_{\text{iter}}+1)\tilde{N}\}$, respectively, for range and Doppler estimations. Thus, Step (4) of Table 6.3 has a CC of $\mathcal{O}\{(2N_{\text{iter}}+1)Px\}$ with $x = \max\{\bar{M},\tilde{N}\}$.

As will be shown in Section 6.4, a small value of N_{iter}, e.g. five, can ensure a near-maximum likelihood (ML) estimation performance of Table 6.4. Moreover, P is typically smaller than $\tilde{N}$ or $\bar{M}$ in practice. In summary, we can assert that the CC of the OTFS sensing scheme is dominated by that of the two-dimensional DFT performed in Step (3) of Table 6.3.

Since the ML-based OTFS sensing method [7] will be employed as a benchmark in the simulations, we provide below its CC as a comparison. As a lower bound on its CC, we only consider the first iteration of the ML method under a single target assumption. From (22) and its context in [7], we can see that the number of range-Doppler grids is $N_{\text{CP}}N$, where N_{CP} is the number of samples in the CP of the underlying OTFS communication system and N is the size of the Doppler dimension. For each grid, the likelihood ratio given in [7, (21)] needs to be calculated, yielding the CC of $\mathcal{O}\{(MN)^2\}$. Thus, a lower bound of the overall CC of the ML method can be given by $\mathcal{O}\{N_{\text{CP}}M^2N^3\}$. Based on (6.3) and the text above (6.1), we have $\bar{M} = \tilde{M} - Q < \tilde{M}$ and $\tilde{N}\bar{M} < \tilde{N}\bar{M} < MN$, and hence $\tilde{N}\bar{M}\log(\bar{M}\tilde{N}) \ll N_{\text{CP}}M^2N^3$. This demonstrates the high-computational efficiency of the OTFS sensing presented in this chapter.

6.3 Optimizing Parameters of Sensing Methods

In this section, we optimize the sensing scheme presented above by deriving the optimal $\tilde{M}$ such that the estimation signal-to-interference-plus-noise ratio (SINR) is maximized. To start with, we derive the respective power expressions of the useful signal, interference, and noise components in the input of the algorithm given in Table 6.3; namely $\tilde{X}_n[l]$ given in (6.17). To do so, we first recover the accurate signal model without neglecting the intrasymbol Doppler impact. By using the accurate signal model in (6.2a), in contrast to using (6.2b) in Section 6.1, the signal in (6.4) can be rewritten as follows:

$$\tilde{x}_n[m] = \sum_{p=0}^{P-1} \overbrace{\tilde{\alpha}_p e^{\mathrm{j}2\pi \nu_p n\tilde{M}T_{\mathrm{s}}}}^{\breve{\alpha}_p} \left(\sum_{l=0}^{\bar{M}-1} \tilde{S}_n[l] e^{-\mathrm{j}\frac{2\pi l l_p}{\bar{M}}} \mathcal{Z}_{\bar{M}}^{-lm} \right) \times e^{\mathrm{j}2\pi \nu_p m T_{\mathrm{s}}} + z_n^{\mathrm{A}}[m] + z_n^{\mathrm{B}}[m] + w_n[m], \tag{6.32}$$

where to account for the ICI, the time-domain signal $\tilde{s}_n[\cdot]$ is replaced by its inverse IDFT, as enclosed in the round brackets. Strictly speaking, $z_n^{\mathrm{A}}[m]$ and $z_n^{\mathrm{B}}[m]$ are different from those given in (6.5) and (6.6), since $\tilde{s}_{n-1}\left[m + \tilde{M} - l_p\right]$ and $\tilde{s}_n\left[m + \bar{M}\right]$ are multiplied by extra exponential terms in a pointwise manner. The multiplication, however, changes neither the white Gaussian nature of $z_n^{\mathrm{A}}[m]$ and $z_n^{\mathrm{B}}[m]$ nor their variances. This can be validated based on the proof of

Lemma 6.1. Thus, we continue using the same symbols here. Based on the new expression of $\tilde{x}_n[m]$, $\tilde{X}_n[l]$, as given in (6.17), can be rewritten as follows:

$$\begin{aligned}\tilde{X}_n[l] &= \frac{\sum_{m=0}^{\bar{M}-1} \tilde{x}_n[m] \mathcal{Z}_{\bar{M}}^{ml}}{\mathrm{k}\tilde{S}_n[l]} \\ &= \underbrace{\sum_{p=0}^{P-1} \frac{\breve{\alpha}_p}{\mathrm{k}} e^{-\mathrm{j}\frac{2\pi l l_p}{\bar{M}}} \frac{1}{\bar{M}} \sum_{m=0}^{\bar{M}-1} e^{\mathrm{j}2\pi \nu_p m T_s}}_{\tilde{X}_n^{\mathrm{S}}[l]} \\ &+ \underbrace{\sum_{p=0}^{P-1} \breve{\alpha}_p \sum_{m=0}^{\bar{M}-1} \sum_{\substack{l'=0 \\ l' \neq l}}^{\bar{M}-1} \frac{\tilde{S}_n[l']}{\mathrm{k}\tilde{S}_n[l]} e^{-\mathrm{j}\frac{2\pi l' l_p}{\bar{M}}} e^{\mathrm{j}2\pi \nu_p m T_s} \frac{\mathcal{Z}_{\bar{M}}^{(l-l')m}}{\sqrt{\bar{M}}}}_{\tilde{X}_n^{\mathrm{I}}[l]} \\ &+ \overbrace{Z_n^{\mathrm{A}}[l]/(\mathrm{k}\tilde{S}_n[l]) + Z_n^{\mathrm{B}}[l]/(\mathrm{k}\tilde{S}_n[l])}^{\tilde{X}_n^{\mathrm{Z}}[l]} + \tilde{W}_n[l],\end{aligned} \tag{6.33}$$

where the coefficient $\frac{1}{\bar{M}}$ in $\tilde{X}_n^{\mathrm{S}}[l]$ comes from the product of $\mathcal{Z}_{\bar{M}}^{-ml} \mathcal{Z}_{\bar{M}}^{ml}$ and the coefficient $\frac{1}{\sqrt{\bar{M}}}$ in $\tilde{X}_n^{\mathrm{I}}[l]$ is likewise produced. For the same reason that $z_n^{\mathrm{A}}[m]$ and $z_n^{\mathrm{B}}[m]$ are reused in (6.32), we continue using $Z_n^{\mathrm{A}}[l]$, $Z_n^{\mathrm{B}}[l]$, and $\tilde{W}_n[l]$ in (6.33). *Note that $\tilde{X}_n^{\mathrm{S}}[l]$ is the useful signal term, $\tilde{X}_n^{\mathrm{I}}[l]$ is the interference caused by the ICI and $\tilde{X}_n^{\mathrm{Z}}[l]$ is the interference introduced when adding VCPs.* Next, we first derive the power expressions of the four components in (6.33) and then study their relations.

6.3.1 Preliminary Results

Some handy features that will be frequently used later are introduced first. We notice that the p-related summation is involved in $\tilde{X}_n^{\mathrm{S}}[l]$ and $\tilde{X}_n^{\mathrm{I}}[l]$. Lemma 6.4 is helpful in simplifying the calculations of their powers. We also notice that the ratio of complex Gaussian variables appears in $\tilde{X}_n^{\mathrm{I}}[l]$ and $\tilde{X}_n^{\mathrm{Z}}[l]$. The finite moments of such a ratio does not exist [1]. Thus, we provide Proposition 6.1 to approximate the variance of the ratio; the proof is given in Appendix A.6.

Lemma 6.4 *Let $\mathcal{X}$ denote a general expression. We have*

$$\mathbb{V}\left\{\sum_{p=0}^{P-1} \alpha_p \mathcal{X}\right\} = \sum_{p=0}^{P-1} \sigma_p^2 \mathbb{E}\left\{\mathcal{X}^* \mathcal{X}\right\},$$

where $\mathbb{V}\{x\}$ denotes the variance of x.

Proof: We consider i.i.d. zero-mean α_p ($\forall p$). Thus, $\mathbb{E}\left\{\sum_{p=0}^{P-1}\alpha_p\mathcal{X}\right\}=0$. Then, the variance can be calculated as follows:

$$\mathbb{V}\left\{\sum_{p=0}^{P-1}\alpha_p\mathcal{X}\right\}=\mathbb{E}\left\{\left(\sum_{p=0}^{P-1}\alpha_p\mathcal{X}\right)\left(\sum_{p=0}^{P-1}\alpha_p\mathcal{X}\right)^*\right\}=\sum_{p=0}^{P-1}\sigma_p^2\mathbb{E}\left\{\mathcal{X}^*\mathcal{X}\right\}, \quad (6.34)$$

where the cross-terms are suppressed due to $\mathbb{E}\left\{\alpha_{p_1}\alpha_{p_2}^*\right\}=0$ given $\forall p_1\neq p_2$.

Proposition 6.1 *Let $x\sim\mathcal{CN}(0,\sigma_x^2)$ and $y\sim\mathcal{CN}(0,\sigma_y^2)$ denote two uncorrelated complex Gaussian variables. Defining $z=\frac{x}{y}$, we have $\mathbb{E}\{z\}=0$. Provided that $\mathbb{P}\{|y|\leq 1\}=\epsilon$, $\sigma_x^2\ll\sigma_y^2$ and $\rho=\sigma_x^2/\sigma_y^2$, the variance of z can be approximated by*

$$\sigma_z^2\approx\beta(\epsilon)\rho,\quad s.t.\ \beta(\epsilon)=2\left(\ln\left(\frac{2(1-\epsilon)}{\sqrt{\epsilon(2-\epsilon)}}\right)-1\right). \quad (6.35)$$

6.3.2 Maximizing SINR for Parameter Estimation

We start with deriving the SINR of the RDM obtained by the sensing framework, i.e. $\tilde{X}_n[l]$ given in (6.33). To do so, the powers of individual terms in $\tilde{X}_n[l]$ need to be calculated first. With the lengthy math encapsulated in Appendix A.7, the powers of $\tilde{X}_n^{\mathrm{S}}[l]$, $\tilde{X}_n^{\mathrm{I}}[l]$, $\tilde{X}_n^{\mathrm{Z}}[l]$, and $\tilde{W}_n[l]$ can be, respectively, given by

$$\sigma_{\mathrm{S}}^2\approx\frac{1}{\mathrm{k}^2}\left(\sigma_P^2+\alpha-\alpha\bar{M}^2\right),\quad \text{s.t. } \alpha=\frac{\pi^2T_{\mathrm{S}}^2}{3}\sum_{p=0}^{P-1}\sigma_p^2\nu_p^2,\quad \sigma_P^2=\sum_{p=0}^{P-1}\sigma_p^2, \quad (6.36)$$

$$\sigma_{\mathrm{I}}^2=\beta(\epsilon)\mathrm{k}^{-2}(\alpha\bar{M}^2-\alpha),\quad \text{s.t. } \alpha=\frac{\pi^2T_{\mathrm{S}}^2}{3}\sum_{p=0}^{P-1}\sigma_p^2\nu_p^2, \quad (6.37)$$

$$\sigma_{\mathrm{Z}}^2\overset{\substack{\mathrm{i}_{\min}=0\\ \mathrm{i}_{\max}=Q}}{\leq}\frac{\beta(\epsilon)Q(P+1)\sigma_P^2}{\mathrm{k}^2P\bar{M}},\quad \text{and} \quad (6.38)$$

$$\sigma_{\mathrm{W}}^2=\mathbb{V}\left\{\tilde{W}_n[l]\right\}=\frac{\beta(\epsilon)\sigma_w^2}{\mathrm{k}^2\sigma_d^2}. \quad (6.39)$$

We are now ready to analyze the overall SINR for the parameter estimation method illustrated in Section 6.2 and derive the optimal $\bar{M}$. Note that a two-dimensional DFT along n and l is taken for $\tilde{X}_n[l]$ before estimating target parameters; see Table 6.3. The maximum SINR improvement brought by the DFT is $\bar{M}\tilde{N}$; see (6.20). This is because $\tilde{X}_n^{\mathrm{S}}[l]$ is coherently accumulated with the maximum amplitude gain of $\bar{M}\tilde{N}$, while $\tilde{X}_n^{\mathrm{I}}[l]$ and $\tilde{X}_n^{\mathrm{Z}}[l]$ are only incoherently accumulated with the power gain of $\bar{M}\tilde{N}$. With the SINR gain of $\bar{M}\tilde{N}$ taken into account, the overall SINR for parameter estimation can be given by

$$\gamma=\frac{f(\bar{M})}{g(\bar{M})}\overset{\Delta}{=}\frac{\bar{M}\tilde{N}\sigma_{\mathrm{S}}^2}{\sigma_{\mathrm{I}}^2+\sigma_{\mathrm{Z}}^2+\sigma_{\mathrm{W}}^2}, \quad (6.40)$$

where σ_S^2, σ_I^2, σ_Z^2, and σ_W^2 are derived in (6.36), (6.37), (6.38), and (6.39), respectively. To avoid overly cumbersome expressions, we denote the numerator and denominator as a different function of $\bar{M}$ and study their monotonic features separately. In particular, the following interesting results are achieved, with the proof given in Appendices A.8 and A.9.

Proposition 6.2 *The numerator function $f(\bar{M})$ in (6.40) is concave w.r.t. $\bar{M}$. Given $\bar{M} \gg 3Q/2$, $f(\bar{M})$ is maximized at*

$$\bar{M} = \bar{M}_f^* \approx \sqrt[3]{(Q\alpha + Q\sigma_P^2)/(2\alpha)}.$$

Proposition 6.3 *The denominator function $g(\bar{M})$ in (6.40) is convex w.r.t. $\bar{M}$ and is minimized at*

$$\bar{M} = \bar{M}_g^* = \sqrt[3]{(Q(P+1)\sigma_P^2)/(2\alpha P)}.$$

From the two propositions, $f(\bar{M})$ and $g(\bar{M})$ seem to achieve the extremum at different values of $\bar{M}$. Nevertheless, taking into account the practical values of the variables in their respective expressions, we have the following result.

Corollary 6.1 *The optimal $\bar{M}$ for the numerator and denominator functions in (6.40) satisfy $\bar{M}_f^* \approx \bar{M}_g^* \approxeq \sqrt[3]{Q\sigma_P^2/(2\alpha)}$, where the equality can be approached as P becomes larger.*

Proof: As defined in (6.36), $\alpha = \frac{\pi^2 T_s^2}{3} \sum_{p=0}^{P-1} \sigma_p^2 \nu_p^2$. Thus, we have the following upper bound of 2α

$$2\alpha = \frac{\pi^2 T_s^2}{3} \sum_{p=0}^{P-1} 2\sigma_p^2 \nu_p^2 \leq \frac{\pi^2 T_s^2}{3} \sum_{p=0}^{P-1} (\sigma_p^2 + \nu_p^2).$$

Applying the above inequality, we have

$$Q\sigma_P^2/(2\alpha) \geq \frac{Q}{\frac{\pi^2 T_s^2}{3}\left(1 + \frac{1}{\sigma_P^2} \sum_{p=0}^{P-1} \nu_p^2\right)} \gg Q/2,$$

where the last result is because T_s can be in the order of 10^{-6}, while ν_p is only in the order of 10^3. Reflecting the above relation in Proposition 6.2, we obtain $\bar{M}_f^* \approx \sqrt[3]{Q\sigma_P^2/(2\alpha)}$. Given a large P, $\bar{M}_g^*$ derive in Proposition 6.3 becomes $\bar{M}_g^* \approxeq \sqrt[3]{Q\sigma_P^2/(2\alpha)} \approx \bar{M}_f^*$.

In summary, we highlight below the key features of the OTFS sensing methods illustrated in Section 6.2. *First*, the actual range and Doppler dimensions, as denoted by $\tilde{M}$ and $\tilde{N}$, respectively, are not limited to the original dimensions of the underlying OTFS system, i.e. M and N. *Second*, the maximum range that can be estimated by the design is not subject to the CP length of the OTFS system and can be flexibly adjusted by changing Q, i.e. the length of VCP. *Third*, the SINR for target parameter estimation can be maximized by setting

$$\tilde{M} = \sqrt[3]{\frac{Q\sigma_P^2}{(2\alpha)}} + Q, \quad \tilde{N} = \left\lceil \frac{MN}{\tilde{M}} \right\rceil, \tag{6.41}$$

where α and σ_P^2 are defined in (6.36). Note that the relation $\tilde{M} = \bar{M} + Q$, as given in (6.3), is used above. Also note that the optimization in this section has included the potential ICI caused by the intrasubblock Doppler impact. Thus, even if a large Q results in a large $\tilde{M}$ according to (6.41), the optimal $\tilde{M}$ is guaranteed to maximize the sensing SINR, whether ICI is present or not.

6.4 Simulation Results

Based on the theory developed above, we now present some simulation results to validate the designs of the presented sensing framework and performance of the algorithms described.

6.4.1 Comparison with Benchmark Method

In the first set of simulations, we compare the design with the state-of-the-art OTFS sensing method [7] that performs the ML estimation. In short, the ML method [7] first transforms $x[i]$ given in (5.6) into the delay-Doppler domain by performing an inverse symplectic Fourier transform, i.e. the inverse transform of that performed in (5.1). The result is stacked into an I-dimensional ($I = MN$) column vector, as denoted by $\mathbf{y}_{\mathrm{ML}}$. Then a multilayer, two-dimensional searching is carried out, using the coefficients $\mathbf{H}_{\mathrm{ML}}^{a}(l_g^a, \nu_g^a)\mathbf{d}$, where $\mathbf{d}$ is obtained by stacking d_{kM+l} into an I-dimensional column vector (k first) and $\mathbf{H}_{\mathrm{ML}}^{a}(l_g^a, \nu_g^a)$ is the input–output response matrix at a delay-Doppler grid (l_g^a, ν_g^a); refer to [7, Eq. (15)] for its detailed expression. The superscript $()^a$ denotes the layer index and the subscript $()_g$ denotes the grid index. For a single target, the 2D searching can be depicted as follows:

$$(\hat{l}^a, \hat{\nu}^a) : \max_{(l_g^a, \nu_g^a) \in S_a} \frac{\left(\mathbf{H}_{\mathrm{ML}}^{a}(l_g^a, \nu_g^a)\mathbf{d}\right)^{\mathrm{H}} \mathbf{y}_{\mathrm{ML}}}{\left(\left(\mathbf{H}_{\mathrm{ML}}^{a}(l_g^a, \nu_g^a)\mathbf{d}\right)^{\mathrm{H}} \left(\mathbf{H}_{\mathrm{ML}}^{a}(l_g^a, \nu_g^a)\mathbf{d}\right)\right)}, \tag{6.42}$$

where $(\hat{l}^a, \hat{\nu}^a)$ is the pair of estimated parameters and S_a is the delay-Doppler region to be searched at layer a.

For simulation efficiency, we take $l_g^a = l^a$ $(\forall a, g)$, where l^a is the true value of sample delay at layer a. Thus, we only perform ML estimation for the target velocity. Moreover, we set the number of grids in each layer as five, i.e. $|S_a| = 5$, and the total number of layers as eight, i.e. $a = 0, 1, \dots, 7$. More specifically, at $a = 0$, we set $v_g^0 = g\Delta_f/4$ $(g = 0, 1, 2, 3, 4)$. Note that Δ_f, the subcarrier interval, is also the unambiguously measurable region of the Doppler frequency for the ML method [7]. For $a \geq 1$, the searching region of layer a is reduced to $\Delta_f/4^a$ centered at $\hat{v}^{a-1}$. In addition, we set a rule that stops the iterative searching at layer a if $\hat{v}_g^a \neq v^a$ and take v^a as the final velocity estimate, where v^a is the true value. For a fair comparison with the ML method [7], the key system parameters set therein are used here only with minor modification. The settings are summarized in Table 6.5.

Figure 6.2 plots the mean squared error (MSE) of the velocity estimation, denoted by $\hat{v}$, against the SNR, denoted by γ_0. Here, $\gamma_0 = (\sigma_0^2/\sigma_w^2)$ is defined based on the sensing-received signal in the time domain, i.e. $x_n[m]$ given in (6.2). We see that, overall, the presented method has comparable performance to the ML method [7]. Interestingly, we see different performance relations between the two methods in three SNR regions:

(1) As γ_0 increases from −10 to −2 dB, the ML outperforms the method presented in this chapter, with the performance gap becoming increasingly smaller. This is because the estimation algorithm, as summarized in Table 6.4, can return

Table 6.5 Simulation parameters.

Variable	Description	Value
f_c	Carrier frequency	5.89 GHz
B	Bandwidth	10 MHz
Δ_f	Subcarrier interval	B/M MHz
M	No. of subcarriers per symbol	25
N	No. of symbols	40
σ_d^2	Power of data symbol d_{kM+l} give in (5.1)	0 dB
σ_0^2	Power of α_p $(p = 0)$	0 dB
σ_w^2	Variance of additive white Gaussian noise (AWGN) $w_n[m]$ given in (6.2)	$-10 : 2 : 10$ dB
r	Target range	30 m
v	Target velocity	80 km/h
N_{iter}	No. of iteration for Table 6.4	5

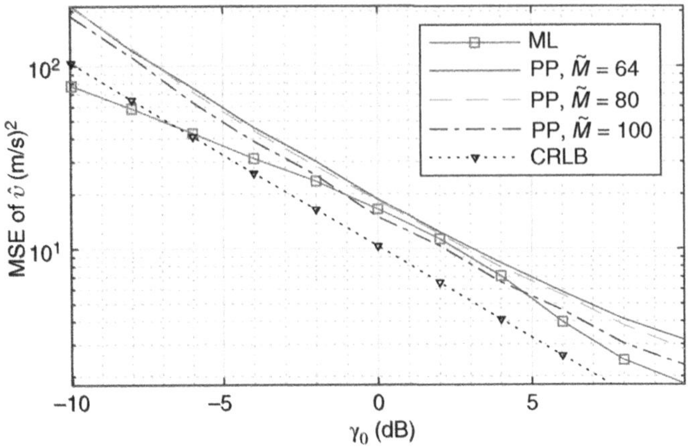

Figure 6.2 MSE of velocity estimation, denoted by $\hat{v}$, vs. $\gamma_0(=\sigma_0^2/\sigma_w^2)$, i.e. the SNR of $x_n[m]$ given in (6.2), where *ML* is for the benchmark method and PP for the design in this chapter. Adapted from [7].

any value in $[0, \frac{1}{\tilde{M}T_s}]$,[3] in the presence of strong noises. In contrast, the way the ML method is simulated makes its velocity estimation error bounded to 80/3.6.

(2) For $\gamma_0 \in [-1,5]$ dB, the presented method is able to outperform the ML method. The reason is that the presented method provides off-grid estimations, while the ML method performs on-grid searching.

(3) As γ_0 increases over 5 dB, an increasing (yet small) performance gap between the presented and the ML methods can be seen. The slight degrading of the presented method is caused by the interference introduced during adding VCP; refer to Figure 6.1. In particular, when the noise power keeps decreasing, the impact of the inter- and intrasymbol interference becomes gradually dominant.

Nevertheless, we see from Figure 6.2 that increasing $\tilde{M}$ can help reduce the impact of the interference. This is expected. Based on the parameter setting in Table 6.5 and Corollary 6.1, we have $\bar{M}_f^* \approx \bar{M}_g^* \approxeq 247.8675$. According to Propositions 6.2 and 6.3, the estimation SINR increases with $\tilde{M}$ when $\tilde{M} \leq \bar{M}_f^*$. That is, a larger $\tilde{M}$ corresponds to a more accurate $\hat{v}$ estimation.

3 In Section 6.2, the estimation of Doppler frequency is turned into estimating $n_p(=\tilde{N}\tilde{M}T_s v_p)$ defined in (6.20). The algorithm for estimating n_p, as summarized in Table 6.4, can return any value in $[0, \tilde{N}]$ as the estimate of n_p in the presence of strong noises. Thus, the estimation of nu_p can take any value in the region of $[0, \frac{1}{\tilde{M}T_s}]$.

Figure 6.2 also plots the CLRB of velocity estimation to validate the high accuracy of the parameter estimation methods developed in Section 6.2. As derived Remark 6.1, the Cramer-Rao lower bound (CRLB) is given by

$$\mathrm{CRLB}\{\hat{v}\} = \frac{\lambda^2}{4}\frac{1}{(\tilde{M}\tilde{N}T_s)^2}\frac{6}{4\pi^4\tilde{N}\tilde{M}\gamma_0/\beta(\epsilon)}, \tag{6.43}$$

where $\beta(\epsilon)$ is given in Proposition 6.1. We see from Figure 6.2 that the asymptotic performance of our estimation is slightly worse than the ML estimation. This owes to the interference introduced when adding VCP. Note that the low-SNR performance of the ML method is not accurate in Figure 6.2. As explained above, this is because the way we simulate the method bounds estimation error.

Remark 6.1 From Section 6.2, we see that the velocity and range estimations are both turned into frequency estimations. Thus, we can use the CRLB of a single-tone frequency estimator to derive the CRLBs of $\hat{\nu}_p$ and $\hat{l}_p$ obtained in (6.30). As the derivations for the two estimates are very similar, we only take $\hat{\nu}_p = (\tilde{n}_p + \hat{\epsilon}_p)/(\tilde{N}\tilde{M}T_s)$ for an illustration. According to [8], the CRLB of $(\tilde{n}_p + \hat{\epsilon}_p)$ can be given by $6/(4\pi^2\tilde{\gamma})$, where $\tilde{\gamma}$ is the SNR of $\check{X}^{(p)}_{\tilde{n}_p^{\pm}}[\tilde{l}_p]$ obtained in (6.28). The SNR $\tilde{\gamma}$ can be approximated by suppressing σ_{I}^2 and σ_{Z}^2 in (6.40), leading to $\tilde{\gamma} = \frac{\tilde{M}\tilde{N}\sigma_{\mathrm{S}}^2}{\sigma_{\mathrm{W}}^2}$, where σ_{S}^2 is given in (6.36) and σ_{W}^2 in (6.39). Given $\hat{\nu}_p = (\tilde{n}_p + \hat{\epsilon}_p)/(\tilde{N}\tilde{M}T_s)$, we have $\mathrm{CRLB}\{\hat{\nu}_p\} = \mathrm{CRLB}\{\tilde{n}_p + \hat{\epsilon}_p\}/(\tilde{N}\tilde{M}T_s)^2$. Since $\nu_p = 2v_p/\lambda$, we further have $\mathrm{CRLB}\{\hat{v}_p\} = \frac{\lambda^2}{4}\,\mathrm{CRLB}\{\hat{\nu}_p\}$. Combining the above analyses gives the CRLB in (6.43).

Figure 6.3 plots the MSE of $\hat{v}$ vs. different values of v. We see that the presented method can achieve comparable performance to the ML method [7] at moderately high or high SNRs. We also see that the rendered method presents an even performance over a large region of v, while the good performance of the ML method can be velocity-selective. From Figures 6.2 and 6.3, we conclude that the presented method can provide an ML-like estimation performance for OTFS-based sensing.

6.4.2 Wide Applicability

Next, we validate the analysis of the impact of $\tilde{M}$ on the presented OTFS sensing framework. To also show the wide applicability of the presented method, we employ another set of system parameters, as summarized in Table 6.6, where $\mathcal{U}_{[a,b]}$ denotes the uniform distribution in $[a, b]$. Note that $P = 4$ means four targets are set. Also, note that the Doppler frequency of 4.63 KHz corresponds to the maximum target velocity of 500 km/h.

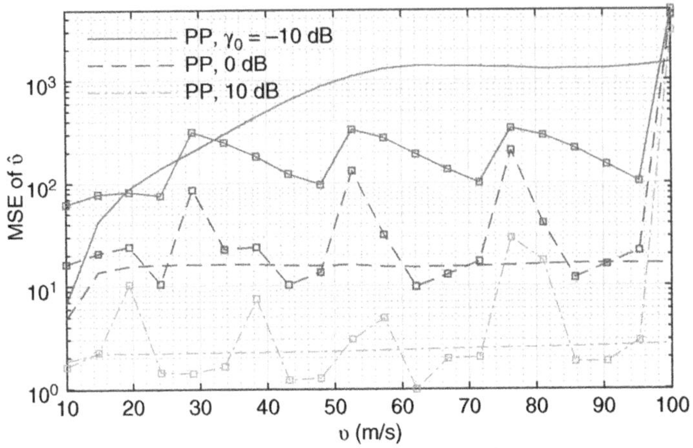

Figure 6.3 MSE of $\hat{\upsilon}$ vs. υ, where $\tilde{M} = 100$ is taken for the method presented in this chapter (PP) and the curves with markers correspond to the ML method. Adapted from [7].

Table 6.6 Simulation parameters.

Variable	Value	Variable	Value	Variable	Value
f_c	5 GHz	B	12 MHz	P	4
M	400	N	100	l_p $(\forall p)$	$\mathcal{U}_{[0,Q-1]}$
σ_w^2	−10 dB	σ_d^2	0 dB	σ_p $(\forall p)$	0 dB
σ_0^2	0 dB	ν_p $(\forall p)$	$\mathcal{U}_{[-4.63\ \text{KHz},4.63\ \text{KHz}]}$		

Figure 6.4 illustrates the impact of $\tilde{M}$ on the powers of the signal and other components in $\tilde{X}_n[l]$ given in (6.33), where the power expressions of σ_S^2, σ_I^2, σ_Z^2, and σ_W^2, as derived in (6.36), (6.37), (6.38), and (6.39), respectively, are plotted as the theoretical references. From Figure 6.4, we see that the derived expressions can precisely depict the actual powers of different signal components. We also see from the left subfigure that the signal power presents an approximate concave relation with $\tilde{M}$, as proved in Proposition 6.2. Moreover, we see from the right subfigure that the overall power of interference plus noises is indeed a convex function of $\tilde{M}$, which validates Proposition 6.3. In addition, jointly comparing the two subfigures, we can see that the extrema of σ_S^2 and $(\sigma_I^2 + \sigma_Z^2 + \sigma_W^2)$ are achieved at approximately identical $\tilde{M}$ for each value of Q. This confirms the analysis in Corollary 6.1.

Figure 6.5 plots the MSE of range and velocity estimations against $\tilde{M}$. We see that, as consistent with the SINR variations reflected in Figure 6.4, the MSEs of

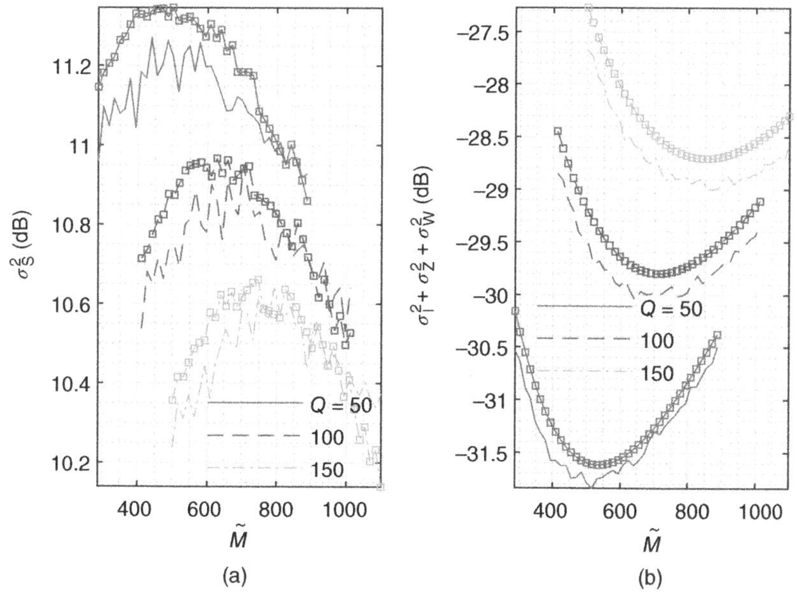

Figure 6.4 Illustrating the impact of $\tilde{M}$ on the powers of signal and other components, as derived in Section 6.3.2, where the curves with markers are theoretical results while the ones without markers are simulated. (a) is for the signal power, and (b) is for the power of interference plus noise.

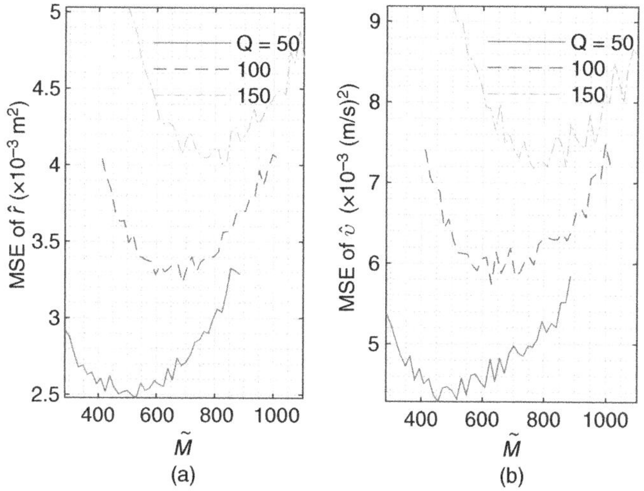

Figure 6.5 MSE of parameter estimations vs. $\tilde{M}$, where the mean is calculated over the squared estimation errors of all the four targets. (a) is for range estimation, and (b) is for velocity estimation.

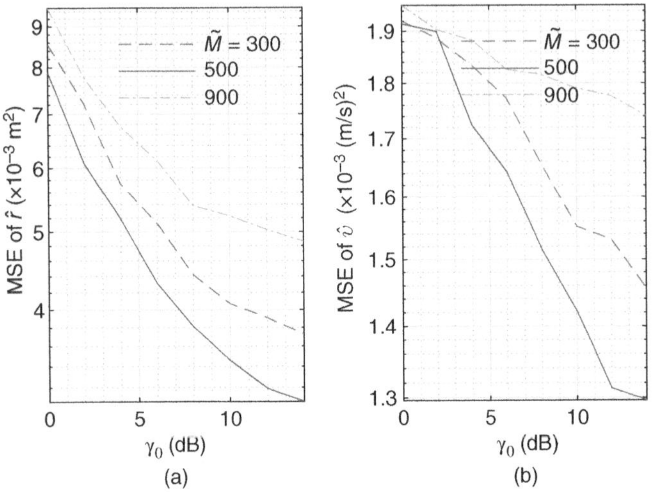

Figure 6.6 MSE of parameter estimations vs. γ_0, where $Q = 50$ and the mean is calculated over the squared estimation errors of all the four targets. (a) is for range estimation, and (b) is for velocity estimation.

the estimations of both parameters are convex functions of $\tilde{M}$ and are minimized at the optimal $\tilde{M}$ derived in Corollary 6.1. We also see that the presented method achieves the high-accuracy estimations of the ranges and velocities of all four targets (as the MSE is calculated over the squared estimation errors of all targets). Moreover, we see from Figure 6.4 that Q also has a nontrivial impact on OTFS sensing performance. In particular, as Q becomes larger, the whole MSE curve shifts upward. This is because a larger Q leads to a smaller number of samples in the essential part of each subblock; see Figure 6.1; as a further result, the coherent accumulation gain of the two-dimensional DFT performed in (6.20) for parameter estimation is also reduced.

Figure 6.6 illustrates the MSE of parameter estimations against γ_0. Based on Figure 6.4, we see that the optimal $\tilde{M}$ is about 500. Thus, we take three values of $\tilde{M}$ in this simulation to further illustrate the impact of $\tilde{M}$ under different γ_0. As consistent with Figures 6.4 and 6.5, the best MSE performance vs. γ_0 is achieved at $\tilde{M} = 500$. Moreover, we also see that the MSE performance under $\tilde{M} = 500$ is better than that under $\tilde{M} = 900$. This is also consistent with what was observed in Figure 6.5. In addition, we see that the impact of $\tilde{M}$ is more prominent at a higher SNR. This is because as noise becomes weaker, the interference component has a more dominating effect, which also makes the impact of $\tilde{M}$ more obvious.

6.5 Conclusions

A low-complexity OTFS sensing scheme with the near-ML performance for OTFS-based JCAS is described in this chapter. The scheme is enabled by a series of waveform preprocessing, which addresses the challenges of ICI and ISI, and successfully removes the impact of communication data symbols in the time-frequency domain without amplifying noises. The scheme also entails a high-accuracy and off-grid method for estimating ranges and velocities of sensing targets; the complexity of the method is only dominated by a two-dimensional DFT. A comprehensive analysis of the SINR for parameter estimation and optimizing a key parameter in the scheme is performed. Numerical simulations are provided, validating the near-ML performance and wide applicability of the presented OTFS sensing scheme and the precision of its SINR analysis.

References

1 M. K. Simon. *Probability Distributions Involving Gaussian Random Variables: A Handbook for Engineers and Scientists*. Springer Science & Business Media, 2007.

2 C. Sturm and W. Wiesbeck. Waveform design and signal processing aspects for fusion of wireless communications and radar sensing. *Proceedings of the IEEE*, 99(7):1236–1259, 2011.

3 S. Ye and E. Aboutanios. Rapid accurate frequency estimation of multiple resolved exponentials in noise. *Signal Processing*, 132:29–39, 2017.

4 A. Serbes and K. Qaraqe. A fast method for estimating frequencies of multiple sinusoidals. *IEEE Signal Processing Letters*, 27:386–390, 2020. doi: 10.1109/LSP.2020.2970837.

5 H. L. Van Trees. *Optimum Array Processing: Part IV of Detection, Estimation, and Modulation Theory*. John Wiley & Sons, 2004.

6 K. Wu, J. A. Zhang, X. Huang, and Y. J. Guo. Accurate frequency estimation with fewer DFT interpolations based on padé approximation. *arXiv preprint arXiv:2105.13567*, 2021.

7 L. Gaudio, M. Kobayashi, G. Caire, and G. Colavolpe. On the effectiveness of OTFS for joint radar parameter estimation and communication. *IEEE Transactions on Wireless Communications*, 19(9):5951–5965, 2020. doi: 10.1109/TWC.2020.2998583.

8 K. Wu, W. Ni, J. A. Zhang, R. P. Liu, and Y. J. Guo. Refinement of optimal interpolation factor for DFT interpolated frequency estimator. *IEEE Communications Letters*, 1, 2020. ISSN 2373-7891. doi: 10.1109/LCOMM.2019.2963871.

Part III

Radar-Centric Joint Communications and Sensing

7

FH-MIMO Dual-Function Radar-Based Communications: Single-Antenna Receiver

Part II is focused on performing sensing based on communication waveforms. In the three chapters of Part III, we consider using radar waveforms for communications. In particular, we base our discussions on the frequency-hopping multiple-input and multiple-output (FH-MIMO) radar. As discussed in Chapter 2, such a radar can relieve the restriction on the communication symbol rate as imposed by the radar repetition frequency (PRF). The signal processing for the radar function has been illustrated in Chapter 2, and hence, we only focus on the communication aspect from now on.

Channel estimation is one of the most important tasks for any communications system. Therefore, in this chapter, we focus on the channel estimation in the FH-MIMO dual-function radar communications (DFRC) system depicted in Figure 7.1a. In such a system, we assume that there are an FH-MIMO radar and a single-antenna user terminal; more complicated systems with multiple antennas at the user terminal will be dealt with in Chapters 8 and 9. In addition to sensing signals to a target for detection, the DFRC system also performs downlink communication with a communication user through a line-of-sight (LoS) channel. A challenging scenario is considered here: the radar hopping frequencies are changed without notifying the communication receiver. To fulfill data communications and reduce overhead, we propose a new scheme in which the packet communications have the signal frame structure depicted in Figure 7.1b. In each such frame, the first two hops are set identical to enable effective estimation of timing offset, carrier frequency offset (CFO), and channel, and the remaining hops are used for data transmission.

Using two identical hops enables simple and effective coarse timing and CFO estimation by performing the conventional energy-based or autocorrelation packet detection [1]. A coarse timing offset, denoted by $\eta(>0)$, can be readily estimated at each frame. When the CFO is not very large, a high-accuracy CFO estimation can be readily achieved based on the well-developed methods in the literature; refer to

Joint Communications and Sensing: From Fundamentals to Advanced Techniques, First Edition.
Kai Wu, J. Andrew Zhang, and Y. Jay Guo.

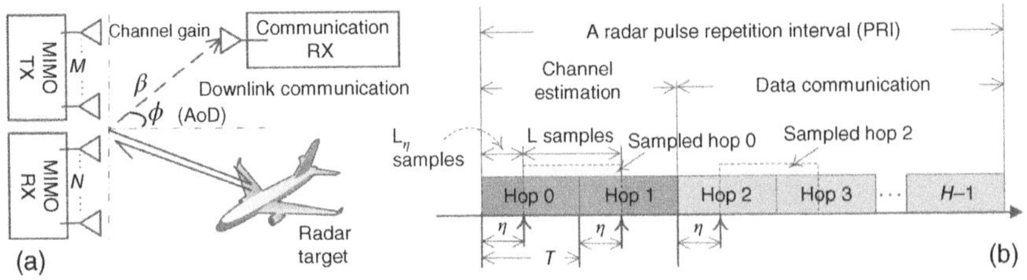

Figure 7.1 (a) Illustration on the system diagram of an FH-MIMO DFRC; (b) The signal frame structure for the downlink communication in Figure 7.1a.

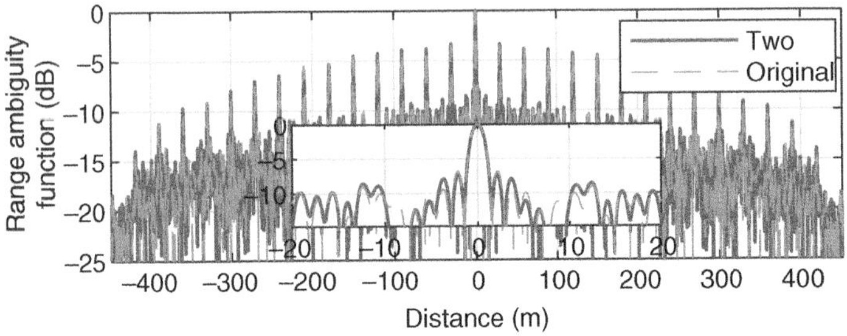

Figure 7.2 The impact of using two identical hops on the range ambiguity function (RAF) of an FH-MIMO radar, where, with reference to [3], the radar is configured as $M = 10$, $K = 20$, $H = 15$, $T = 0.2$ μs, $f_L = 8$ GHz, and $B = 100$ MHz. For the original RAF, the hopping frequencies, f_{hm} $\forall h, m$, are randomly selected from $\{f_k (= kB/K)\ \forall k\}$.

[2, Chapter 5.4] for a review. Thus, we proceed with the assumption of zero CFO and a nonzero timing offset of $\eta(> 0)$.

It should be pointed out that using two identical hops at a radar pulse can affect the range ambiguity function but only slightly. Figure 7.2 compares the range ambiguity function of an FH-MIMO radar having two identical hops at the beginning or not, where the suboptimal hopping frequency sequence, denoted by the $M \times 1$ vector $\mathbf{f}^*$, are used. ($\mathbf{f}^*$ will be designed in Section 7.3.3). We see from Figure 7.2 that only slight changes are incurred to the sidelobes of the range ambiguity function with the mainlobe unaffected by using $\mathbf{f}^*$ at the first two hops.

7.1 Problem Statement

Following the notations in Section 2.3.3, we use ϕ to denote the angle-of-departure (AoD) of the LoS path with respect to (w.r.t.) the radar transmitter array and β to denote the LoS path gain. Moreover, given the sampling time T_s, the number of samples in each hop is $L (= \frac{T}{T_s})$, where T denotes the hop duration. Under a perfect timing, the signal received by the communication receiver at hop h can be given by

$$y_h(t) = \beta \sum_{m=0}^{M-1} e^{-j\pi m \sin\phi} F_{hm} e^{-j2\pi f_{hm} t} + \xi(t), \tag{7.1}$$

where $F_{hm} = e^{j\varpi_{hm}}$ and $\xi(t)$ is an additive white Gaussian noise (AWGN).

Affected by the timing offset η, the initial sampling point of each hop is delayed by $L_\eta (= \lfloor \frac{\eta}{T_s} \rceil)$ samples, where $\lfloor x \rceil$ rounds x to the nearest integers. Based on (7.1), the ith ($i = 0, 1, \dots, L-1$) digitized communication signal at hop h, denoted by $y_h(i)$, is given by

$$y_h(i) = \beta \begin{cases} \sum_{m=0}^{M-1} F_{hm} e^{-j\pi m \sin\phi} e^{-j2\pi \frac{k_{hm} B}{K}(iT_s + \eta)} + \xi(i), \\ \times \text{ for } i = 0, 1, \dots, L - L_\eta - 1, \\ \sum_{m=0}^{M-1} F_{(h+1)m} e^{-j\pi m \sin\phi} e^{-j2\pi \frac{k_{(h+1)m} B}{K}((i-L)T_s + \eta)} + \xi(i), \\ \times \text{ for } i = L - L_\eta, \dots, L-1, \end{cases} \tag{7.2}$$

where $F_{hm} = e^{j\varpi_{hm}}$ denotes a PSK modulation symbol, and $k_{hm} (\in \{0, 1, \dots, K-1\})$ denotes the selected subband for antenna m at hop h.

We see from (7.2) that extracting F_{hm} for communication demodulation can be nontrivial due to the disturbing phases caused by k_{hm}, η, ϕ, and β. Although the M hopping frequencies at hop h can be estimated from the discrete Fourier transform (DFT) of $y_h(i)$, determining k_{hm} requires the pairing between the hopping frequencies and antennas. Acquiring the pairing information and updating the information as frequently as the primary radar does (i.e. H times per radar pulse repetition interval [PRI]) can be challenging.

We also see from (7.2) that the phases incurred by ϕ and η are coupled in a multiplicative manner for each antenna m. This is drastically different from the conventional communications with a constant η across antennas and, hence, invalidates conventional methods for estimating η and ϕ, e.g. in [2]. The coupling relation destroys the linear phase of $e^{-j\pi m \sin\phi}$ (across m), as required for ϕ estimation [4]. On the other hand, due to the random, independent frequency hopping across hops, the exponential term $e^{-j2\pi \frac{k_{hm} B\eta}{K}}$ is also random across hops. These factors make the joint estimation of η, ϕ, and β challenging.

7.2 Waveform Design for FH-MIMO DFRC

To solve the issues mentioned above, we need to design the waveform first. Then we present a channel estimation scheme based on the waveform.

7.2.1 FH-MIMO Radar Waveform

Due to the waveform orthogonality given in (2.10), the M signals transmitted from the M radar antennas have different center frequencies. This indicates that the M signals can be differentiated in the frequency domain. Thanks to the two identical hops at the beginning of each PRI (see Figure 7.1b), the waveform orthogonality condition given in (2.10) can be ensured, as analyzed below. Taking the L-point DFT of the digitized samples of hop h, i.e. $y_h(i)$ $(i = 0, 1, \dots, L-1)$ given in (7.2), the frequency-domain received signal at the lth discrete frequency $\frac{l}{LT_s}$, denoted by $Y_h(l)$, is

$$Y_h(l) = \sum_{i=0}^{L-1} y_h(i) e^{-j\frac{2\pi il}{L}} = L\beta \sum_{m=0}^{M-1} e^{-j\pi m \sin\phi} e^{-j2\pi \frac{k_{hm} B\eta}{K}} \times e^{-j\frac{\pi(L-1)l}{L}} \delta\left(l - \left(L - k_{hm}BT/K\right)\right) e^{-j\frac{\pi(L-1)k_{hm}BT}{KL}} + \Xi(l), \tag{7.3}$$

where $\delta(l)$ denotes the Dirac delta function and $\Xi(l)$ is the DFT of the AWGN $\xi(i)$. Note that the summation term in (7.3), as indexed by m, is solely related to the signal transmitted by radar antenna m. These terms will be separated and used for the estimation methods to be developed in Sections 7.3 and 7.4.

Due to the delta function in (7.3), M peaks of $|Y_h(l)|$ can be detected at $(L - k_{hm}BT/K)$ $(m = 0, 1, \dots, M-1)$. By identifying the M largest peaks of $|Y_h(l)|$, the set of $\{k_{hm}\ \forall m\}$ can be obtained. However, we cannot determine the pairing between the hopping frequencies and the radar transmitter antennas, as $k_{hm}\ \forall m$ can take $\forall k (\in [0, K-1])$ in conventional FH-MIMO radars [5, 6]. The waveform designed in [7] can solve this problem and will be called the new waveform below for convenience (in contrast to the conventional waveform illustrated in Section 2.3.3). The new waveform introduces a re-ordering of hopping frequencies at any hop in an ascending order,[1] as given by

$$\tilde{s}_{hm}(t) = e^{j2\pi(f_L + \tilde{k}_{hm}B/K)t}\ \forall h, \quad 0 \le t - hT \le T$$
$$\text{s.t. } \tilde{k}_{h0} < \tilde{k}_{h1} < \cdots < \tilde{k}_{h(M-1)}, \tag{7.4a}$$
$$\left\{\tilde{k}_{hm}\ \forall m\right\} = \left\{k_{hm}\ \forall m\right\}, \tag{7.4b}$$

where both $\tilde{k}_{hm}$ and k_{hm} denote the subband index of the radar-transmitted signal from antenna m at hop h, and the former is for the new waveform while the latter is for the conventional waveform.

An illustration of the new waveform is provided in Figure 7.3b, where the hopping frequencies across antennas and hops are displayed in scaled gray colors, and the hopping frequencies of a conventional FH-MIMO waveform are given in Figure 7.3a for reference.

---

1 It can also be a descending order, which does not affect the property of the waveform to be unveiled in Proposition 7.1 and the estimation methods to be presented in Sections 7.3 and 7.4.1.

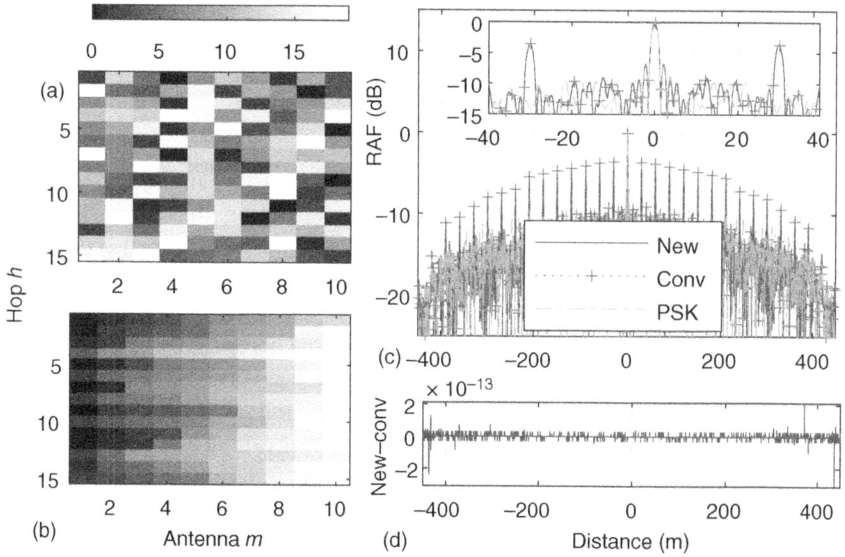

Figure 7.3 (a) The hopping frequencies of a conventional FH-MIMO radar waveform given in (2.11); (b) The hopping frequencies of the new waveform given by (7.4); (c) Comparison of the range ambiguity function (RAF) of the FH-MIMO radar using the conventional, new waveforms, and PSK-embedded new waveform; and (d) Difference between the first two RAFs from Figure 7.3c. The same radar configuration as in Figure 7.2 is used here.

Enabled by the new waveform, we can now determine the pairing between the hopping frequencies and radar transmitter antennas. Let l_m^* denote the index of the mth peak of $|Y_h(l)|$, satisfying

$$0 \le l_0^* < l_1^* < \cdots < l_{M-1}^* \le L-1. \tag{7.5}$$

From the parameter of the delta function in (7.3), we see that a smaller k_{hm} corresponds to a larger index of the peak. Based on this observation, (7.4a) and (7.5), we can estimate $\tilde{k}_{hm}$ as

$$\hat{k}_{hm} = \left(\frac{L - l_{M-1-m}^*}{LT_s}\right) \Big/ \left(\frac{B}{K}\right) = \frac{K(L - l_{M-1-m}^*)}{BT}. \tag{7.6}$$

The above estimation is achieved without degrading radar ranging due to the following property.

Proposition 7.1 *The novel FH-MIMO radar waveform, $\tilde{s}_{hm}(t)$, has the same range ambiguity function as the original FH-MIMO radar based on $s_{hm}(t)$ given in (2.11).*

Proof: The proof can be established by analyzing the range ambiguity function of the FH-MIMO radar. Based on [6, Eq. (27)], the range ambiguity function of the radar, denoted by $R(\tau)$, can be expressed as follows:

$$R(\tau) = \left| \sum_{m=0}^{M-1} \sum_{m'=0}^{M-1} \sum_{h,h'=0}^{H-1} \underbrace{\chi(\tilde{\tau}, \nu) e^{j2\pi\nu hT}}_{\mathcal{B}} \underbrace{e^{j2\pi f_{h'm'}\tau}}_{\mathcal{D}} \right|, \tag{7.7}$$

where $\tilde{\tau} = \tau - T(h' - h)$, $\nu = f_{hm} - f_{h'm'}$, and $\chi(x, y)$ is the ambiguity function of a standard rectangular pulse with x and y spanning range and Doppler domains, respectively. According to [6, Eq. (26)], we have $\chi(x, y) = (T - |x|)\, S\,(y(T - |x|))\, e^{j\pi y(x+T)}$, if $|x| < T$, and otherwise, $\chi(x, y) = 0$, where $S(\alpha) = \frac{\sin(\pi\alpha)}{\pi\alpha}$ is the sinc function.

As hopping frequencies are independently selected across hops, any change of f_{hm} at hop h has no impact on $f_{h'm'}$ at hop h', and vice versa. Therefore, we can claim that the set of combinations of $(\nu, f_{h'm'})$ remain the same given any ordering of hopping frequencies at hops h and h'. The underlying principle is that the overall combinations of $(\nu, f_{h'm'})$ are independent of element orderings [8].

Moreover, we see from (7.7) that the combinations of $(\mathcal{B}, \mathcal{D})$ are uniquely determined by the combinations of $(\nu, f_{h'm'})$, since the other two parameters, $\tilde{\tau}$ and τ, are independent of m or m'. Given the independence of the set, $\{(\nu, f_{h'm'})\ \forall h, h', m, m'\}$, on the ordering of hopping frequencies, we conclude that the range ambiguity function, $R(\tau)$, is unaffected by the reordering introduced in (7.4).

Figure 7.3c compares the range ambiguity functions of an FH-MIMO radar, where the conventional range ambiguity function is calculated by substituting the hopping frequencies shown in Figure 7.3a into (7.7), and the new range ambiguity function is calculated based on the hopping frequencies shown in Figure 7.3b. We see from Figure 7.3c that the new range ambiguity function overlaps with the conventional one. This is further validated by Figure 7.3d. Figure 7.3c also plots the range ambiguity function using the PSK-embedded new waveform, where $(M \times H)$ number of randomly generated binary phase shift keying (BPSK) symbols are multiplied onto $\tilde{s}_{hm}$ $\forall h, m$, one for each. Same as the conventional FH-MIMO waveform [5], the new waveform, when combined with PSK, can have range sidelobe spikes suppressed. This is because incoherent PSK phases can prevent periodic energy accumulations in range sidelobes; refer to [5] for an in-depth analysis of the spike suppression.

7.2.2 Overall Channel Estimation Scheme

Based on the FH-MIMO radar waveform described in Section 7.2.1, a two-step channel estimation scheme is introduced in this section. The estimation methods can be performed based on a single hop, i.e. the sampled hop 0 as highlighted in Figure 7.1. Hence, we drop the subscript "$(\cdot)_h$," unless otherwise specified.

7.2.2.1 Estimate Timing Offset

Substituting (7.5) and (7.6) into (7.3), the signal from the mth radar transmitter antenna can be extracted, as given by

$$Y_m = Y(l^*_{M-1-m}) = \tilde{\beta} e^{-j\frac{2\pi m u}{M}} \omega_\eta^{\hat{k}_m} + \Xi(l^*_{M-1-m}), \tag{7.8}$$

where the intermediate variables, $\tilde{\beta}$, u, and ω_η, are defined as follows:

$$\tilde{\beta} \triangleq L\beta e^{-j\pi(L-1)}, \quad u \triangleq M \sin\phi/2, \quad \omega_\eta \triangleq e^{-j2\pi\frac{B\eta}{K}}. \tag{7.9}$$

The way we define u is to enable accurate estimation of ϕ, as will be designed in Section 7.4.1. Note that $Y(l^*_{M-1-m})$ takes the DFT output at the l^*_{M-1-m}th discrete frequency, while the DFT output given in (7.3) is calculated based on the L signal samples in the first single hop received by the communication receiver.

From (7.8), we see that $\omega_\eta^{\hat{k}_m}$ $\forall m$ is multiplied to $e^{-j\frac{2\pi m u}{M}}$ $\forall m$ pointwise. Due to the phase disturbance caused by $\omega_\eta^{\hat{k}_m}$, the linear phase relation in $e^{-j\frac{2\pi m u}{M}}$, which is the key for angle estimation [4], is scrambled. Nevertheless, we notice that the linear phase relation in $e^{-j\frac{2\pi m u}{M}}$ can be exploited to suppress the impact of u on η estimation. This enables us to estimate ω_η unambiguously, as to be designed in Section 7.3. We estimate ω_η rather than η due to the nontrivial phase ambiguity issue in η estimation. This will be clear in Section 7.3. It is noteworthy that using the estimate of ω_η is sufficient to suppress the impact of η on ϕ estimation and communication decoding.

7.2.2.2 Estimate Channel Parameters

Given the estimate of ω_η, we can remove $\omega_\eta^{\hat{k}_m}$ $\forall m$ in (7.8) to further estimate ϕ and β. It is nontrivial to estimate ϕ based on the single-hop signals Y_m $\forall m$. The spatial searching methods developed in [9, 10] can be performed using a single snapshot. However, these methods [9, 10] can be time-consuming in achieving a satisfactory estimation accuracy, since more searching grids are required for a better angle resolution. We have illustrated in Chapter 3 two low-complexity frequency estimators. In Section 7.4.1, we will apply the first estimator, detailed in Section 3.2, to efficiently estimate ϕ and β.

7.3 Estimating Timing Offset

As illustrated in Section 7.2.2, the estimation of ω_η is key to the overall channel estimation. In this section, we first present two estimators for ω_η, then derive their mean squared error lower bounds (MSELBs), and moreover, design a suboptimal hopping frequency sequence.

7.3.1 Two Estimation Methods

To estimate ω_η based on Y_m given in (7.8), we need to suppress the impact of $\tilde{\beta}$ and u. Since $\tilde{\beta}$ is independent of m, we can suppress $\tilde{\beta}$ by taking the ratio of adjacent Y_m,

$$\breve{Y}_m = \frac{Y_m}{Y_{m+1}} = e^{\mathrm{j}\frac{2\pi u}{M}} \omega_\eta^{\hat{k}_m - \hat{k}_{(m+1)}}, \quad m = 0, 1, \dots, M-2. \tag{7.10}$$

We see from (7.10) that the u-related term is now independent of m, and hence, by taking the ratio of adjacent $\breve{Y}_m$, the impact of u can be suppressed, i.e.

$$\begin{aligned} \bar{Y}_m &= \frac{\breve{Y}_m}{\breve{Y}_{m+1}} = \omega_\eta^{\kappa_m} = e^{\mathrm{j}\kappa_m \angle\omega_\eta}, \quad m = 0, \dots, M-3, \\ \text{s.t. } \kappa_m &\triangleq \hat{k}_m - 2\hat{k}_{(m+1)} + \hat{k}_{(m+2)}, \end{aligned} \tag{7.11}$$

where $\angle\omega_\eta$ takes the phase of ω_η in the interval of $[-\pi, \pi]$. By estimating $\angle\omega_\eta$, ω_η can be determined.

We see from (7.11) that $\angle\omega_\eta$ can be estimated by taking the phase of $\bar{Y}_m$, which, however, requires $\kappa_m \neq 0$. Based on (7.6), we can identify the set of antenna indexes satisfying

$$\mathcal{M} = \{\forall \tilde{m}\}, \quad \text{s.t. } \kappa_{\tilde{m}} \neq 0. \tag{7.12}$$

We also see from (7.11) that directly taking the angle of $\bar{Y}_m$ can lead to phase ambiguity due to potential cases of $|\kappa_m \angle\omega_\eta| > \pi$. Hence, we have two possible estimates of $\angle\omega_\eta$, i.e.

$$\measuredangle\omega_\eta(d_{\tilde{m}}) = \begin{cases} \kappa_{\tilde{m}} \measuredangle\bar{Y}_{\tilde{m}} \ (d_{\tilde{m}} = 0), & \text{if } |\kappa_{\tilde{m}}| = 1, \\ \frac{\measuredangle\bar{Y}_{\tilde{m}} + 2d_{\tilde{m}}\pi}{\kappa_{\tilde{m}}} \ (d_{\tilde{m}} = 0, \pm 1, \dots), & \text{otherwise,} \end{cases} \tag{7.13}$$

where $d_{\tilde{m}}$ is the ambiguity degree.

From (7.13), we see that, if $|\kappa_{\tilde{m}}| = 1$ holds for some $\tilde{m}$, we obtain the unambiguous estimate of $\angle\omega_\eta$ directly. Let $\bar{\mathcal{M}} \subseteq \mathcal{M}$ denote the set of $\bar{m}$ such that $|\kappa_{\bar{m}}| = 1$, i.e.

$$\bar{\mathcal{M}} = \{\forall \bar{m}\} \subseteq \mathcal{M}, \quad \text{s.t. } |\kappa_{\bar{m}}| = 1 \ \forall \bar{m} \in \mathcal{M}. \tag{7.14}$$

Substituting (7.14) into (7.13), we can accumulate $\bar{Y}_{\bar{m}}$ coherently across $\bar{m}$ and then take the angle for $\angle\omega_\eta$ estimation. This leads to the first estimator of $\angle\omega_\eta$, referred to as the coherent accumulation estimator (CAE):

$$\text{CAE: } \measuredangle\bar{\omega}_\eta = \angle\left(\frac{1}{\bar{M}} \sum_{\bar{m} \in \bar{\mathcal{M}}} \left(\Re\{\bar{Y}_{\bar{m}}\} + \mathrm{j}\kappa_{\bar{m}} \Im\{\bar{Y}_{\bar{m}}\} \right) \right), \tag{7.15}$$

where $\bar{M}$ is the dimension of the set $\bar{\mathcal{M}}$. Note that $\kappa_{\bar{m}}$ is multiplied to $\Im\{\bar{Y}_{\bar{m}}\}$, as $\kappa_{\bar{m}} = -1$ can happen for some $\bar{m}$.

Depending on the hopping frequencies, we can have $\bar{\mathcal{M}} = \emptyset$, which invalidates CAE given in (7.15). In this case, we can still estimate $\angle\omega_\eta$ by removing the estimation ambiguity in (7.13). Specifically, we can exploit the Chinese remainder theorem [11] to suppress the ambiguity in (7.13). To do this, we need to identify the set $\breve{\mathcal{M}}$ rendering $\{|\kappa_{\breve{m}}|\ \forall \breve{m} \in \breve{\mathcal{M}}\}$ coprime with at least two elements, i.e.

$$\breve{\mathcal{M}} = \{\forall \breve{m}\}, \quad \text{s.t. } \mathcal{G}\{|\kappa_{\breve{m}}|(\neq 1)\ \forall \breve{m} \in \mathcal{M}\} = 1, \quad \breve{M} \geq 2, \tag{7.16}$$

where $\mathcal{G}\{\cdot\}$ takes greatest common divisor (GCD), and $\breve{M}$ denotes the dimension of the set $\breve{\mathcal{M}}$. By identifying the ambiguity degree $d^*_{\breve{m}}\ \forall \breve{m}$ such that the estimates $\measuredangle\omega_\eta(d^*_{\breve{m}})\ \forall \breve{m} \in \breve{\mathcal{M}}$ are identical, the second estimator for $\angle\omega_\eta$, referred to as the Chinese remainder theorem estimator (CRE), is achieved:

$$\text{CRE: } \measuredangle\breve{\omega}_\eta = \frac{1}{\breve{M}} \sum_{\breve{m}\in\breve{\mathcal{M}}} \measuredangle\omega_\eta(d^*_{\breve{m}}). \tag{7.17}$$

Note that CAE and CRE have their favorable working conditions and correspondingly different estimation accuracy, as analyzed below.

7.3.2 Performance Analysis and Comparison of the Estimators

To compare the two estimators, we first derive their MSELBs. In the following, the signal-to-noise ratio (SNR), denoted by γ, refers to the ratio between the received signal power and the communication receiver noise power. Based on (7.2), we have $\gamma = |\beta|^2/\sigma_n^2$, where $\sigma_n^2 = \mathbb{E}\{|\xi(i)|^2\}$ is the noise variance of the AWGN $\xi(i)$. The high-SNR MSELBs of CAE and CRE are derived as follows:[2]

Proposition 7.2 *At high SNR, the MSELBs of CAE and CRE are (7.18) and (7.19), respectively.*

$$\bar{\sigma}_\eta^2 = 3/(\bar{M}L\gamma). \tag{7.18}$$

$$\breve{\sigma}_\eta^2 = \frac{1}{\breve{M}^2} \sum_{\breve{m}\in\breve{\mathcal{M}}} 3/(\kappa_{\breve{m}}^2 L\gamma). \tag{7.19}$$

Proof: The proof is established by first proving that the estimators given in (7.15) and (7.17) are maximum-likelihood estimators (MLEs), and then evaluating the estimation SNRs for the two estimators to derive their MSELBs. Let Ξ_m denote the noise term $\Xi(l^*_{M-1-m})$ in (7.8). Here, Ξ_m, as the DFT of the AWGN $\xi(i)$, is still an AWGN; refer to (7.3). (The underlying principle is that linear calculations involved

2 Note that Cramer–Rao lower bound (CRLB) is the lower limit of the MSELB derived here; while, according to [12], CRLB does not apply to estimators, like CAE and CRE, which estimate a random phase with finite support $[-\pi, \pi)$.

in DFT do not change the statistic distribution of AWGNs [13].) Substituting (7.8) into (7.10), the noise term added to $\check{Y}_m$ can be given by

$$\begin{aligned}\check{\Xi}_m &= \frac{Y_m + \Xi_m}{Y_{m+1} + \Xi_{m+1}} - \check{Y}_m = \frac{\check{Y}_m + \frac{\Xi_m}{Y_{m+1}}}{1 + \frac{\Xi_{m+1}}{Y_{m+1}}} - \check{Y}_m \\ &\approx \check{Y}_m \left(1 - \frac{\Xi_{m+1}}{Y_{m+1}}\right) + \frac{\Xi_m}{Y_{m+1}} \left(1 - \frac{\Xi_{m+1}}{Y_{m+1}}\right) - \check{Y}_m \\ &= -\frac{\check{Y}_m \Xi_{m+1}}{Y_{m+1}} + \frac{\Xi_m}{Y_{m+1}} - \frac{\Xi_m}{Y_{m+1}} \frac{\Xi_{m+1}}{Y_{m+1}},\end{aligned} \tag{7.20}$$

where the approximation is based on the Taylor series of $\frac{1}{1+x} = 1 - x \;\forall x \ll 1$ [14].

Similarly, by substituting (7.10) and (7.20) into (7.11), the noise term added to $\bar{Y}_m$ becomes

$$\begin{aligned}\bar{\Xi}_m &= \frac{\check{Y}_m + \check{\Xi}_m}{\check{Y}_{m+1} + \check{\Xi}_{m+1}} - \bar{Y}_m = \frac{\bar{Y}_m + \frac{\check{\Xi}_m}{\check{Y}_{m+1}}}{1 + \frac{\check{\Xi}_{m+1}}{\check{Y}_{m+1}}} - \bar{Y}_m \\ &\approx \frac{\bar{Y}_m \Xi_{m+2}}{Y_{m+2}} - \frac{\bar{Y}_m \Xi_{m+1}}{\check{Y}_{m+1} Y_{m+2}} - \frac{\check{Y}_m \Xi_{m+1}}{\check{Y}_{m+1} Y_{m+1}} + \frac{\Xi_m}{\check{Y}_{m+1} Y_{m+1}} \\ &\quad + \frac{\bar{Y}_m \Xi_{m+1} \Xi_{m+2}}{\check{Y}_{m+1} Y_{m+2}^2} - \frac{\Xi_{m+1} \Xi_m}{\check{Y}_{m+1} Y_{m+1}^2} - \frac{\check{\Xi}_m}{\check{Y}_{m+1}} \frac{\check{\Xi}_{m+1}}{\check{Y}_{m+1}},\end{aligned} \tag{7.21}$$

where we have used the same mathematical manipulations as those applied in (7.20). From the most right-hand side (RHS) of (7.21), we see that only the first four terms dominate the statistical distribution of $\bar{\Xi}_m$, since the other terms have the products of at least two independent AWGNs. Thus, we obtain that the additive noise to $\bar{Y}_m$ approaches an AWGN in high SNR regions.

In the background of $\bar{\Xi}_{\tilde{m}}$ ($\forall \tilde{m} \in \mathcal{M} \subseteq \{\bar{Y}_m \forall m\}$), $\measuredangle \bar{Y}_{\tilde{m}}$ is an MLE of the phase of $\bar{Y}_{\tilde{m}}$, since the angle estimate of a complex number corrupted by AWGN is an unbiased MLE. Accordingly, $\measuredangle \omega_\eta(d_{\tilde{m}})$ obtained in (7.13) is an MLE due to the linear relation between $\measuredangle \omega_\eta(d_{\tilde{m}})$ and $\measuredangle \bar{Y}_{\tilde{m}}$. By comparing (7.13) and (7.15), we also see that $\measuredangle \bar{\omega}_\eta$ is an MLE, since the summation does not change the AWGN distribution. Moreover, by substituting (7.13) into (7.17), we further see that $\measuredangle \breve{\omega}_\eta$ is an MLE, since $\measuredangle \breve{\omega}_\eta$ is a constant-scaled sum of the MLEs $\measuredangle \omega_\eta(d_{\tilde{m}})$. Being MLEs, the high-SNR MSELB of $\measuredangle \bar{Y}_{\tilde{m}}$, $\measuredangle \bar{\omega}_\eta$ and $\measuredangle \breve{\omega}_\eta$ can be approximated as the reciprocal of the doubled estimation SNRs [12]. The estimation SNRs of $\measuredangle \bar{\omega}_\eta$ and $\measuredangle \breve{\omega}_\eta$ are calculated below.

We start by deriving the estimation SNR of $\measuredangle \bar{Y}_{\tilde{m}}$, denoted by Υ, since Υ is the basis of the estimation SNRs of $\measuredangle \bar{\omega}_\eta$ and $\measuredangle \breve{\omega}_\eta$; see (7.15) and (7.17). Seen from (7.11), the signal power of $\bar{Y}_{\tilde{m}}$ is unit one; hence, we have $\Upsilon = \frac{1}{\bar{\sigma}_n^2}$, where $\bar{\sigma}_n^2$ is the noise variance of $\bar{\Xi}_m$. Based on (7.10) and (7.11), we notice that the second and third terms in (7.21) are identical. Therefore, we have $\bar{\sigma}_n^2 = \frac{6\tilde{\sigma}_n^2}{|\tilde{\beta}|^2}$, where $\tilde{\sigma}_n^2$ denotes the noise variance of Ξ_m $\forall m$ and $|\tilde{\beta}|^2$ is the power of the signal component in (7.8). Based on (7.9), we have $|\tilde{\beta}|^2 = L^2|\beta|^2$. Based on (7.3) and (7.8), we obtain $\tilde{\sigma}_n^2 = L\sigma_n^2$, where L is the number of DFT points, and σ_n^2 denotes the noise variance of the time-domain AWGN $\xi(i)$; see (7.2). The above calculations lead to $\Upsilon = \frac{L|\beta|^2}{6\sigma_n^2}$.

As the summation in (7.15) is a coherent accumulation, we have $\bar{\Upsilon} = \bar{M}\Upsilon = \frac{\bar{M}L|\beta|^2}{6\sigma_n^2}$ with $\bar{\Upsilon}$ denoting the estimation SNR for $\measuredangle\bar{\omega}_\eta$. By calculating $\frac{1}{2\bar{\Upsilon}}$, the MSELB of $\measuredangle\bar{\omega}_\eta$ is achieved in (7.18). Based on (7.17), the variance of $\measuredangle\breve{\omega}_\eta$ can be approximated as $\mathbb{E}\left\{\left(\frac{1}{\breve{M}}\sum_{\breve{m}\in\breve{\mathcal{M}}}\measuredangle\omega_\eta(d_{\breve{m}}^*) - \angle\omega_\eta\right)^2\right\}$. Here, $\angle\omega_\eta$ is used as the mean of $\measuredangle\breve{\omega}_\eta$, since $\measuredangle\omega_\eta(d_{\breve{m}}^*)$ is unbiased in the sense of $\mathbb{E}\{\measuredangle\omega_\eta(d_{\breve{m}}^*)\} = \angle\omega_\eta$ [12]. By suppressing the cross-terms, the above variance is lower bounded by $\frac{1}{\breve{M}^2}\sum_{\breve{m}\in\breve{\mathcal{M}}}\mathbb{E}\left\{\left(\measuredangle\omega_\eta(d_{\breve{m}}^*) - \angle\omega_\eta\right)^2\right\} = \frac{1}{\breve{M}^2}\sum_{\breve{m}\in\breve{\mathcal{M}}}\sigma_{\breve{m}}^2$, where $\sigma_{\breve{m}}^2$ denotes the MSELB of $\measuredangle\omega_\eta(d_{\breve{m}}^*)$. From (7.17), we obtain $\sigma_{\breve{m}}^2 = \frac{1}{2\Upsilon\kappa_{\breve{m}}^2}$, which gives (7.19).

We see from (7.18) and (7.19) that the accuracy of both estimators are dependent on hopping frequencies. Specifically, $\bar{\sigma}_\eta^2$ decreases when the number of ones in $\bar{\mathcal{M}}$ (i.e. $\bar{M}$) increases. Thus, the MSELB for CAE has a lower limit, i.e. $\underline{\bar{\sigma}_\eta^2} = \frac{3}{(M-2)L\gamma} \leq \bar{\sigma}_\eta^2$, where $(M-2)$ is the maximum value that $\bar{M}$ can take. In contrast, the accuracy of CRE depends on the number and the values of the coprime elements in $\breve{\mathcal{M}}$. Given (7.16), $\breve{\mathcal{M}} = \{2,3\}$ with $\breve{M} = 2$ is the smallest set with the minimum coprime numbers. Substituting $\breve{\mathcal{M}} = \{2,3\}$ into (7.19), we obtain the upper limit of $\breve{\sigma}_\eta^2$, as given by $\overline{\breve{\sigma}_\eta^2} = \frac{2}{L\gamma} \times \frac{\frac{1}{4}+\frac{1}{9}}{4} = \frac{3}{L\gamma} \times \frac{13}{144} \geq \breve{\sigma}_\eta^2$. Note that if $\underline{\bar{\sigma}_\eta^2} > \overline{\breve{\sigma}_\eta^2}$ then $\bar{\sigma}_\eta^2 > \breve{\sigma}_\eta^2$ is assured. Moreover, $\underline{\bar{\sigma}_\eta^2} > \overline{\breve{\sigma}_\eta^2}$ leads to $\frac{1}{M-2} > \frac{13}{144}$, and further $M \leq 13$. Thus, we have the following corollary:

Corollary 7.1 *For a uniform linear array with the antenna spacing of half a wavelength, provided the number of antennas at the radar transmitter satisfies $M \leq 13$, CRE always has a better asymptotic performance than CAE, i.e. $\bar{\sigma}_\eta^2 > \breve{\sigma}_\eta^2$.*

Remark 7.1 Corollary 7.1 compares the asymptotic performance of the two estimates in high-SNR regions, where the ambiguity degree $d_{\breve{m}}^*$ required for CRE can be reliably identified; see (7.17). In low-SNR regions, however, the correct identification of $d_{\breve{m}}^*$ cannot be ensured, which degrades the estimation accuracy of CRE. This issue does not exist for CAE which does not have estimation ambiguity. Moreover, when $\bar{M}$ is large, the coherent accumulation in (7.15) can help improve the estimation SNR of CAE. CAE is more suited for low-SNR regions than CRE in this sense. As to be observed from simulation in Section 7.7, there is an SNR threshold of γ, denoted by γ_{T}, satisfying: if $\gamma > \gamma_{\mathrm{T}}$, CRE is more accurate than CAE; otherwise, CAE is better.

7.3.3 Design of a Suboptimal Hopping Frequency Sequence

At the communication receiver, γ can be estimated. By comparing γ and γ_{T}, the receiver can choose which estimator to use between CAE and CRE. However, this does not apply to the radar transmitter with no *a-priori* information on γ. To this end, it is necessary to design a hopping frequency sequence that renders both estimators applicable at the communication receiver. Such a sequence is optimal when the MSELB of CAE, $\bar{\sigma}_\eta^2$, and that of CRE, $\breve{\sigma}_\eta^2$, are minimized simultaneously. The optimality, however, cannot be achieved, since $\bar{M} = (M-2)$ is required to minimize $\bar{\sigma}_\eta^2$; whereas $\bar{M}$ can take no greater than $(M-4)$ to ensure at least two elements in $\breve{\mathcal{M}}$ for CRE. Next, we illustrate a suboptimal design of hopping frequency sequence that ensures the largest coherent accumulation gain of $(M-4)$ in low-SNR regions and accordingly minimizes the MSELB of CRE. Let $\mathbf{f}^*$ denote the suboptimal hopping frequency sequence to be designed. Since $\mathbf{f}^* = f_{\mathrm{L}} + B\mathbf{k}^*/K$, where $\mathbf{k}^* = [k_0^*, k_1^*, \dots, k_{M-1}^*]^{\mathrm{T}}$, we can design $\mathbf{k}^*$ equivalently.

(1) *Minimizing* $\bar{\sigma}_\eta^2$: The minimization of the MSELB of CAE, $\bar{\sigma}_\eta^2$, can be achieved at $\bar{M} = (M-4)$, i.e. having $(M-4)$ elements in $\bar{\mathcal{M}}$. This requires at least $(M-2)$ hopping frequencies, since according to (7.11), one element of $\bar{\mathcal{M}}$ is calculated using three hopping frequencies. Based on (7.11), we design the following recursive calculation of k_m^* to ensure $\bar{M} = M-4$,

$$k_{m+2}^* = 2k_{m+1}^* - k_m^* \pm 1, \tag{7.22a}$$

$$\text{s.t. } k_{m+2}^* > k_{m+1}^*, \quad m = 0, 1, \dots, M-5, \tag{7.22b}$$

$$k_0^* = 0, \quad k_1^* = 1, \tag{7.22c}$$

where the constraint (7.22b) complies with the constraint (7.4a) of the new FH-MIMO radar waveform; and (7.22c) initializes the first two hopping frequencies associated with antennas $m = 0$ and 1. Given the recursive calculation in (7.22a), taking the minimum values for k_0^* and k_1^* also minimizes k_{M-3}^*. The minimization of k_{M-3}^* is important for the design of the remaining two elements to minimize the MSELB of CRE, $\breve{\sigma}_\eta^2$, as elaborated on below.

(2) *Minimizing* $\breve{\sigma}_\eta^2$: By solving (7.22), the first $(M-2)$ elements in $\mathbf{k}^*$ are determined, which leaves k_{M-2}^* and k_{M-1}^* to be designed for minimizing $\breve{\sigma}_\eta^2$. Moreover, k_{M-2}^* and k_{M-1}^* can only be selected from $\mathcal{K} = \{k_{M-3}^* + 1, k_{M-3}^* + 2, \dots, K-1\}$. According to (7.19), the problem of minimizing $\breve{\sigma}_\eta^2$ is turned into: the selection of two elements from $\mathcal{K}$ as the last two elements of $\mathbf{k}^*$, so that the last four elements of $\mathbf{k}^*$ can produce two coprime numbers to

minimize $\rho = \frac{1}{(\breve{M})^2}\sum_{\breve{m}\in\breve{\mathcal{M}}}\frac{1}{\kappa_{\breve{m}}^2}$ and hence $\breve{\sigma}_{\eta}^2$. Let $\{\mathbf{k}_b,\ b = 0, 1, \ldots, C_{|\mathcal{K}|}^2 - 1\}$ denote the set for the combinations of selecting two elements from $\mathcal{K}$, where $|\mathcal{K}|$ is the cardinality of $\mathcal{K}$. By substituting $[k_{M-4}^*, k_{M-3}^*, \mathbf{k}_b^{\mathrm{T}}]^{\mathrm{T}}$ into (7.11) and (7.16), the obtained set of coprime numbers is denoted by $\breve{\mathcal{M}}_b$. Its dimension is $\breve{M}_b$. Thus, $\breve{\sigma}_{\eta}^2$ can be minimized via solving

$$\{\mathbf{k}_{b^*}, \rho_{b^*}\} : \quad \min_{b\in\{0,1,\ldots,C_{|\mathcal{K}|}^2-1\}} \rho_b = \frac{1}{\breve{M}_b^2}\sum_{\breve{m}\in\breve{\mathcal{M}}_b}\frac{1}{\kappa_{\breve{m}}^2}. \tag{7.23}$$

Based on (7.22) and (7.23), the suboptimal hopping frequency sequence is obtained as $\mathbf{f}^* = f_{\mathrm{L}} + B\mathbf{k}^*/K$, where

$$\mathbf{k}^* = \left(\overbrace{k_0^*, k_1^*, \ldots, k_{M-4}^*, k_{M-3}^*}^{(7.22):\ \text{minimizing } \breve{\sigma}_\eta^2},\ \underbrace{\ \mathbf{k}_{b^*}^{\mathrm{T}}}_{(7.23):\ \text{minimizing } \breve{\sigma}_\eta^2}\right)^{\mathrm{T}}. \tag{7.24}$$

7.4 Estimating Channel Response

With ω_η estimated, we proceed to estimate ϕ and $\tilde{\beta}$. Then, a complexity analysis is provided for the overall channel estimation scheme.

7.4.1 Estimation Method

Based on the estimation obtained in (7.15) or (7.17), we obtain the estimate of ω_η as $\hat{\omega}_\eta = e^{\mathrm{j}\measuredangle\tilde{\omega}_\eta}$ or $e^{\mathrm{j}\measuredangle\breve{\omega}_\eta}$. Dividing both sides of (7.8) by $\hat{\omega}_\eta^{\hat{k}_m}$ leads to

$$Z_m = Y_m / \hat{\omega}_\eta^{\hat{k}_m} = \tilde{\beta}e^{-\mathrm{j}\frac{2\pi m u}{M}}, \tag{7.25}$$

where $\tilde{\beta}$ and the ϕ-related variable, u, are defined in (7.9) and the noise term is dropped to focus on algorithm illustration. Note in (7.25) that $\omega_\eta^{\hat{k}_m}$ is assumed to be fully suppressed so that we can focus on formulating the estimation method for u and $\tilde{\beta}$. (The impact of the ω_η estimation error on the estimations of u and $\tilde{\beta}$ will be illustrated in Section 7.7.)

We see from (7.25) that u can be regarded as a discrete frequency, and hence, u estimation is turned into the frequency estimation of a sinusoidal signal Z_m $(m = 0, 1, \ldots, M-1)$. The frequency estimator presented in Section 3.2 is applied here for u estimation. In overall, the estimator first searches for the DFT peak of the sinusoidal signal Z_m to obtain a coarse estimation of u, and then interpolates the DFT coefficients around the peak to refine the estimation. Taking the DFT of Z_m w.r.t. m leads to $z_{m'} = \sum_{m=0}^{M-1} Z_m e^{-j\frac{2\pi mm'}{M}}$. By identifying the peak of $|z_{m'}|$, a coarse estimation of u can be obtained as $\frac{\tilde{m}}{M}$, where $\tilde{m}$ is the index of the peak. The true value of u can be written as $\frac{\tilde{m}+\delta}{M}$ with $\delta(\in[-0.5, 0.5])$ being a fractional frequency residual. By estimating δ, the coarse u estimate can be refined, as developed below.

We can estimate δ recursively from the interpolated DFT coefficients around $\tilde{m}$. Initially, we set $\delta = 0$, and calculate the interpolated DFT at the discrete frequency $\tilde{m} \pm \epsilon + \delta$, where $\epsilon = \min\{M^{-\frac{1}{3}}, 0.32\}$ [15, Eq. (23)] is an auxiliary variable of the u estimation algorithm. It has been proved in [14] that the above value of ϵ leads to an efficient estimator in the sense of approaching CRLB. The interpolated DFT coefficients, denoted by $z_{\pm}$, can be calculated as $z_{\pm} = \sum_{m=0}^{M-1} Z_m e^{-j\frac{2\pi m(\tilde{m}+\delta\pm\epsilon)}{M}}$. An update of δ can be obtained using $z_{\pm}$, i.e.

$$\delta = \frac{\epsilon\cos^2(\pi\epsilon)}{1-\pi\epsilon\cot(\pi\epsilon)} \times \Re\{\zeta\} + \delta, \tag{7.26}$$

where δ on the RHS is the old value and $\zeta = \frac{z_+ - z_-}{z_+ + z_-}$. Use the new value of δ to update $z_{\pm}$ which is then used, as above, for δ update. By updating δ three times in overall, the algorithm can generally converge [14]. The final estimate of u is obtained as $\hat{u} = (\tilde{m}+\delta)/M$. Substituting $\hat{u}$ into (7.9) and (7.25), the ϕ and $\tilde{\beta}$ estimations are

$$\hat{\phi} = \arcsin \hat{u}\lambda/(Md), \quad \hat{\tilde{\beta}} = \frac{1}{M}\sum_{m=0}^{M-1} Z_m e^{j\frac{2\pi m\hat{u}}{M}}. \tag{7.27}$$

7.4.2 Complexity Analysis

In overall, the channel estimation scheme has low computational complexity since no computationally intensive operations (e.g. matrix inversion/decomposition) are required. As a matter of fact, the major computations involved in the scheme are an L-dimensional DFT, an M-dimensional DFT and N_{iter} numbers of M-dimensional complex vector operations. The first DFT is used for identifying hopping frequencies and extracting the M peaks that are used for channel estimation; see Sections 7.2.2 and 7.3.1. The second DFT is used for obtaining the coarse estimation of u; see Section 7.4.1. The third vector operation is used for refining u estimation through N_{iter} iterations. Thus, the computational complexity of the scheme is given by

$$\mathcal{O}\left(L'\log L' + M'\log M' + M'N_{\text{iter}}\right) \overset{L\gg M}{\approx} \mathcal{O}\left(L'\log L'\right),$$

where the fast Fourier transform is used for calculating the two DFTs; $L' = 2^{\lceil \log_2 L\rceil}$ and $M' = 2^{\lceil \log_2 M\rceil}$; and the last approximation is established as N_{iter} is small. As illustrated in Section 7.4.1, $N_{\text{iter}} = 3$ is generally sufficient for the convergence of the u estimation method.

7.5 Using Estimations in Data Communications

In this section, we illustrate how to apply the estimations of ω_η, ϕ, and β to perform data communications. To improve the data rate, we combine PSK and frequency-hopping code selection (FHCS) into a new constellation, referred to as PSK FHCS (PFHCS). To perform PFHCS modulation at the radar transmitter, we only need to multiply the modulation term $F_{hm}(t) = e^{j\varpi_{hm}}$ onto radar waveform, as with the sole PSK. Due to the use of PSK, PFHCS can also suppress the sidelobe spikes in the range ambiguity function of an FH-MIMO radar [5], as illustrated in Figure 7.3c. At the communication receiver, the PFHCS demodulation can be performed by first demodulating the FHCS subsymbol and then the PSK subsymbol.

To demodulate an FHCS symbol, we need to extract the hopping frequencies, i.e. k_{hm}. The subscript $(\cdot)_h$ is re-added here to differentiate the $(H-2)$ hops used for data communication. Referring to Figure 7.1, we see that the sample shift L_η caused by the timing offset η needs to be compensated to recover a complete data hop. The value of L_η can be extracted from the estimation of $\angle\omega_\eta$ obtained in Section 7.3. Let $\measuredangle\omega_\eta$ denote the $\angle\omega_\eta$ estimation which can be either (7.15) or (7.17). Based on the definition of ω_η given in (7.9), the η estimation can be extracted from $\measuredangle\omega_\eta$ as follows:

$$\hat{\eta}_d = (K\measuredangle\omega_\eta + 2d\pi)/(2\pi B), \quad \text{s.t. } 0 < \hat{\eta}_d < T,$$

where $d(=0, \pm 1, \ldots)$ is the ambiguity degree. Note that the constraint can make the number of η estimates limited. Using $\hat{\eta}_d$, the sample shift L_η can be estimated as $\hat{L}_\eta^{(d)} = \lfloor \hat{\eta}_d / T_s \rceil$, $d = 0, \pm 1, \ldots$. Given the waveform orthogonality; see (2.10), we can remove the estimation ambiguity in $\hat{L}_\eta^{(d)}$.

Reconstruct the hth $(h = 1, \ldots, H-1)$ sampled hop as follows:

$$\mathbf{y}_h^{(d)} = \left[y_{h-1}(L - \hat{L}_\eta^{(d)}), \ldots, y_{h-1}(L-1),\ y_h(0), \ldots, y_h(L - \hat{L}_\eta^{(d)} - 1) \right]^{\mathrm{T}}. \quad (7.28)$$

By calculating the L-point DFT of $\mathbf{y}_h^{(d)}$ and searching for the spectrum peaks as done in (7.3) and (7.8), the mth peak is denoted by $Y_{hm}^{(d)}$. Similarly, we can calculate the DFT of $\mathbf{y}_h^{(d)}$ at the l_h^*th discrete frequency, leading to $\tilde{Y}_{hm}^{(d)}$. Here, l_h^* is taken such that $l_h^* \notin \{l_{h(M-1-m)}^*\ \forall m\}$, where $l_{h(M-1-m)}^*$ is defined similar to l_{M-1-m}^* given in (7.5). Provided that $\mathbf{y}_h^{(d)}$ is correctly reconstructed at $d = d^*$, we have $\sum_{m=0}^{M-1} |Y_{hm}^{(d^*)}| = ML$ and $\sum_{m=0}^{M-1} |\tilde{Y}_{hm}^{(d^*)}| = 0$ in the absence of noises. The two equations are due to the waveform orthogonality given in (2.10). Considering inevitable noises, we can identify d^* robustly via

$$d^* : \max_{\left\{ \substack{d=0,\\ \pm 1, \ldots} \right\}} \sum_{h=2}^{H-1} \left(\sum_{m=0}^{M-1} \left| Y_{hm}^{(d)} \right| \Big/ \sum_{m=0}^{M-1} \left| \tilde{Y}_{hm}^{(d)} \right| \right). \quad (7.29)$$

The PFHCS demodulation is summarized below.

(a) *FHCS subsymbol*: After identifying d^*, substitute the indexes of the M DFT peaks into (7.6), producing $\hat{k}_{hm}$ and the estimate of hopping frequency $\hat{f}_{hm}(=\hat{k}_{hm}B/K)$. Comparing $\{\hat{f}_{hm}\ \forall m\}$ with the FHCS constellations, the FHCS subsymbol is demodulated.
(b) *PSK subsymbol*: Using the identified k_{hm} and the estimations of channel parameters, we can estimate ϖ_{hm} based on (7.8), as given by $\hat{\varpi}_{hm} = \angle\left(Y_{hm}^{(d^*)}\hat{\beta}^{\dagger}e^{\mathrm{j}\frac{2\pi m\hat{u}}{M}}e^{-\mathrm{j}\measuredangle\omega_\eta\hat{k}_{hm}}\right)$, where $()^{\dagger}$ takes conjugate. Comparing $\hat{\varpi}_{hm}$ with the PSK constellation, the PSK subsymbol is demodulated.

7.6 Extensions to Multipath Cases

Next, we extend the methods presented in Sections 7.3 and 7.4 to multipath scenarios. The two estimators, CAE and CRE, can be extended to flat Rician fading channels, where the radar-transmitted signal arrives at the communication receiver through several nonline-of-sight (NLoS) paths in addition to the LoS. By considering a quasistatic flat-fading channel, the delay spread can be confined within a radar snapshot [1]. Thus, the waveform orthogonality defined in (2.10) is preserved, and the method developed in Section 7.2.1 can still be used to estimate the hopping frequencies at the communication receiver, leading to the same result in (7.6). As in the LoS case, we use the frequency-domain signals, i.e. $Y_m(l^*_{M-1-m})$ in (7.8), to estimate communication channel.

Let $p \in [0, P-1]$ denote path index. The multipath version of (7.8) can be given by

$$Y_m = Y(l^*_{M-1-m}) = \rho_m \omega_\eta^{\hat{k}_m} + \Xi_m, \quad \rho_m = \sum_{p=0}^{P-1} \tilde{\beta}_p e^{-\mathrm{j}\frac{2\pi m u_p}{M}}, \tag{7.30}$$

where variables are defined in the same way as in (7.8) and $\Xi(l^*_{M-1-m})$ therein is shortened into Ξ_m. Compared with (7.8), the linear phase relation of the coefficients of $\omega_\eta^{\hat{k}_m}$ has been destroyed in (7.30) by NLoS components. To apply the estimators CAE and CRE developed in Section 7.3, we estimate and suppress ρ_m first.

From (7.30), we observe that *if the hopping frequency attached to the mth antenna is zero, i.e.* $\hat{k}_m = 0$, *then* $Y_m = \rho_m + \Xi_m$ *becomes an estimate of* ρ_m. Thus, ρ_0 can be estimated as Y_0, since $k_0 = 0$ is ensured in the suboptimal hop sequence designed in Section 7.3.3. To estimate ρ_m $(m > 0)$, we resort to the hops originally assigned for data communication; see Figure 7.1b. In specific, we require that for antenna m

at hop $(m+2)$, the hopping frequency is zero and no information bit is embedded, i.e.

$$k_{hm} = 0, \quad F_{hm} = 1, \quad h = m+2, \quad m = 0, 1, \ldots, M-1, \tag{7.31}$$

where $F_{hm} = 1$ denotes no information modulation.

Referring to Figure 7.1b, the sampled hop $(m+2)$ spans across hops $(m+2)$ and $(m+3)$. To avoid interhop interference, we take an L-dimensional yet $\frac{L}{2}$-point DFT on the sampled hop $(m+2)$. This leads to

$$\begin{aligned} Y_h(l) &= \sum_{i=0}^{L/2-1} y_h(i) e^{-\mathrm{j}\frac{2\pi il}{L}} = \sum_{m=0}^{M-1} \rho_m e^{\mathrm{j}\pi(L-1)} \omega_\eta^{k_{hm}} \\ &\times \frac{\sin\frac{\pi}{2}\left(\frac{k_{hm}BT}{K} + l\right)}{\sin\frac{\pi}{L}\left(\frac{k_{hm}BT}{K} + l\right)} e^{-\mathrm{j}\frac{\pi(L-2)\left(\frac{k_{hm}BT}{K}+l\right)}{2L}} + \Xi(l), \end{aligned} \tag{7.32}$$

where ρ_m is given in (7.30), u and ω_η are defined in (7.9), and $\Xi(l)$ denotes the DFT of the AWGN $\xi(i)$. Based on (7.31) and (7.32), we have

$$\begin{aligned} Y_{m+2}(0) &= \frac{L}{2}\rho_m e^{\mathrm{j}\pi(L-1)} + \sum_{\substack{m'=0 \\ m''=\neq m}}^{M-1} \rho_{m'} e^{\mathrm{j}\pi(L-1)} \omega_\eta^{k_{hm'}} \\ &\times \frac{\sin\frac{k_{hm'}BT\pi}{2K}}{\sin\frac{k_{hm'}BT\pi}{LK}} e^{-\mathrm{j}\frac{\pi(L-2)\frac{k_{hm'}BT}{K}}{2L}} + \Xi(0). \end{aligned} \tag{7.33}$$

We see from (7.33) that taking $\frac{k_{hm'}BT}{2K}$ as an integer makes the sine function in the numerator become zero, hence, avoiding interantenna interference. This can be achieved by configuring the hopping frequencies as follows:

$$k_{hm'} = \begin{cases} \forall k \in [1, K-1], & \text{if } BT/K \text{is even}, \\ \forall k \in \{2, 4, 6, \ldots\}, & \text{otherwise}, \end{cases} \tag{7.34}$$

Based on the above analyses and derivations, estimating ω_η in Rician channels is summarized in the following proposition:

Proposition 7.3 *Provided that (7.31) and (7.34) are satisfied, CAE given in (7.15) and CRE given in (7.17) are capable of estimating ω_η in Rician channels, by replacing Y_m $\forall m$ in (7.10) with $Y_m/\hat{\rho}_m$, where*

$$\hat{\rho}_m = \begin{cases} Y_0 \text{ based on (7.30)}, & \text{at } m = 0, \\ \frac{2}{L} Y_{m+2}(0) e^{-\mathrm{j}\pi(L-1)} \text{ based on (7.33)}, & \text{elsewhere}, \end{cases}$$

With ω_η estimated, two options are available for the remaining processing: (i) We can proceed to estimate u_0 and β_0 using the method developed in Section 7.4.1, and then perform data communication as illustrated in Section 7.5; (ii) We can suppress timing offset using the method provided in Section 7.5, and then divide Y_m by $\hat{\rho}_m$ $\forall m$ to remove the impact of other channel parameters. The benefit of the first option is that, with the path AoD estimated, beamforming can be performed to enhance LoS and suppress NLoS paths, provided that multiple antennas are equipped. In contrast, the second option can be more efficient in NLoS scenarios when a single antenna is available.

7.7 Simulation Results

In what follows, we shall provide some simulation results to validate the high accuracy of the channel estimation methods presented in this chapter. Unless otherwise specified, the FH-MIMO radar is configured as follows: $M = 10, K = 20, B = 100\,\text{MHz}, f_\text{L} = 8\,\text{GHz}, f_\text{s} = 2B, T = 0.8\,\mu\text{s}$, and $d = \frac{\lambda}{2}$. Both LoS and Rician channels are simulated, where $\eta \sim \mathcal{U}_{[0.05\,\mu\text{s},0.35\,\mu\text{s}]}$, $\phi_0 = 20^\circ$, $\phi_p \sim \mathcal{U}_{[-90^\circ,90^\circ]}(\forall p > 0)$, $\beta_0 = e^{j\angle\beta_0}$ with $\angle\beta_0 \sim \mathcal{U}_{[0^\circ,360^\circ]}$, β_p $(\forall p > 0) \sim \mathcal{CN}(0, -5\text{dB})$ (the Rician factor is 5 dB), and $P = 4$ NLoS paths are added for Rician channels. By taking $p = 0$, LoS channels are obtained. Here, $\mathcal{U}_{[\cdot,\cdot]}$ stands for the uniform distribution in the subscript region. Based on the above parameters, the suboptimal hopping frequency sequence can be calculated as in Section 7.3.3, leading to

$$\text{suboptimal: } \mathbf{f}^* = f_\text{L} + [0, 1, 3, 4, 6, 7, 9, 10, 17, 19]^\text{T} \times B/K.$$

In addition, exploiting (7.11) and (7.14), we can identify the hopping frequency sequence satisfying $\bar{M} = (M-2)$, i.e. CAE: $\bar{\mathbf{f}} = f_\text{L} + [0, 1, 3, 4, 6, 7, 9, 10, 12, 13]^\text{T} \times B/K$. When applying $\bar{\mathbf{f}}$, only CAE is applicable with its MSELB minimized. Similarly, substituting the above parameters into (7.11) and (7.16), the hopping frequency sequence leading to the minimum $\breve{\sigma}_\eta^2$ can be obtained, as given by CRE: $\breve{\mathbf{f}} = f_\text{L} + [0, 1, 2, 3, 4, 5, 6, 7, 17, 19]^\text{T} \times B/K$. Note that $\breve{\mathbf{f}}$ leads to $\bar{\mathcal{M}} = \emptyset$, and hence, only CRE is applicable. The above three sequences of hopping frequencies are adopted and compared in the following simulations:

Figure 7.4a plots the MSE of $\angle\omega_\eta$ estimation against the received SNR γ at the communication receiver in LoS and Rician channels. We first analyze the estimation results for LoS channels. We see from the figure that CRE has a much better high-SNR performance than CAE, whereas CAE outperforms CRE in low-SNR regions. This validates the analysis in Remark 7.1. We also see that CRE and CAE can asymptotically approach their MSELBs derived in Proposition 7.2. This validates the analysis in the proof of Proposition 7.2. By comparing CRE and CAE, the SNR threshold $\gamma_\text{T} = 18\,\text{dB}$ can be obtained from the zoomed-in turning point. As analyzed in Remark 7.1, we perform CAE and CRE below and above γ_T, respectively, meanwhile exploiting the suboptimal $\mathbf{f}^*$. We see from

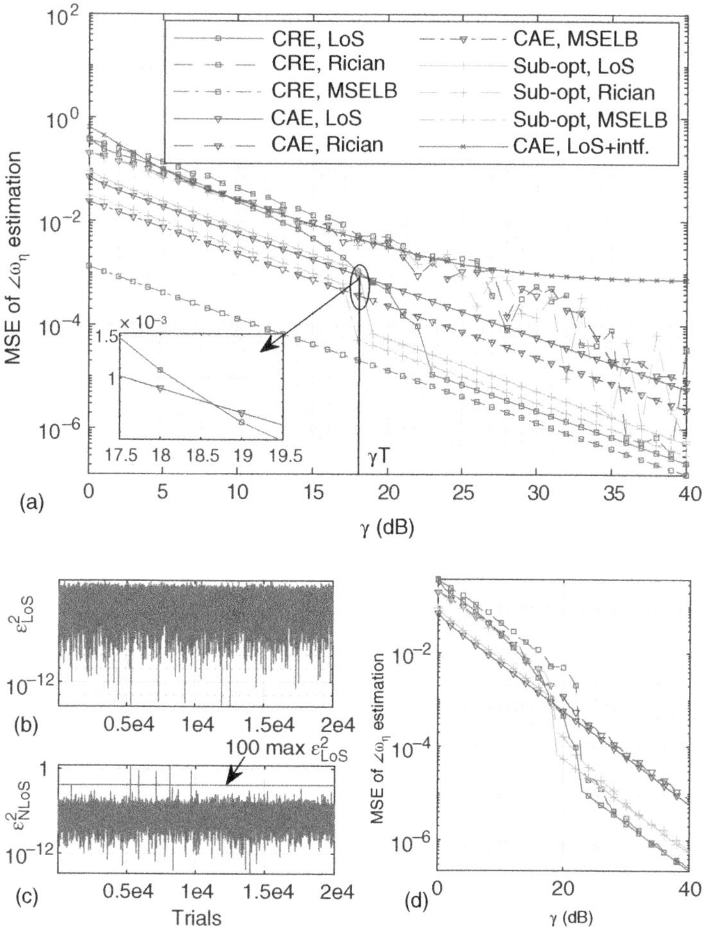

Figure 7.4 (a) MSE of $\angle\omega_\eta$ estimation against $\gamma(= |\beta|^2/\sigma_n^2)$, where "intf." is short for interference, and σ_n^2 is the power of $\xi(i)$; see (7.2) for β and $\xi(i)$. (b, c) are obtained at $\gamma = 30$ dB, where ϵ^2_{LoS} and ϵ^2_{NLoS} denote the squared estimation errors for LoS and NLoS scenarios, respectively; and (d) MSE of $\angle\omega_\eta$ estimation in NLoS scenarios with abnormal trials removed, where an abnormal trial has $\epsilon^2_{\text{NLoS}} > 100\max\{\epsilon^2_{\text{LoS}}\}$.

Figure 7.4 that the suboptimal $\mathbf{f}^*$ provides the suboptimal estimation accuracy in the whole SNR regions. Nevertheless, we see that the suboptimal $\mathbf{f}^*$ improves the estimation accuracy obviously over CRE in low-SNR regions ($\gamma \leq \gamma_T$), and substantially outperforms CAE in high-SNR regions ($\gamma > \gamma_T$). It is noteworthy that the improvement achieved by the suboptimal $\mathbf{f}^*$ across the whole SNR region is based on a single hop.

For Rician channels, we see from Figure 7.4a that the estimation performance improves with γ in overall. This validates the effective extension of the methods to multipath channels, as elaborated on in Section 7.6. We also see oscillations in the MSE results, particularly in high-SNR regions. The oscillations are caused by a few trials whose estimation errors are abnormally larger than the overall MSE. (Note that the few abnormal estimations are caused by signal canceling among multiple paths in Rician channels.) This can be validated by jointly observing Figure 7.4b,c. By removing the abnormal trials, the estimation performance under Rician channels approach that under LoS channels, as demonstrated in Figure 7.4d.

Figure 7.4a also shows the robustness of CAE against interference, where we consider a −5 dB interference signal from another nonsynchronized FH-MIMO radar. The interference radar is configured the same as the target radar, except that its hopping frequencies are randomly taken (leaving the first three subbands used by CAE uninterfered). We see from Figure 7.4a that CAE is robust to interference, particularly in low-SNR regions. This validates the conditional robustness of the method to interference, as discussed in Section 7.6. We also see that, unlike in interference-free scenarios, the MSE of CAE converges to about 9×10^{-4} as SNR increases. As expected, interference, rather than AWGN, dominates the estimation performance at high SNRs.

Figure 7.5 observes the u estimation accuracy against γ, where the $\angle\omega_\eta$ estimations obtained from Figure 7.4 are used for calculating Z_m in (7.25). We see that the u estimation accuracy is closely dependent on the $\angle\omega_\eta$ estimations. In particular, we see a fast decay from 22 to 23 dB, when the $\angle\omega_\eta$ estimates obtained by CRE are used for suppressing the timing offset. This is because the performance of CRE improves by an order of magnitude over the same SNR region; see Figure 7.4a. We also see that the suboptimal $\mathbf{f}^*$ enables the MSE of u estimation to outperform those achieved by CRE and CAE in low- and high-SNR regions, respectively. On the one hand, this validates the superiority of the suboptimal $\mathbf{f}^*$ over $\bar{\mathbf{f}}$ and $\check{\mathbf{f}}$, as consistent with Figure 7.4; and on the other

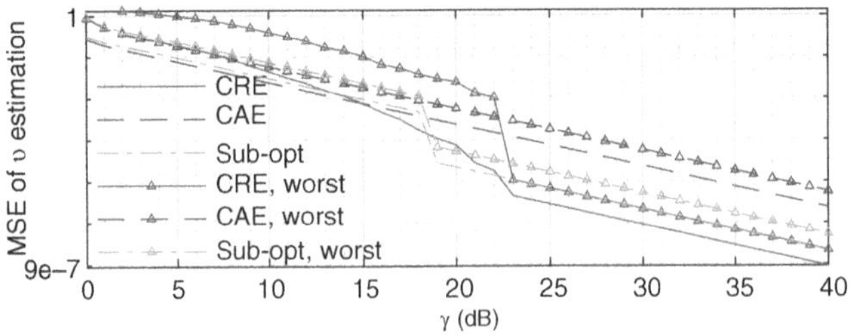

Figure 7.5 MSE of u estimation against γ, where the $\angle\omega_\eta$ estimations from Figure 7.4 are used to calculate Z_m in (7.25) (as required for u estimation). The worst case adds the mean and standard deviation of squared estimation errors.

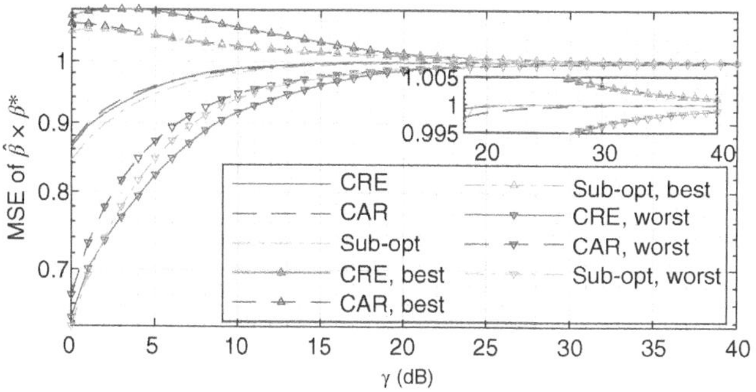

Figure 7.6 MSE of $\hat{\tilde{\beta}} \times \tilde{\beta}^*$ against γ, where the u estimations in Figure 7.5 are used to estimate $\hat{\tilde{\beta}}$ as given in (7.27) and $\tilde{\beta}^*$ is the conjugate of $\tilde{\beta}$. The best case subtracts the standard deviation of squared estimation errors from the MSE.

hand, this demonstrates the robustness of u estimation to the estimation error of CAE and CRE, even in low-SNR regions.

Figure 7.6 plots the MSE of $\hat{\tilde{\beta}} \times \tilde{\beta}^*$ against γ, where the u estimations obtained in Figure 7.5 are used in (7.27). We see that, owing to the high accuracy of $\angle\omega_\eta$ and u, the estimate $\hat{\tilde{\beta}}$ is very close to the true value. From the zoomed-in subfigure, we see that at $\gamma = 20$ dB the MSE of $\hat{\tilde{\beta}} \times \tilde{\beta}^*$ approaches to 1 with an error of less than 0.001. We also see that, due to the different performance index from that used in Figures 7.4 and 7.5, the MSEs achieved by the three estimators are not as differentiable as they are in the previous two figures. Nevertheless, from the deviations of their squared errors, we see that CAE and the suboptimal $\mathbf{f}^*$ can produce better $\tilde{\beta}$ estimations compared with CRE (particularly in low-SNR regions).

We proceed to demonstrate the efficacy of applying the channel estimations for data communications in FH-MIMO DFRC. Two existing constellations BPSK [16] and FHCS [3], and the new constellation PFHCS are evaluated based on the ideal and estimated channels. For a fair comparison with BPSK [16] and FHCS [3], channel coding is not considered here. Figure 7.7 illustrates the achievable data rate against the communication SNR. The estimations obtained at $\gamma = 15$ dB from Figures 7.4–7.6 are used to perform data communications. We see that the new constellation PFHCS improves the data rate substantially over BPSK and FHCS. In particular, the converging data rate of PFHCS is $170\%(= \frac{33.75-12.5}{12.5})$ and $58.82\%(= \frac{33.75-21.25}{21.25})$ higher than that of BPSK and FHCS, respectively. We also see that the estimated parameters enable the data rate to tightly approach that corresponds to the ideally known parameters. From the right y-axis, we see that the suboptimal estimator and CAE produce much smaller achievable rate difference across the whole communication SNR region, compared with CRE. This is consistent with the channel estimation accuracy shown in Figures 7.4–7.6.

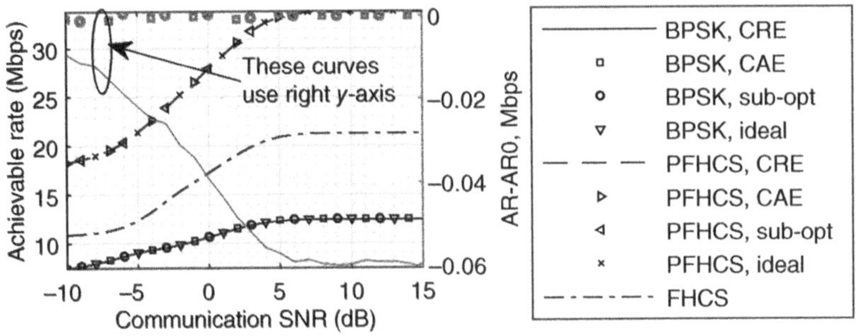

Figure 7.7 Achievable data rate vs. communication SNR, where AR0 is obtained based on the ideal channels.

Figure 7.7 compares the abbtextsymbol error rates (SERs) achieved based on the ideal timing offset and channel parameters and the estimated ones, where the estimations obtained at $\gamma = 15\,\text{dB}$ from Figures 7.4–7.6 are used. We see that due to the larger number of symbol bits of the new constellation PFHCS, its SER vs. Eb/N0 is improved substantially, compared with that achieved by BPSK and FHCS. We also see that the channel estimation error incurred by CRE makes the SERs of BPSK and PFHCS converge to 10^{-2}, whereas the smaller estimation error of the suboptimal estimator and CAE produces the continuously decreasing SERs against Eb/N0. Interestingly, we see from Figure 7.8 that using channel estimations obtained under Rician channels can achieve better SER performance compared with using the estimations under LoS channels. The main reason is that

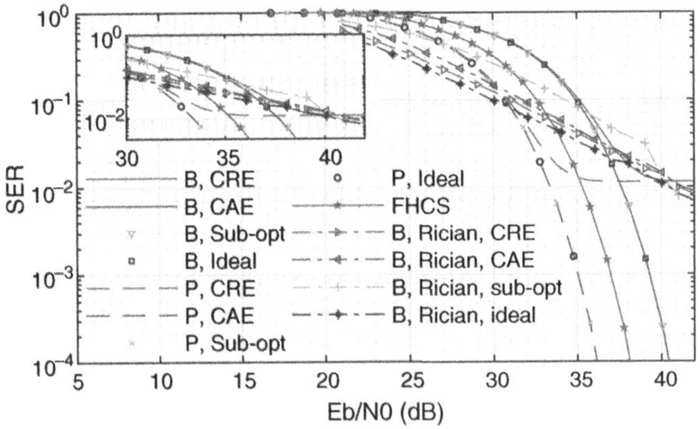

Figure 7.8 SER vs. Eb/N0, where Eb/N0 is the energy per bit to noise power density ratio and is calculated as $L\gamma_{\text{com}}BT/E$, where γ_{com} denotes communication SNR and E is the number of bits conveyed per radar hop. In the legend, B and P are short for BPSK and PFHCS, respectively.

multiple paths can constructively enhance LoS signals and improve the detecting probability of identifying hopping frequencies; c.f., the destructive signal canceling leading to the abnormal ω_η estimations in Figure 7.4c.

7.8 Conclusions

To enable communications in radar systems, it is critically important to have a modified waveform and, importantly, the associated channel estimation method. In this chapter, a novel FH-MIMO radar waveform which enables a communication receiver to estimate, rather than acquiring from radar, the hopping frequency associated with each radar transmitter antenna, is presented. Some accurate methods to estimate timing offset and channel for FH-MIMO DFRC are developed. In particular, two new estimators for timing offset suited for different hopping frequency sequences and an accurate channel estimation method are described. This is further fulfilled by effectively extending the methods to multipath scenarios. Simulation results validate the high accuracy of the estimation methods, and the high communication performance attained using estimated channels. A single-antenna communications receiver is considered in this work for a single-path communications channel. More complicated scenarios, e.g. multiple antennas and paths, will be discussed in subsequent chapters.

References

1 F. Liu, A. Garcia-Rodriguez, C. Masouros, and G. Geraci. Interfering channel estimation in radar-cellular coexistence: How much information do we need? *IEEE Transactions on Wireless Communications*, 18(9):4238–4253, 2019.

2 Y. S. Cho, J. Kim, W. Y. Yang, and C. G. Kang. *MIMO-OFDM Wireless Communications with MATLAB*. John Wiley & Sons, 2010.

3 W. Baxter, E. Aboutanios, and A. Hassanien. Dual-function MIMO radar-communications via frequency-hopping code selection. In *2018 52nd Asilomar Conference on Signals, Systems, and Computers*, pages 1126–1130, October 2018. doi: 10.1109/ACSSC.2018.8645212.

4 L. C. Godara. Application of antenna arrays to mobile communications. II. Beam-forming and direction-of-arrival considerations. *Proceedings of the IEEE*, 85(8):1195–1245, 1997.

5 I. P. Eedara, A. Hassanien, M. G. Amin, and B. D. Rigling. Ambiguity function analysis for dual-function radar communications using PSK signaling. In *52nd Asilomar Conference on Signals, Systems, and Computers*, pages 900–904, October 2018. doi: 10.1109/ACSSC.2018.8645328.

6 C. Chen and P. P. Vaidyanathan. MIMO radar ambiguity properties and optimization using frequency-hopping waveforms. *IEEE Transactions on Signal Processing*, 56(12):5926–5936, December 2008. ISSN 1941-0476. doi: 10.1109/TSP.2008.929658.

7 K. Wu, J. A. Zhang, X. Huang, Y. J. Guo, and R. W. Heath. Waveform design and accurate channel estimation for frequency-hopping MIMO radar-based communications. *IEEE Transactions on Communications*, pages 1, 2020. doi: 10.1109/TCOMM.2020.3034357.

8 K. P. Bogart. *Introductory Combinatorics*, Chapter 2.3 Combinations (subsets) of sets. Academic Press, Inc., 2004.

9 Q. Li, T. Su, and K. Wu. Accurate DOA estimation for large-scale uniform circular array using a single snapshot.*IEEE Communications Letters*, 23(2):302–305, February 2019. ISSN 2373-7891. doi: 10.1109/LCOMM.2018.2889855.

10 R. Cao, B. Liu, F. Gao, and X. Zhang. A low-complex one-snapshot DOA estimation algorithm with massive ULA. *IEEE Communications Letters*, 21(5):1071–1074, May 2017. ISSN 2373-7891. doi: 10.1109/LCOMM.2017.2652442.

11 C. Wang, Q. Yin, and H. Chen. Robust Chinese remainder theorem ranging method based on dual-frequency measurements. *IEEE Transactions on Vehicular Technology*, 60(8):4094–4099, October 2011. ISSN 1939-9359. doi: 10.1109/TVT.2011.2167690.

12 R. Reggiannini. A fundamental lower bound to the performance of phase estimators over Rician-fading channels. *IEEE Transactions on Communications*, 45(7):775–778, July 1997. ISSN 1558-0857. doi: 10.1109/26.602582.

13 A. V. Oppenheim. *Discrete-Time Signal Processing*. Pearson Education India, 1999.

14 A. Serbes. Fast and efficient sinusoidal frequency estimation by using the DFT coefficients. *IEEE Transactions on Communications*, 67(3):2333–2342, March 2019. ISSN 0090-6778. doi: 10.1109/TCOMM.2018.2886355.

15 K. Wu, W. Ni, J. A. Zhang, R. P. Liu, and Y. J. Guo. Refinement of optimal interpolation factor for DFT interpolated frequency estimator. *IEEE Communications Letters*, 1, 2020. ISSN 2373-7891. doi: 10.1109/LCOMM.2019.2963871.

16 I. P. Eedara, M. G. Amin, and A. Hassanien. Analysis of communication symbol embedding in FH MIMO radar platforms. In *2019 IEEE Radar Conference (RadarConf)*, pages 1–6, April 2019. doi: 10.1109/RADAR.2019.8835532.

8

Frequency-Hopping MIMO Radar-Based Communications with Multiantenna Receiver

Frequency-hopping (FH) multiple-input and multiple-output (MIMO) radar is currently an attractive underlying system for realizing dual-function radar-communication (DFRC) systems. It has the potential to achieve high communication symbol rates. In Chapter 7, we present a channel estimation scheme for FH-MIMO DFRC but with a single-antenna communication receiver. Considering the popularity of antenna arrays in modern electronic systems, we present in this chapter a multiantenna receiver-based downlink communication scheme for the FH-MIMO DFRC. In particular, we will unveil some unique features of the FH-MIMO radar waveform and show how they can help suppress interantenna and interhop interference. We will also introduce accurate estimation methods for timing offset and channel parameters. These methods are further employed to design reliable demodulation methods for communications.

8.1 Signal Model

The FH-MIMO DFRC system considered in this chapter is illustrated in Figure 8.1a, where an FH-MIMO radar senses a target while communicating with user equipment (UE) through flat-fading multipath channels. The radar and UE are equipped with a uniform linear array whose antenna spacing is half a wavelength. In this section, we provide the signal models of the radar and DFRC.

Based on the signal model given in Section 2.3.3, at hop h, the mth antenna of the radar transmitter transmits the following signal:

$$s_{hm}(t) = e^{j2\pi f_{hm} t} g_{\text{rect}}(t - hT), \quad hT \leq t \leq (h+1)T, \tag{8.1}$$

where $f_{hm} = k_{hm}B/K$, $k_{hm} \in \{0, 1, \dots, K-1\}$, and $g_{\text{rect}}(t)$ denotes the rectangular function taking unit one if $0 \leq t - hT \leq T$ and, otherwise, zero. Refer to Section 2.3.3 for an elaboration on FH-MIMO radar signal processing.

Joint Communications and Sensing: From Fundamentals to Advanced Techniques, First Edition.
Kai Wu, J. Andrew Zhang, and Y. Jay Guo.

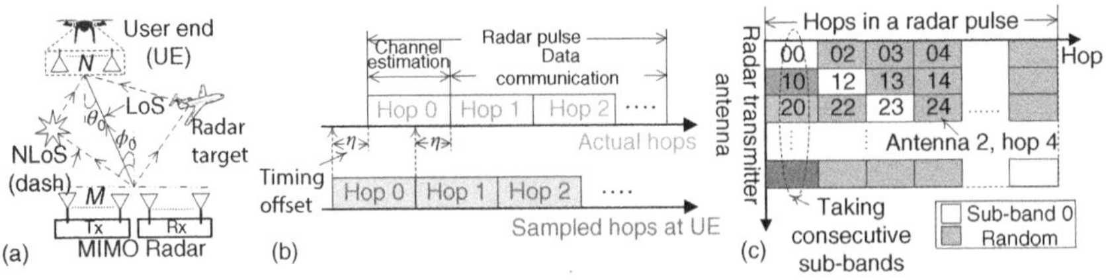

Figure 8.1 (a) Illustration of the system diagram of an FH-MIMO DFRC; (b) the signal frame structure devised for the downlink communication, where the timing offset η is smaller than half the time duration of a hop; and (c) illustration of the arrangement of hopping frequencies over antennas and hops.

Consider P paths between the radar and UE, each having the angle of departure (AoD) of ϕ_p, the angle of arrival (AoA) of θ_p, and the complex path gain of β_p. Assuming quasistatic channels in a short radar pulse, it follows that ϕ_p, θ_p, and β_p are independent of h [1]. The steering vectors of radar and UE arrays are $\mathbf{a}_M(u_p)$ and $\mathbf{a}_N(v_p)$, respectively. They are given by

$$\begin{aligned}\mathbf{a}_M(u_p) &= [1, e^{\mathrm{j}u_p}, \dots, e^{\mathrm{j}(M-1)u_p}]^{\mathrm{T}}, \quad u_p = \pi \sin \phi_p;\\ \mathbf{a}_N(v_p) &= [1, e^{\mathrm{j}v_p}, \dots, e^{\mathrm{j}(N-1)v_p}]^{\mathrm{T}}, \quad v_p = \pi \sin \theta_p,\end{aligned} \tag{8.2}$$

where N is the number of antennas in the UE array. To perform data communication, a phase shift keying (PSK) modulation signal,[1] as given by

$$F_{hm} = e^{\mathrm{j}\varpi_{hm}}, \quad \varpi_{hm} \in \Omega_{J_{\mathrm{PSK}}} \ (J_{\mathrm{PSK}} \geq 1), \quad \Omega_{J_{\mathrm{PSK}}} = \left\{0, \dots, \frac{2\pi(2^{J_{\mathrm{PSK}}} - 1)}{2^{J_{\mathrm{PSK}}}}\right\} \tag{8.3}$$

can be multiplied with $s_{hm}(t)$ to embed information bits in radar signal [2], where $\Omega_{J_{\mathrm{PSK}}}$ is a J_{PSK}-bit PSK constellation.

We consider the packet communications and employ the signal frame structure to fulfill data communications and reduce overhead, as shown in Figure 8.1b. In each frame, the first radar hop is assigned for estimating timing offset and channel, and the remaining hops are used for data transmission. Moreover, we consider more practical scenarios that the UE performs a coarse timing based on the conventional energy-based or auto-correlation-based packet detection [5], resulting in a timing offset η. In general, $\eta < \frac{T}{2}$ can be ensured [6], which is assumed in the following. Let $f_s(\geq 2B)$ denote the sampling frequency; the

1 For ease of exposition, we employ PSK [2] to develop the receiving scheme, given the much simpler signal models than those of differential PSK (DPSK) [3] and continuous phase modulation (CPM) [4]. Note that the methods can be readily applied to other modulations exploiting the methods, as will be developed in Section 8.3.

sampling interval is then $T_s = 1/f_s$. The number of samples in T and η are given by $L = \left\lceil \frac{T}{T_s} \right\rceil$ and $L_\eta = \left\lceil \frac{\eta}{T_s} \right\rceil$, respectively.

Based on (8.1) and (8.2), the UE-received baseband signal at hop h can be given by the following $N \times 1$ vector,

$$\mathbf{y}_h(i) = \sum_{p=0}^{P-1} \left(\beta_p \sum_{m=0}^{M-1} \tilde{s}_{hm}(i) e^{jmu_p} \right) \mathbf{a}_N(\theta_p) + \mathbf{n}_h(i), \tag{8.4}$$

where i is the sample index and $\mathbf{n}_h(i) \in \mathbb{C}^{N\times 1}$ collects N independent and identically distributed (IID) additive white Gaussian noises (AWGNs) on the UE antennas. In (8.4), $\tilde{s}_{hm}(i)$ is the sampled signal of $F_{hm}s_{hm}(t)$ at hop h. Given $\eta > 0$, $\tilde{s}_{hm}(i)$ can span over two actual radar hops, as illustrated in Figure 8.1b. Taking the sampled hop 1 for an illustration, we see from Figure 8.1b that the sampled hop consists of the actual hops 0 and 1. The interference caused by hop 0 to hop 1 is referred to as the *interhop interference*. Considering that the starting point of sampled hop h is earlier than the actual point, $\tilde{s}_{hm}(i)$ is given by

$$\tilde{s}_{hm}(i) = \begin{cases} F_{(h-1)m}s_{(h-1)m}(i + L - L_\eta), & i \in \mathcal{I}_1, \\ F_{hm}s_{hm}(i - L_\eta), & i \in \mathcal{I}_2, \end{cases}$$
$$\text{s.t. } \mathcal{I}_1 = \{0, 1, \dots, L_\eta - 1\}; \quad \mathcal{I}_2 = \{L_\eta, \dots, L-1\}. \tag{8.5}$$

In (8.5), $s_{hm}(i)$ is the digitized $s_{hm}(t)$ given in (8.1), specifically

$$s_{hm}(i) = e^{j\frac{2\pi k_{hm} B}{K} iT_s} = e^{j\frac{2\pi k_{hm} \Delta}{L} i}, \tag{8.6}$$

where $\Delta \triangleq BT/K$. Note that B is the bandwidth, T is the hop duration, and K is the total number of subbands. Based on (8.5), we can validate that the inner product between $\tilde{s}_{hm}(i)$ and $\tilde{s}^*_{hm'}(i)$ ($\forall m' \neq m$) is nonzero, i.e. $\sum_{i=0}^{L-1} \tilde{s}^*_{hm}(i)\tilde{s}_{hm'}(i) \neq 0$. To this end, the signal transmitted by antenna m causes interference to that by antenna m', or vice versa. This interference is referred to as the *interantenna interference* which prevents UE from extracting the signal component related to a single antenna (which is achievable without interantenna interference). Note that accurate signal extraction is critical to information demodulation, as seen in Section 8.3.3.

8.2 The DFRC Signal Mode

In this section, we first simplify the DFRC signal model presented above to reveal some of the hidden features and then provide some insights into the simplified signal model, which will be useful in developing the receiver scheme. From (8.4), we notice that the signals transmitted by all the radar transmitter antennas are superimposed on the UE receiver. Assuming perfect timing, we can employ the waveform orthogonality, as given in (2.10), to separate the signals in the frequency domain [7]. However, in the presence of a nonzero timing offset η, as illustrated in Figure 8.1b, interhop interference is introduced, and the waveform orthogonality is lost. We notice that the unique signal structure of the FH-MIMO radar can be exploited to combat the adverse consequence of η.

In particular, we can take the L-dimensional yet $\frac{L}{2}$-point discrete Fourier transform (DFT) of $y_{hn}(i)$ $(i = L/2, \dots, L-1)$ to avoid interhop interference, where $y_{hn}(i)$, $\forall i$, denotes the nth entry of $\mathbf{y}_h(i)$, $\forall i$, as given in (8.4). This leads to

$$Y_{hn}(l) = \frac{2}{L}\sum_{i=L/2}^{L-1} y_{hn}(i)e^{-j\frac{2\pi il}{L}} = \sum_{p=0}^{P-1}\beta_p \sum_{m=0}^{M-1} S_{hm}(l)e^{jmu_p}e^{jnv_p} + \Xi_{hn}(l), \tag{8.7}$$

where $\Xi_{hn}(l) = \frac{2}{L}\sum_{i=L/2}^{L-1}[\mathbf{n}_h(i)]_n e^{-j\frac{2\pi il}{L}}$ and $[\mathbf{n}_h(i)]_n$ denotes the nth entry of $\mathbf{n}_h(i)$ given in (8.4). In (8.7), $S_{hm}(l)$ is the DFT of $\tilde{s}_{hm}(i)$, as given by

$$\begin{aligned} S_{hm}(l) &= \frac{2}{L}\sum_{i=L/2}^{L-1} F_{hm}s_{hm}(i-L_\eta)e^{-j\frac{2\pi il}{L}} \\ &= \frac{F_{hm}\sin\frac{\pi}{2}\left(l-k_{hm}\Delta\right)}{\frac{L}{2}\sin\frac{\pi}{L}\left(l-k_{hm}\Delta\right)} e^{j\frac{\pi(3L-2)(k_{hm}\Delta-l)}{2L}} e^{-j\frac{2\pi k_{hm}\Delta L_\eta}{L}}, \end{aligned} \tag{8.8}$$

where $\tilde{s}_{hm}(i)$ is replaced with $F_{hm}s_{hm}(i-L_\eta)$ based on (8.5).

Based on the results of the special DFT, we can also extract the signal transmitted by each antenna without interantenna interference, as developed in the following proposition.

Proposition 8.1 *Provided that the following condition holds,*

$$k_{hm}\Delta/2 \in \{0, \mathbb{I}_+\}, \quad \forall m \tag{8.9}$$

the interantenna interference is suppressed in the interhop interference-free signal obtained in (8.7). This enables the extraction of the information-bearing signal that is solely related to the mth radar transmitter antenna at any hop h, given by

$$Y_{hn}(l_{hm}^*) = \rho_{nm}F_{hm}e^{-j\frac{2\pi L_\eta l_{hm}^*}{L}} + \Xi_{hn}(l_{hm}^*), \text{ s.t. } \rho_{nm} = \sum_{p=0}^{P-1}\beta_p e^{jmu_p}e^{jnv_p};\ l_{hm}^* = k_{hm}\Delta, \tag{8.10}$$

where Δ is given in (2.10), and ρ_{nm} depicts the channel coefficient between the mth radar transmitter antenna and the nth UE receiver antenna.

Proof: Referring to Figure 8.1b, we notice that the second half of each sampled hop is free of interhop interference. Since the special DFT performed in (8.7) only relies on the second half of each sampled hop, $Y_{hn}(l)$ obtained in (8.7) and $S_{hm}(l)$ calculated in (8.8) are also free of interhop interference. To this end, we focus on proving the suppression of interantenna interference in the rest of the proof.

Note that both sine functions in (8.8) take zero at $l = k_{hm}\Delta$. Thus, applying the L'Hospital's rule, we obtain,

$$S_{hm}(l^*_{hm}) = F_{hm} e^{-\mathrm{j}\frac{2\pi L_\eta l^*_{hm}}{L}}, \quad \text{s.t. } l^*_{hm} = k_{hm}\Delta. \tag{8.11}$$

Also note from (8.8) that the signal transmitted by antenna m', $\forall m' \neq m$, has the following DFT magnitude at $l = l^*_{hm}$,

$$|S_{hm'}(l^*_{hm})| = \frac{|\sin\frac{\pi}{2}\left(l^*_{hm} - k_{hm'}\Delta\right)|}{\frac{L}{2}|\sin\frac{\pi}{L}\left(l^*_{hm} - k_{hm'}\Delta\right)|}, \quad \forall m' \neq m. \tag{8.12}$$

To avoid interantenna interference, we expect the numerator on the RHS of the equation given above is zero, while the denominator is nonzero. This leads to the condition given in (8.9). Applying the condition in (8.12), we obtain $|S_{hm'}(l^*_{hm})| = 0$, since its numerator is zero, while the denominator is nonzero. Thus, under the condition (8.9), substituting (8.11) and (8.12) into (8.7) leads to (8.10).

Given k_{hm} and Δ in (2.10), we can have the condition in (8.9) satisfied by configuring one of the following:

(1) Take even-indexed subbands as hopping frequencies at hop h, i.e. $k_{hm} \in \{0, 2, 4, \ldots\}$;
(2) Properly design T and K such that $\Delta/2 \in \mathbb{I}_+$.

Configuration (1) restricts the hopping frequencies available in a certain number of radar hops (which is M in the methods to be presented). On the other hand, (2) can increase the hop duration and reduce the number of radar hops per pulse. More in-depth analysis of the impact incurred by (8.9) on an FH-MIMO radar can be carried out employing the range ambiguity functions derived in [2, 8]. We proceed with the assumption that (8.9) is satisfied to focus on developing the multiantenna receiving scheme for FH-MIMO DFRC.

Remark 8.1 ***(Insights Drawn from Proposition 8.1)*** Several insights can be drawn from Proposition 8.1. *First*, interhop and interantenna interferences can be suppressed at a UE even in the presence of timing offset, which owes to the special DFT performed in (8.8) and the condition on radar waveform imposed in (8.9). *Second*, a clearer signal structure is obtained in (8.10), as compared with (8.4). Specifically, the channel of FH-MIMO DFRC is now depicted by $\rho_{nm} e^{-\mathrm{j}\frac{2\pi L_\eta l^*_{hm}}{L}}$, which becomes an $M \times N$ flat-fading MIMO channel without the exponential term. Given a nonzero L_η, the channel $\rho_{nm} e^{-\mathrm{j}\frac{2\pi L_\eta l^*_{hm}}{L}}$ varies fast over hops and antennas, since l^*_{hm} randomly changes over h and m. *Third*, noticing the nontrivial impact of l^*_{hm} on the time-varying channel, we can explore the degrees of freedom in the FH-MIMO radar waveform for the efficient estimations of L_η and ρ_{nm}. This will be developed next.

8.3 A Multiantenna Receiving Scheme

In light of the insights drawn in Remark 8.1, a multiantenna receiving scheme is devised for FH-MIMO DFRC with the following tasks performed in sequence: (i) Estimate the channel coefficients ρ_{nm}; see (8.10); (ii) Estimate the timing offset L_η; see also (8.10); (iii) Apply the estimates for information demodulation. The three tasks will be accomplished in Sections 8.3.1–8.3.3, respectively.

8.3.1 Estimating Channel Response

At this stage, the timing offset still exists. Thus, the signal derived in (8.10) is used for estimating ρ_{nm}. As seen therein, $F_{hm}e^{-j\frac{2\pi L_\eta l^*_{hm}}{L}}$ is a disturbance term. Nevertheless, we notice that by taking $k_{hm} = 0$, the exponential term turns into unit, since $l^*_{hm} = k_{hm}\Delta = 0$ according to (8.10). Accordingly, we can estimate ρ_{nm} as follows:

Lemma 8.1 *By designating the zeroth subband as a hopping frequency in the following manner:*

$$k_{hm} = 0, \quad h = m, \quad m = 0, 1, \dots, M-1, \tag{8.13}$$

and correspondingly setting $F_{hm} = 1$, the channel coefficient between radar transmitter antenna m and UE antenna n can be estimated as follows:

$$\hat{\rho}_{nm} = Y_{hn}(l^*_{hm}), \quad h = m, \quad \forall m, n, \tag{8.14}$$

where $Y_{hn}(l^*_{hm})$ is given in (8.10).

Proof: Substituting (8.13) into (8.10) yields $l^*_{hm} = 0$ and $Y_{hn}(l^*_{hm}) = \rho_{nm} + \Xi_{hn}(l^*_{hm})$ for $h = m$ ($\forall m \in [0, M-1]$). Thus, $Y_{hn}(l^*_{hm})$ can be used as an estimate of ρ_{nm}.

The scheduling of the zeroth hopping frequency, as given in (8.13), is also illustrated in Figure 8.1c. We remark that the constraint may only involve altering the pairing between hopping frequencies and antennas, and it incurs no change to the range ambiguity function according to Proposition 7.1. To be more specific, if there are M hops in a radar pulse having the zeroth subband selected originally, then we can reassign the zeroth subband in such a manner that the M radar transmitter antennas can each take the zeroth subband over the M

(nonconsecutive) hops. If there are less than M hops selecting the zeroth subband as hopping frequency, we need to insert the zeroth subband to some hops to make the number of such hops become M. Given that an FH-MIMO radar randomly selects hopping frequencies to use [8], the insertion can have a negligible impact on the range ambiguity function of the radar.

8.3.2 Estimating Timing Offset

Using the estimate $\hat{\rho}_{nm}$, we can obtain equivalent channel coefficients from (8.10), leading to

$$\tilde{Y}_{hnm} = \frac{Y_{hn}(l^*_{hm})}{\hat{\rho}_{nm}} = e^{-\mathrm{j}\frac{2\pi L_\eta l^*_{hm}}{L}} + \tilde{\Xi}_{hnm}, \tag{8.15}$$

where $\tilde{\Xi}_{hnm} = \Xi_{hn}(l^*_{hm})/\hat{\rho}_{nm}$. Note that the exponential term becomes the same for all the UE antennas, and hence, we can coherently combine $\tilde{Y}_{hnm}$ over n to improve the estimation signal-to-noise ratio (SNR). This yields

$$\bar{Y}_{hm} = \frac{1}{N}\sum_{h=0}^{N-1}\tilde{Y}_{hnm} = e^{-\mathrm{j}\frac{2\pi L_\eta \Delta k_{hm}}{L}} + \bar{\Xi}_{hm}, \tag{8.16}$$

where l^*_{hm} is replaced by $k_{hm}\Delta$ given their relation in (8.10), and $\bar{\Xi}_{hm} = \frac{1}{N}\sum_{n=0}^{N-1}\tilde{\Xi}_{hnm}$. To estimate L_η, we transform the timing offset estimation problem as follows:

Lemma 8.2 *Design the hopping frequencies at hop $h = 0$ such that they take continuous values, e.g.*

$$k_{0m} = k_0 + m, \quad m = 0, 1, \ldots, M-1, \tag{8.17}$$

where k_0 can take 0 or any integer from $\{1, \ldots, K-M\}$. Then, we can turn the L_η estimation into a more tractable problem which estimates the frequency, denoted by Υ, of a discrete single-tone signal,

$$\bar{Y}_{0m} = e^{-\mathrm{j}\frac{2\pi\Upsilon k_0}{M}} e^{-\mathrm{j}\frac{2\pi\Upsilon m}{M}} + \bar{\Xi}_{0m}, \quad \text{s.t. } \Upsilon = L_\eta \Delta M/L. \tag{8.18}$$

The proof can be established by substituting (8.17) into (8.16) and reshaping the exponent of the exponential term. Clearly, the single-tone frequency estimation problem, as formulated in Lemma 8.2 is more tractable, compared with estimating L_η based on (8.16). However, there can be an estimation ambiguity problem, as elaborated on below. According to the definition of Υ given in (8.18), we have $\Upsilon \in [0, \Delta M/2]$. That is, $\Upsilon > M$ can happen if $\Delta > 2$. Given the up limit of Υ, we have $\Upsilon/M = \Delta/2$. Thus, we can express Υ as follows:

$$\Upsilon = (\Upsilon)_M + dM, \quad \text{s.t. } d = 0, 1, \ldots, \lceil \Delta/2 \rceil - 1, \tag{8.19}$$

where $(\cdot)_M$ denotes modulo-M and d is referred to as the ambiguity degree. Substituting (8.19) into (8.18), we obtain

$$\bar{Y}_{0m} = e^{-\mathrm{j}\frac{2\pi \Upsilon k_0}{M}} e^{-\mathrm{j}\frac{2\pi (\Upsilon)_M m}{M}} + \bar{\Xi}_{0m}. \tag{8.20}$$

We see that only $(\Upsilon)_M$ can be estimated from $\bar{Y}_{0m}$. In other words, Υ can be estimated from $\bar{Y}_{0m}$ with an ambiguity degree of d. The two estimation methods developed in Section 3 can be employed for estimating $(\Upsilon)_M$ from $\bar{Y}_{0m}$. Next, we take the first method in Section 3.2 for an illustration. We will also design the method to recover Υ from $(\Upsilon)_M$.

8.3.2.1 Estimating L_η

A coarse estimate of Υ can be obtained from the DFT of $\bar{Y}_{0m}$ w.r.t. m which is given by

$$\mathsf{Y}_m = \frac{1}{M}\sum_{m'=0}^{M-1} \bar{Y}_{0m'} e^{-\mathrm{j}\frac{2\pi m' m}{M}}.$$

Let $\tilde{m}$ denote the peak index of $|\mathsf{Y}_m|$, and then a coarse estimate of $(\Upsilon)_M$ is given by $\frac{2\pi\tilde{m}}{M}$. While the actual value of $(\Upsilon)_M$ can be expressed as $\frac{2\pi(\tilde{m}+\delta)}{M}$, with $\delta \in [-0.5, 0.5]$ denoting the fractional part of $(\Upsilon)_M$.

Next, we iteratively interpolate the DFT coefficients around $\tilde{m}$ to estimate δ [9].

Step 1: Compute the interpolated DFTs as follows:

$$\mathsf{Y}_\pm = \frac{1}{M}\sum_{m'=0}^{M-1} Y_{0m'} e^{-\mathrm{j}\frac{2\pi m'(\tilde{m}+\delta\pm\epsilon)}{M}}, \tag{8.21}$$

where δ (to be updated) takes zero initially, i.e. $\delta = 0$, and ϵ is a constant intermediate variable which takes $\epsilon = \min\{M^{-\frac{1}{3}}, 0.32\}$ [10, Eq. (23)].

Step 2: Set $\delta_{\text{old}} = \delta$ and update δ as follows [9]

$$\delta = \frac{\epsilon\cos^2(\pi\epsilon)}{1 - \pi\epsilon\cot(\pi\epsilon)} \times \Re\{\zeta\} + \delta_{\text{old}}, \tag{8.22}$$

where $\zeta = \frac{\mathsf{Y}_+ - \mathsf{Y}_-}{\mathsf{Y}_+ + \mathsf{Y}_-}$ is calculated based on (8.21).

Step 3: Iteratively perform the above two steps for up to N_{iter} rounds. The final δ is given by $\delta^* = \delta_{N_{\text{iter}}}$

In general, $N_{\text{iter}} = 3$ enables the algorithm to converge [9, 10], with the convergent performance approaching the Cramer–Rao lower bound (CRLB) to be derived in Section 8.4.2. The estimate of $(\Upsilon)_M$ is given by $(\tilde{m} + \delta^*)$, thus, based on (8.19), the final estimate of Υ is

$$\hat{\Upsilon} = \tilde{m} + \delta^* + dM, \quad d = 0, 1, \dots, \lceil \Delta/2 \rceil - 1. \tag{8.23}$$

Substituting (8.23) into (8.18), we obtain the estimate of L_η, as given by

$$\hat{L}_\eta^{(d)} = L(\tilde{m} + \delta^*)/(M\Delta) + Ld/\Delta. \tag{8.24}$$

8.3.2.2 Removing Estimation Ambiguity

We enumerate the possible estimates and compare the SNR in the recovered hops since the ambiguity degree is generally limited given the potentially small value of Δ. Besides, we use the first hop to fulfill the task, given that it has a known feature; see Lemma 8.2. Based on (8.4) and (8.5), the first "complete" hop, recovered using $\hat{L}_\eta^{(d)}$, is given by

$$y_{0n}^{(d)}(i) = \left[y_{0n}(\hat{L}_\eta^{(d)}), \dots, y_{0n}(L-1),\ y_{1n}(0), \dots, y_{1n}(\hat{L}_\eta^{(d)} - 1)\right]^{\mathrm{T}}, \tag{8.25}$$

where $y_{hn}(i)$ $(h = 0, 1)$ is the nth entry of $\mathbf{y}_h(i)$ given in (8.4) and $y_{0n}^{(d)}(i)$ denotes the recovered first hop. Take the L-point DFT of $y_{0n}^{(d)}(i)$, leading to

$$X_{0n}^{(d)}(l) = \frac{1}{L}\sum_{i=0}^{L-1} y_{0n}^{(d)}(i) e^{-\mathrm{j}\frac{2\pi i l}{L}}. \tag{8.26}$$

Provided that d^* is the correct ambiguity degree and $\hat{L}_\eta^{(d^*)}$ is sufficiently accurate, $X_{0n}^{(d^*)}(l)$ becomes

$$X_{0n}^{(d^*)}(l) = \sum_{p=0}^{P-1} \beta_p e^{\mathrm{j}n\nu_p} \sum_{m=0}^{M-1} e^{\mathrm{j}m u_p} \delta\left(l - l_{0m}^*\right) + \Pi_{hn}(l).$$

As a result, there are only M effective values in $X_{0n}^{(d^*)}(l)$ as given by $X_{0n}^{(d^*)}(l_{0m}^*) = \rho_{nm} + \Pi_{0n}(l_{0m}^*)$ $(m = 0, 1, \dots, M-1)$, where ρ_{nm} is given in (8.10). The rest of $X_{0n}^{(d^*)}(l)$ $(l \neq l_{0m}^*)$ are noises. Therefore, the SNR of a correctly recovered hop can be given by

$$\gamma_{X_0}^{(d^*)} = \frac{\frac{1}{MN}\sum_{m=0}^{M-1}\sum_{n=0}^{N-1} |X_{0n}^{(d^*)}(l_{0m}^*)|^2}{\frac{1}{N(L-M)}\sum_{\substack{l=0 \\ l \neq l_{0m}^*}}^{L-1}\sum_{n=0}^{N-1} |X_{0n}^{(d^*)}(l)|^2}. \tag{8.27}$$

For $d \neq d^*$, part of the signal power will leak to the sidelobe region, incurring a decrease of the numerator and an increase of the denominator in $\gamma_X^{(d^*)}$. Therefore, the correct ambiguity degree can be identified by solving the following problem:

$$d^* = \operatorname{argmax}_{d=0,1,\dots,\lceil\Delta/2\rceil-1}\ \gamma_{X_0}^{(d)}, \tag{8.28}$$

where $\gamma_{X_0}^{(d)}$ is calculated by replacing $X_{0n}^{(d^*)}(l)$ in (8.27) with $X_{0n}^{(d)}(l)$ obtained in (8.26).

Remark 8.2 We remark that the overall complexity of the presented methods is generally low. The complexity of the estimation of ρ_{nm} developed in Section 8.3.1

is dominated by computing $Y_{hn}(l)$, as given in (8.7). The complexity of this computation can be given by $\mathcal{O}(MNL^2)$, where the factor M is because the computation happens over M hops; see Lemma 8.2, and N is because the computation needs to be performed for each of the N receiver antennas. The complexity of estimating L_η, as developed in Section 8.3.2, is dominated by the computations in (8.15), (8.21), and (8.28). Their complexities can be, respectively, given by $\mathcal{O}(MN)$, $\mathcal{O}(2I_{\text{iter}}M)$, and $\mathcal{O}(\lceil \Delta/2 \rceil L \log_2 L)$, where I_{iter} is the number of iterations performed for estimating $(\Upsilon)_M$ from $\bar{Y}_{0m}$. As mentioned earlier, $I_{\text{iter}} = 3$ ensures asymptotic convergence to the CRLB. Jointly considering the complexity in the four $\mathcal{O}(\cdot)$ expressions given above, the overall computational complexity of the scheme is in the order of $\mathcal{O}(MNL^2)$.

8.3.3 Information Demodulation

With L_η estimated and compensated, we can take the standard L-dimensional DFT of $y_{hn}(i)$ ($\forall h \geq 2$) given in (8.4), leading to

$$\begin{cases} Z_{hn}(l) = \frac{1}{L}\sum_{i=0}^{L-1} y_{hn}(i) = \sum_{p=0}^{P-1} \beta_p e^{\mathrm{j}nv_p} \times \sum_{m=0}^{M-1} e^{\mathrm{j}mu_p} \delta\left(l - l^*_{hm}\right) F_{hm} + \Pi_{hn}(l); \\ Z_{hn}(l^*_{hm}) = \rho_{nm} F_{hm} + \Pi_{hn}(l^*_{hm}), \end{cases} \tag{8.29}$$

where $\Pi_{hn}(l) = \frac{1}{L}\sum_{i=0}^{L-1} [\mathbf{n}_h(i)]_n e^{-\mathrm{j}\frac{2\pi il}{L}}$. To demodulate F_{hm}, we need to extract the signal component solely associated with the mth radar transmitter antenna, i.e. $Z_{hn}(l^*_{hm})$ given in (8.29). However, the index $l^*_{hm} = k_{hm}\Delta$ needs to estimated, which essentially requires the estimation of k_{hm} (the hopping frequency used by antenna m at hop h).

8.3.3.1 Estimating k_{hm}

Conventional FH-MIMO radars can randomly change hopping frequencies over hops [8]. To make it possible for the UE to estimate k_{hm}, we re-pair the hopping frequencies with radar transmitter antennas at each hop. Specifically, after hopping frequencies are randomly selected for a hop, the frequencies are sorted in an ascending order before being assigned to the antennas. This leads to

$$\begin{cases} k_{hm} < k_{h(m+1)} & h \in [1, M-1], \quad \forall m, \quad m \neq h, \\ k_{hm} < k_{h(m+1)} & h \in [M, H-1], \quad \forall m, \end{cases} \tag{8.30}$$

where $m \neq h$ in the first row is to enforce condition (8.13) required in Lemma 8.1. As assured by [7, Proposition 1], the above re-pairing does not incur any change to the radar range ambiguity function. Enabled by (8.30), we can identify the hopping frequencies, as summarized in Algorithm 8.1.

Algorithm 8.1 Estimating k_{hm}

(1) Use $\hat{L}_{\eta}^{(d^*)}$ to remove the timing offset and reconstruct hop h ($\forall h > 0$), as done in (8.25), where the subscripts 0 and 1 are replaced by h and $(h+1)$, respectively;

(2) Take the standard L-point DFT of the reconstructed hops, leading to $Z_{hn}(l)$ given in (8.29);

(3) To exploit the antenna diversity, we can combine $Z_{hn}(l)$ over $\forall n$, which gives $\bar{Z}_h(l) = \frac{1}{N}\sum_{n=0}^{N-1}|Z_{hn}(l)|^2$; refer to Remark 8.3 for the illustration of diversity gain.

(4) Identify the strongest M peaks from $\bar{Z}_h(l)$ with the peak indexes denoted by $l_{m'}$ ($l_0 < l_1 < \cdots < l_{(M-1)}$);

(5) Associate the peaks with the radar transmitter antennas based on (8.30), leading to the estimates of k_{hm} and $l_{hm}^*(= k_{hm}\Delta)$, denoted by $\hat{k}_{hm}$ and $\hat{l}_{hm}^*$, respectively.

Remark 8.3 The information-bearing signal from the mth radar transmitter antenna can be severely attenuated when arriving at the UE. This can be seen from the complex summation of multiple paths, i.e. $\sum_{p=0}^{P-1}\beta_p e^{jmu_p}e^{jnv_p}F_{hm}$ given in (8.10) and (8.29). The phenomenon, referred to as *signal cancelation*, can severely degrade the detection performance of $\hat{k}_{hm}$. As demonstrated in Figure 8.2a, the fourth DFT peak is attenuated severely such that it is missed. From (8.10) and (8.29), we notice that if $\sum_{p=0}^{P-1}\beta_p e^{jmu_p}e^{jnv_p}F_{hm}$ approaches zero, it is likely

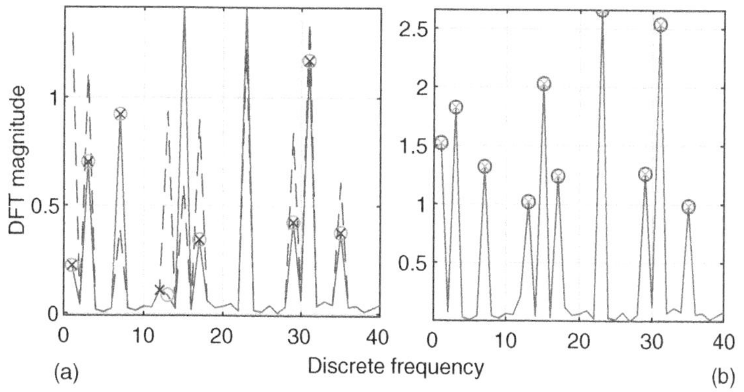

Figure 8.2 Illustration of signal cancelation. (a) The solid curve is for $|Z_{h0}(l)|$ and the dash for $|Z_{h1}(l)|$. (b) the solid curve plots $|\bar{Z}_h(l)|$. In both figures, circular markers indicate the set hopping frequencies and cross markers the detected. Here, $N = 2$ and the other parameters are the same as those used in Section 8.5.

that $\sum_{p=0}^{P-1} \beta_p e^{jmu_p} e^{jn'\nu_p} F_{hm} \gg 0$ for some $n' \neq n$ since the phase $n\nu_p$ ($\forall \nu_p \neq 0$) varies with n. The antenna diversity can be exploited to reduce the effect of signal cancelation, as performed in Step 3 of Algorithm 8.1. The efficacy of the combining in Step 3 is demonstrated by Figure 8.2b. We see that the magnitudes of DFT peaks are substantially enhanced, with only two antennas at the UE.

8.3.3.2 FHCS Demodulation

Note that the combinations of k_{hm} $\forall m$ at any hop can be used for information modulation, which is referred to as frequency hopping code selection (FHCS) [11]. Given K available subbands and M transmitter antennas at the radar, there can be C_K^M combinations of hopping frequencies. We can select $2^{J_{\text{FHCS}}}$ combinations as constellation points which can convey $J_{\text{FHCS}} = \lfloor \log_2 (C_K^M) \rfloor$ bits per radar hop. Comparing $\{\hat{k}_{hm}\ \forall m\}$ with the constellation points can demodulate the FHCS symbol, where $\hat{k}_{hm}$ $\forall m$ is obtained from Step 5 in Algorithm 8.1.

8.3.3.3 PSK Demodulation

With $\hat{l}_{hm}^*$ $\forall m$ estimated using the method given in Algorithm 8.1, we can extract the M values of $Z_{hn}(l)$ at $\hat{l}_{hm}^*$, attaining $Z_{hn}(\hat{l}_{hm}^*)$ which has the same structure as $Z_{hn}(l_{hm}^*)$. Then we use the ρ_{nm} estimates obtained in Lemma 8.1 to perform maximum-ratio combining (MRC) over UE antennas. This yields

$$\tilde{Z}_{hm} = \left(\sum_{n=0}^{N-1} \hat{\rho}_{nm}^* Z_{hn}(\hat{l}_{hm}^*) \right) / \left(\sum_{n=0}^{N-1} |\hat{\rho}_{nm}|^2 \right). \tag{8.31}$$

Finally, $\tilde{Z}_{hm}$ can be used for demodulating F_{hm}. We remark that the enabling factors for exploiting antenna diversity, as done in Step 3 of Algorithm 8.1 and (8.31), are nontrivial. In particular, they are enabled by the accurate estimations of channel responses and timing offset obtained in Sections 8.3.1 and 8.3.2, respectively. Moreover, these estimation methods are developed based on our insights into the UE-received signals in the FH-MIMO DFRC, as illustrated in Section 8.2.

8.4 Performance Analysis

The performance of the estimation and demodulation methods presented above is analyzed.

8.4.1 Performance of Channel Coefficient Estimation

We first analyze the performance of the estimation method for channel coefficients. In particular, we derive the following analytical mean squared error (MSE) of $\hat{\rho}_{nm}$ achieved in Lemma 8.1.

Corollary 8.1 *The analytical MSE of $\hat{\rho}_{nm}$ is given by*

$$\sigma^2_{\hat{\rho}_{nm}} = 2\sigma_0^2/L, \quad \forall n, m, \tag{8.32}$$

where σ_0^2 is the power of each entry of $\mathbf{n}_h(i)$ given in (8.4).

Proof: Given (8.13) and $F_{hm} = 1$, $Y_{hn}(l^*_{hm})$ degenerates into ρ_{nm} plus an AWGN $\Xi_{hn}(l^*_{hm})$. Therefore, the estimation error of ρ_{nm} is $\Xi_{hn}(l^*_{hm})$, and the noise power is the estimation MSE. Given the relation between $\mathbf{n}_h(i)$ and $\Xi_{hn}(l)$; see below (8.4), the noise power of the later is $2\sigma_0^2/L$, so is the MSE of $\hat{\rho}_{nm}$.

From Corollary 8.1, we see that the estimation performance of ρ_{nm} is solely related to the number of samples in a radar hop, i.e. L, and the variance of the antenna-level noise, i.e. σ_0^2. This manifests the applicability of the channel estimation method in various channel models. It is worth pointing out that the flexible applicability is achieved by the novel waveform design, as given in (8.13).

8.4.2 Performance of Timing Offset Estimation

To evaluate the performance of the estimation method for timing offset, we derive the CRLB of $L_\eta^{(d^*)}$.

Proposition 8.2 *Provided that NLoS paths are negligible compared with the LoS path and $\hat{\rho}_{nm}$ approaches its true value ρ_{nm}, the CRLB of $L_\eta^{(d^*)}$ is given by*

$$\text{CRLB}\left\{L_\eta^{(d^*)}\right\} = \frac{3L}{\pi^2 M^3 N \Delta^2 \gamma_0}, \tag{8.33}$$

where $\gamma_0 = \frac{|\beta_0|^2}{\sigma_0^2}$, β_0 is the complex gain of the LoS path and σ_0^2 is given in Corollary 8.1.

Proof: According to [9], the CRLB of estimating Υ from $\gamma_{\bar{Y}}$ given in (8.18) can be expressed as $\frac{6}{4\pi^2 M \gamma_{\bar{Y}}}$, where $\gamma_{\bar{Y}}$ is the SNR in $\bar{Y}_{0m}$. Based on relation between Υ and L_η illustrated in (8.18), the CRLB of L_η estimate can be given by

$$\text{CRLB}\left\{L_\eta^{(d^*)}\right\} = \frac{6L^2}{4\pi^2 M^3 \Delta^2 \gamma_{\bar{Y}}}. \tag{8.34}$$

Next, we derive $\gamma_{\bar{Y}}$ for the CRLB.

Given (8.18), we have $\gamma_{\bar{Y}} = 1/\sigma^2_{\bar{\Xi}}$, where $\sigma^2_{\bar{\Xi}}$ denotes the noise variance of $\bar{\Xi}_{0m}$. Based on (8.16), we obtain $\sigma^2_{\bar{\Xi}} = \frac{1}{N}\sigma^2_{\tilde{\Xi}}$, where $\sigma^2_{\tilde{\Xi}}$ is the noise variance of $\tilde{\Xi}_{0nm}$ give in (8.15). We can derive from (8.15) that $\sigma^2_{\tilde{\Xi}} = \frac{1}{|\hat{\rho}_{nm}|^2}\sigma^2_{\Xi} \approx \frac{1}{|\beta_0|^2}\sigma^2_{\Xi}$, where the last approximation is based on the claimed conditions of Proposition 8.2, and σ^2_{Ξ} is the noise power of $\Xi_{0n}(l^*_{0m})$ given in (8.7). As derived in the Corollary 8.1, $\sigma^2_{\Xi} = 2\sigma_0^2/L$. Backtracking the above changes of noise components, we obtain $\gamma_{\bar{Y}} = \frac{LN\gamma_0}{2}$. Substituting this into (8.34) leads to (8.33).

From Proposition 8.2, we see that the performance of the timing offset estimation method is inversely proportional to the third power of M and the first power of N. This indicates that increasing M, rather than N, can be more effective in achieving better L_η estimation (and hence better communication performance). Note that the L-dependence shown in (8.33) is not straightforward, since Δ in the denominator is also related to L. Taking $f_s = 2B$, we have $\Delta = BT/K = L/(2K)$. Substituting this into (8.33) leads to

$$\text{CRLB}\left\{L_\eta^{(d^*)}\right\} = \frac{12K^2}{\pi^2 M^3 N L \gamma_0}, \tag{8.35}$$

where we see that the estimation performance can be improved by increasing L. This complies with the fact that a larger L corresponds to a greater DFT dimension and hence a larger estimation SNR, as analyzed in the proof of Proposition 8.2.

8.4.3 Communication Performance

We proceed to analyze the communication performance of FH-MIMO DFRC achieved by the receiving scheme.

8.4.3.1 Achievable Rate

The symbol rate of FH-MIMO DFRC is given by H/T_{PRT} (symbol/second), where T_{PRT} denotes the pulse repetition interval of the FH-MIMO radar. Thus, the achievable rate of FH-MIMO DFRC is given by

$$\mathsf{R}_{\text{DFRC}} = \frac{HJ}{T_{\text{PRT}}} = \mathsf{r}J/T, \tag{8.36}$$

where J denotes the number of bits conveyed in each radar hop, r the duty cycle and T the hop duration. Based on (8.3), we have $J = MJ_{\text{PSK}}$ and hence, $\mathsf{R}_{\text{PSK}} = \mathsf{r}MJ_{\text{PSK}}/T$. Based on Section 8.3.3.2, we have $J = J_{\text{FHCS}} = \left\lfloor \log_2\left(C_K^M\right)\right\rfloor$ and hence, $\mathsf{R}_{\text{FHCS}} = \mathsf{r}\left\lfloor \log_2\left(C_K^M\right)\right\rfloor/T$.

From (8.36), we see that reducing T can help improve the achievable rate. This, however, may not stand for the FHCS-based FH-MIMO DFRC, since J_{FHCS} is related to T. According to (2.10), a smaller T results in a smaller $\Delta(= BT/K)$. To keep (8.9) satisfied at a fixed B, we need to reduce K. Moreover, a smaller K can yield a smaller $\left\lfloor \log_2\left(C_K^M\right)\right\rfloor$ given a fixed M. Therefore, reducing T can lead to the decreases of both the numerator and denominator of R_{FHCS}. Note that the system parameters can be holistically designed to optimize radar and communications performance (trade-off) under the receiving scheme. This is left for future work.

8.4.3.2 SER of PSK-Based FH-MIMO DFRC

Note that a DFRC symbol consists of M PSK subsymbols, i.e. F_{hm}, $m = 0, 1, \dots, M-1$, each from a radar transmitter antenna. To this end, the hth DFRC symbol is correctly demodulated only when each $F_{hm}\ \forall m$ is correctly demodulated. Consider the Rician channel here. The resultant performance analysis can be readily extended to Rayleigh or AWGN channel by varying the Rician factor K. According to Section 8.3.3.3, $F_{hm}\ \forall m$ is independently demodulated over identically distributed Rician channels, the symbol error rate (SER) of demodulating $F_{hm}\ \forall m$ is identical and denoted by $\tilde{\mathsf{P}}_{\text{PSK}}$.

From (8.31), we see that $F_{hm}\ \forall m$ is demodulated from the MRC over N antennas with the spacing of half a wavelength. Thus, the identically distributed Rician channels over the N antennas can be highly correlated. To avoid the complexity of analysis incurred by the channel coherence [12], we proceed to derive the lower bound of $\tilde{\mathsf{P}}_{\text{PSK}}$ by treating the MRC result as a single Rician channel with the identical distribution yet an N-fold SNR improvement, compared with the Rician channel on a single UE receiver antenna. Applying the SER of PSK demodulation in a Rician channel, as derived in [13], we obtain

$$\tilde{\underline{\mathsf{P}}}_{\text{PSK}} = \int_0^{\frac{\pi}{2}} \frac{e^{-\mathsf{K}+\frac{\mathsf{K}}{\Omega(\theta)}}}{\pi\Omega(\theta)}\, d\theta + \int_0^{\frac{\pi}{2}-\frac{\pi}{J_{\text{PSK}}}} \frac{e^{-\mathsf{K}+\frac{\mathsf{K}}{\Omega(\theta)}}}{\pi\Omega(\theta)}\, d\theta,$$
$$\text{s.t. } \Omega(\theta) = 1 + \frac{LN\gamma_0 \sin^2\frac{\pi}{J_{\text{PSK}}}}{\mathsf{K}\cos^2\theta}, \tag{8.37}$$

where γ_0 is the antenna-level SNR defined in Proposition 8.2. Accordingly, the SER of demodulating M PSK subsymbols is lower-bounded by

$$\underline{\mathsf{P}}_{\text{PSK}} = 1 - \left(1 - \tilde{\underline{\mathsf{P}}}_{\text{PSK}}\right)^M, \tag{8.38}$$

where $(1 - \tilde{\underline{\mathsf{P}}}_{\text{PSK}})^M$ gives the probability of correctly detecting the M PSK subsymbols in a radar hop.

8.5 Simulations

Next, we validate the presented design through extensive simulations. Unless otherwise specified, we employ the following parameters, where the FH-MIMO radar is configured with reference to [11]: $\mathsf{r} = 0.2$, $B = 100$ MHz, $f_s = 2B$, $K = 20$, $M = H = 10$, $N = 6$, $T = 0.4\ \mu$s (hence $L = f_sT = 80$ and $\Delta = BT/K = 2$), and $L_\eta \in \mathcal{U}_{[0,L/2]}$ with $\mathcal{U}_{[a,b]}$ denoting the uniform distribution in $[a, b]$. Applying the parameters in Section 8.4.3.1, we obtain

$$\mathsf{R}_{\text{PSK}} = 0.2 \times 10/(0.4 \times 10^{-6}) = 5 \text{ Mbps};$$

$$\mathsf{R}_{\text{FHCS}} = 0.2 \times \left\lfloor \log_2\left(C_{20}^{10}\right)\right\rfloor /(0.4 \times 10^{-6}) = 8.5 \text{ Mbps}.$$

As for communication scenarios, the Rician channel with an LoS and three NLoS paths is simulated, i.e. $P = 4$, where $\theta_p \in \mathcal{U}_{[-90°,90°]}\ \forall p$, $\phi_p \in \mathcal{U}_{[-90°,90°]}\ \forall p$,

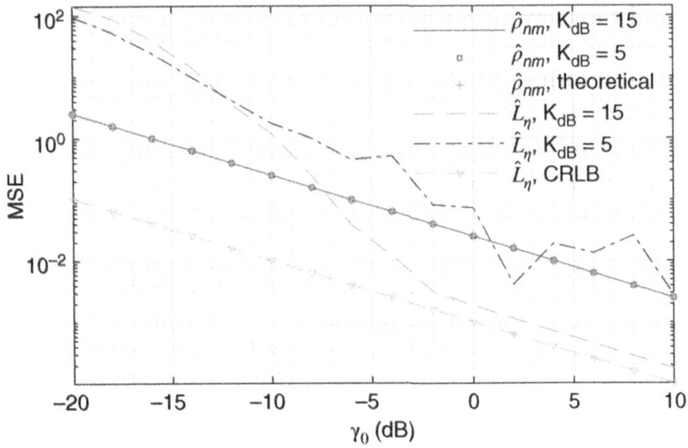

Figure 8.3 MSE of $\hat{\rho}_{nm}$ obtained in (8.14) and $\hat{L}_\eta^{(d^*)}$ given in (8.24).

$\beta_0 = e^{jx}$ with $x \in \mathcal{U}_{[0,2\pi]}$, and $\beta_p \sim \mathcal{CN}(0, \mathrm{K}_{\mathrm{dB}})$ ($\forall p \in [1, P-1]$) with $\mathrm{K}_{\mathrm{dB}} = -\log_{10}\frac{\mathrm{K}}{P-1}$ denoting the Rician factor in decibel. Below, $\gamma_0 = \frac{|\beta_0|}{\sigma_0^2}$ represents the antenna-level SNR at the UE.

Figure 8.3 plots the MSE of $\hat{\rho}_{nm}$ and $\hat{L}_\eta$ against γ_0, where both $\mathrm{K}_{\mathrm{dB}} = 15$ dB and 5 dB are considered. In terms of $\hat{\rho}_{nm}$, we see that the simulated MSE of $\hat{\rho}_{nm}$ monotonically decreases with γ_0 and coincides with the analytical MSE derived in Corollary 8.1. This validates the ρ_{nm} estimation method presented in Section 8.3.1. We also see that the MSE of $\hat{\rho}_{nm}$ is invariant under different values of K_{dB}. This is because we can fully decouple the estimations of ρ_{nm} and L_η, making the estimation error of $\hat{\rho}_{nm}$ solely dependent on background noises; see Section 8.3.1.

As for $\hat{L}_\eta$, we see from Figure 8.3 that, at $\mathrm{K}_{\mathrm{dB}} = 15$ dB, the MSE of $\hat{L}_\eta$ approaches the CRLB derived in Proposition 8.2 in the region of $\gamma_0 > -2$ dB. This validates the high accuracy of L_η estimation, particularly given a large K_{dB}. We also see from the figure that the MSE of $\hat{L}_\eta$ achieved at $\mathrm{K}_{\mathrm{dB}} = 5$ dB is not as low as that at $\mathrm{K}_{\mathrm{dB}} = 15$ dB, and fluctuates in high-SNR regions. This is because the multipath components lead to uneven noise variance across the M signals constructed for L_η estimation; see (8.15). Despite the slight MSE fluctuation, the overall performance of $\hat{L}_\eta$ at $\mathrm{K}_{\mathrm{dB}} = 5$ dB improves with γ_0, which is adequate for high-communication performance, as illustrated subsequently.

Figure 8.4 plots the SER and achievable rate of FHCS-based FH-MIMO DFRC against γ_0, where $\hat{L}_\eta^{(d^*)}$ obtained for Figure 8.3 is used to remove the timing offset in the UE-received signals, as elaborated on in Section 8.3.3. We see that the SER curves achieved by the estimated and actual values of L_η are nearly overlapped in the observed SNR regions. This validates the multiantenna receiving scheme, as developed in Section 8.3, and manifests its robustness against the estimation error

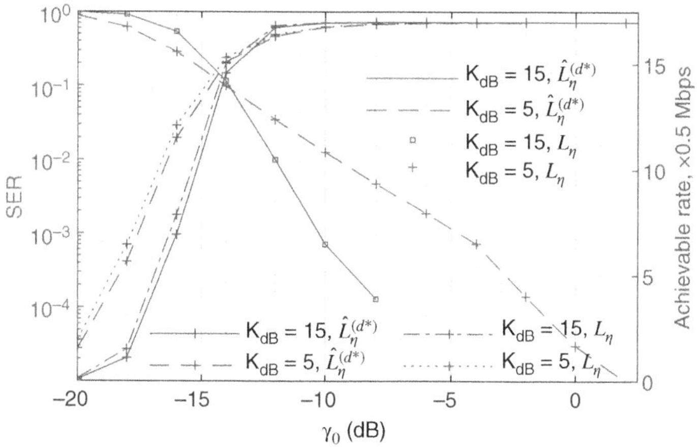

Figure 8.4 Communication performance of FHCS-based FH-MIMO DFRC against γ_0, where the curves with "+" markers use the y-axis on the right.

of $\hat{L}_\eta$. We also see from Figure 8.4 that, enabled by the presented methods, the SNR required for the achievable rate of FHCS to approach R_{PSK} with the SER of 10^{-4} is as low as $\gamma_0 = -7$ dB given $K_{dB} = 15$ dB, whereas the required SNR increases to -1 dB for $K_{dB} = 5$ dB. The moderate increase is caused by the signal cancellation illustrated in Remark 8.3.

Figure 8.5 plots the SER and achievable rate of binary PSK (BPSK)-based FH-MIMO DFRC against γ_0 in the case of $K_{dB} = 15$ dB. The multiantenna

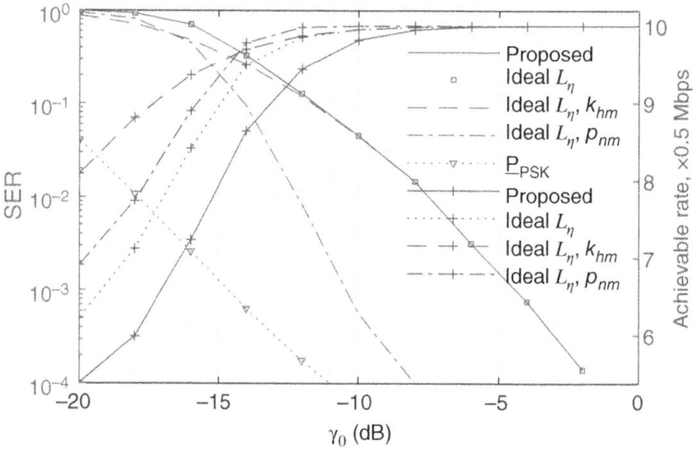

Figure 8.5 Communication performance of BPSK-based FH-MIMO DFRC, with $K_{dB} = 15$ dB and the estimated parameters used unless otherwise specified. The curves with "+" markers use the y-axis on the right. The SER lower bound $\underline{P}_{PSK}$ is given in (8.38).

receiving scheme enables the SER of BPSK to decrease consistently over the observed region of γ_0. In particular, the SER decreases to nearly 10^{-4} at a low SNR of $\gamma_0 = -2$ dB. We also see that in the region of $\gamma_0 \leq -16$ dB, k_{hm} has a more dominant impact on the SER than ρ_{nm} and the reverse is seen in the region of $\gamma_0 > -16$ dB. From the achievable rate curves, we see that the design enables BPSK-based FH-MIMO DFRC to approach R_{PSK} once $\gamma_0 \geq -6$ dB. Moreover, the SNR required for 10^{-4} SER is as low as -1 dB.

Figure 8.6 plots the SER and achievable rate of BPSK-based FH-MIMO DFRC against γ_0 in the case of $\mathsf{K}_{\text{dB}} = 5$ dB. We see a decreasing SER w.r.t. γ_0. However, the decreasing rate is smaller as compared with the large-K_{dB} results given in Figure 8.5. The main reason for this change is the SNR fluctuation across UE antennas, where the fluctuation is further caused by the different extents of signal cancellation presented on UE antennas. As for the achievable rate, we see from Figure 8.6 that the SNR required for approaching 10 Mbps is -1 dB, which is 5 dB higher than observed in Figure 8.5.

Figures 8.5 and 8.6 also plot the derived SER lower bounds for BPSK-based FH-MIMO DFRC. We see that the bound can precisely depict the actual SER performance, particularly when K is small. This is because the Rician channel can degenerate into the LoS-dominant AWGN channel for large values of K. Jointly observing the two figures, we further see that the SER is more prone to multipath fading than the achievable rate between the two performance metrics. Precoding can be designed at the radar to enhance the SER performance. This is left for future work.

Figure 8.7 compares the presented multiantenna receiving scheme with two single-antenna schemes designed in [11] and [2]. For a fair comparison, the same

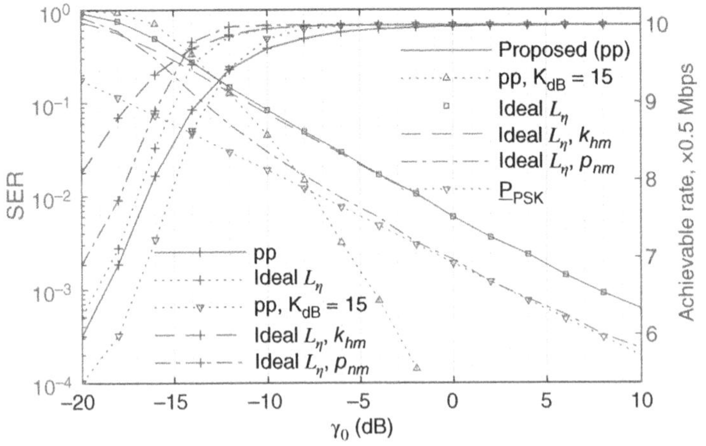

Figure 8.6 The same simulation as done in Figure 8.5 with $\mathsf{K}_{\text{dB}} = 5$ dB here.

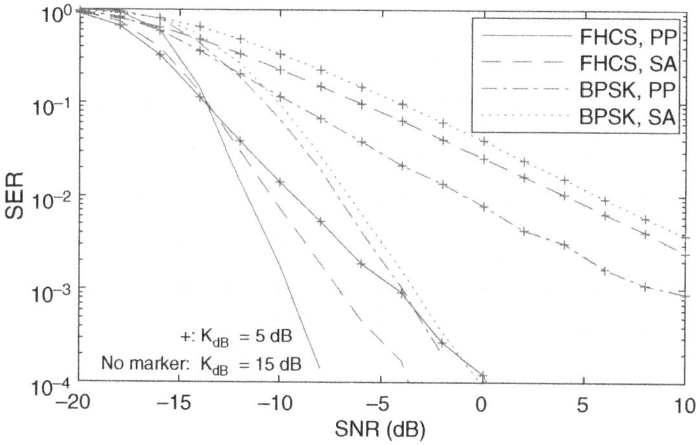

Figure 8.7 Comparing SER between the presented (PP) multiantenna receiving scheme and the single-antenna (SA) schemes, where "FHCS, SA" and "BPSK, SA" are obtained by the methods designed in [11] and [2], respectively. PP, proposed.

multipath fading and total transmission power are set for all schemes. Since the benchmark methods use a single antenna, their SNR is thus $10\log_{10}N$ dB higher than the presented scheme at each receiving antenna. For example, at the SNR of 0 dB in the figure, the SNR for the two benchmark schemes is $10\log_{10}N$ dB. From Figure 8.7, the presented receiving scheme substantially outperforms the benchmark schemes [2, 11], particularly when FHCS modulation is performed. Specifically, we see that to achieve a SER of 10^{-3} for FHCS, the presented scheme only requires an SNR of −4 dB, while the scheme [11] would require an SNR larger than 10 dB. In addition, we see that the SER curves achieved by the presented scheme are much steeper than those by the benchmark schemes, which shows the diversity gain achieved by the presented scheme.

Figure 8.8 plots the SER performance of FHCS-/BPSK-based FH-MIMO DFRC against N. We see that both FHCS and BPSK have better SER performance as N increases. Particularly from $N = 1$ to $N = 2$, the SER decrease of FHCS can be orders of magnitude. This improvement is because the demodulation SNRs of FHCS and BPSK increase with N; see Section 8.3.3. We also see that N has a more evident impact on FHCS. This is because FHCS is more prone to the signal cancelation illustrated in Remark 8.3; hence, increasing N can introduce more diversity over UE antennas. We conclude from Figure 8.8 that furnishing the UE with a multiantenna receiver can dramatically improve the performance of FH-MIMO DFRC. It is worth mentioning that the high-communication performance is achieved by the presented receiving scheme and estimation methods for channel parameters.

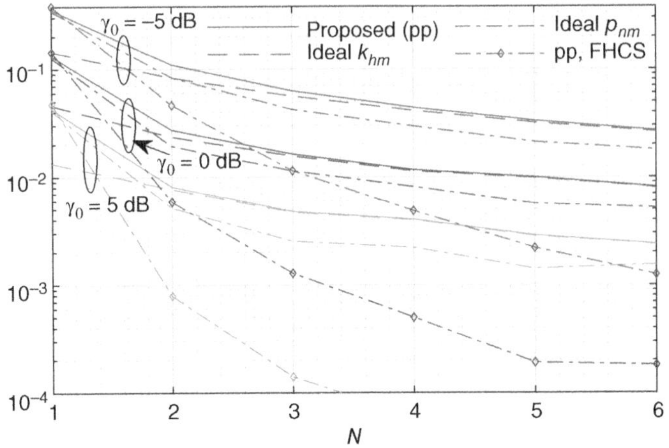

Figure 8.8 SER of FH-MIMO DFRC against N, where, unless otherwise specified, $K_{dB} = 5$ dB, and we use BPSK and the estimated parameters.

8.6 Conclusions

This chapter has developed a high-performance multiantenna-receiving scheme for FH-MIMO DFRC in multipath channels. This is achieved by a special time-frequency transformation that suppresses interantenna and interhop interference and substantially simplifies the DFRC signal model. This is also accomplished by devising new minimal constraints on radar waveform, decoupling the estimations of channel coefficients and timing offset. This is further fulfilled by newly proposed estimation methods for channel parameters. Validated by extensive simulations, superior communication performance is attained by the presented receiving scheme, tightly approaching the derived analytical bound. For the interest of researchers, we notice that a holistic system parameter design, which considers the performances of parameter estimation, data communication, and radar sensing, is still missing in the open literature. This can be an interesting, important future work.

References

1 M. Wang, F. Gao, S. Jin, and H. Lin. An overview of enhanced massive MIMO with array signal processing techniques. *IEEE Journal on Selected Topics in Signal Processing*, 13(5):886–901, 2019.

2 I. P. Eedara, A. Hassanien, M. G. Amin, and B. D. Rigling. Ambiguity function analysis for dual-function radar communications using PSK signaling. In *52nd*

Asilomar Conference on Signals, Systems, and Computers, pages 900–904, October 2018. doi: 10.1109/ACSSC.2018.8645328.

3 I. P. Eedara and M. G. Amin. Dual function FH MIMO radar system with DPSK signal embedding. In *2019 27th European Signal Processing Conference (EUSIPCO)*, pages 1–5, September 2019. doi: 10.23919/EUSIPCO.2019.8902743.

4 I. P. Eedara, M. G. Amin, and A. Hassanien. Controlling clutter modulation in frequency hopping MIMO dual-function radar communication systems. In *IEEE International Radar Conference (RADAR)*, pages 466–471, 2020.

5 F. Liu, A. Garcia-Rodriguez, C. Masouros, and G. Geraci. Interfering channel estimation in radar-cellular coexistence: How much information do we need? *IEEE Transactions on Wireless Communications*, 18(9):4238–4253, 2019.

6 Y. S. Cho, J. Kim, W. Y. Yang, and C. G. Kang. *MIMO-OFDM Wireless Communications with MATLAB*. John Wiley & Sons, 2010.

7 K. Wu, Y. J. Guo, X. Huang, and R. W. Heath. Accurate channel estimation for frequency-hopping dual-function radar communications. In *2020 IEEE International Conference on Communications Workshops (ICC Workshops)*, pages 1–6, 2020.

8 C. Chen and P. P. Vaidyanathan. MIMO radar ambiguity properties and optimization using frequency-hopping waveforms. *IEEE Transactions on Signal Processing*, 56(12):5926–5936, December 2008. ISSN 1941-0476. doi: 10.1109/TSP.2008.929658.

9 A. Serbes. Fast and efficient sinusoidal frequency estimation by using the DFT coefficients. *IEEE Transactions on Communications*, 67(3):2333–2342, March 2019. ISSN 0090-6778. doi: 10.1109/TCOMM.2018.2886355.

10 K. Wu, W. Ni, J. A. Zhang, R. P. Liu, and Y. J. Guo. Refinement of optimal interpolation factor for DFT interpolated frequency estimator. *IEEE Communications Letters*, 1, 2020. ISSN 2373-7891. doi: 10.1109/LCOMM.2019.2963871.

11 W. Baxter, E. Aboutanios, and A. Hassanien. Dual-function MIMO radar-communications via frequency-hopping code selection. In *2018 52nd Asilomar Conference on Signals, Systems, and Computers*, pages 1126–1130, October 2018. doi: 10.1109/ACSSC.2018.8645212.

12 M. K. Simon and M.-S. Alouini. *Digital Communication Over Fading Channels*, volume 95. John Wiley & Sons, 2005.

13 M. G. Shayesteh and A. Aghamohammadi. On the error probability of linearly modulated signals on frequency-flat Ricean, Rayleigh, and AWGN channels. *IEEE Transactions on Communications*, 43(2/3/4):1454–1466, 1995.

9

Integrating Secure Communications into Frequency Hopping MIMO Radar with Improved Data Rates

As illustrated in Chapters 7 and 8, dual-function radar-communication (DFRC) based on frequency hopping (FH) multiple-input and multiple-output (MIMO) radar can achieve symbol rates much higher than the radar pulse repetition frequency. Such DFRC, however, is prone to eavesdropping due to the spatially widespread of an FH-MIMO radar signal. This chapter reveals the potential of using permutations of hopping frequencies to achieve both secure and high data rate FH-MIMO DFRC. Specifically, we highlight the angle-dependent issue in detecting permutations and develop an elementwise phase compensation to solve the issue for a legitimate user (Bob). Elementwise phase compensation makes the demodulation at an eavesdropper (Eve) conditioned on knowing the angle-of-departure (AoD) of Bob. We also present the random sign reversal technique which randomly selects several antennas over hops and reverses their signs. Owing to elementwise phase compensation, there is a sign rule available for Bob. We show that the rule can be employed for developing a low-complexity algorithm to remove random sign reversal at Bob. We further provide theoretical analyses, illustrating that, given the same signal-to-noise ratio (SNR), random sign reversal plus elementwise phase compensation make the demodulation performance of Eve inferior to that of Bob in most angular regions.

9.1 Signal Models and Overall Design

We first present the system structure and signal model of FH-MIMO DFRC. Different from the system structure in in Chapters 7 and 8 with a single communication user involved, and we have two here, i.e. Bob and Eve mentioned above. Thus, the previous signal model is extended to the two-user signal model. Moreover, we also highlight the difference between the signal models of Bob and Eve. Figure 9.1 illustrates the overall system block diagram. The system consists of an FH-MIMO radar,

Joint Communications and Sensing: From Fundamentals to Advanced Techniques, First Edition.
Kai Wu, J. Andrew Zhang, and Y. Jay Guo.

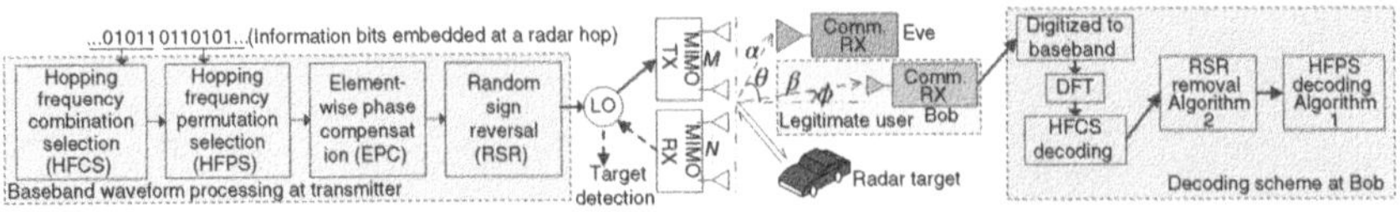

Figure 9.1 System block diagram of an FH-MIMO DFRC, where radar, besides detecting targets, also performs downlink communication through an LoS link with a legitimate user named Bob. Meanwhile, an unintended user, Eve, eavesdrops on the communication between radar and Bob. The presented baseband waveform processing methods highlighted on the left improve both communication secrecy and data rate. As highlighted on the right, the presented demodulation scheme enables Bob to demodulate with low complexity.

a single-antenna communication user called Bob, and a potential eavesdropper called Eve. The radar is equipped with colocated uniform linear arrays as transmitter and receiver. In addition to target detection, the radar also performs downlink data transmission to Bob through a line-of-sight (LoS) channel. This chapter focuses on developing information embedding and demodulation schemes to realize secure and high data rate communications between the DFRC transmitter and Bob. Thus, we assume that the AoD of Bob is available at the radar. Also, we consider a practical scenario that the channel information of Eve is unknown to either the radar or Bob.

9.1.1 Signal Model of Bob

For ease of reading, we briefly recapitulate the signal model of the FH-MIMO radar illustrated in Section 2.3.3. Assume that the radar has M transmitter antennas and N receiver antennas. Each radar pulse is divided into H subpulses, also known as hops. Each hop has a time duration of T. The radar frequency band with bandwidth B is divided evenly into K subbands. The kth ($k = 0, 1, \dots, K-1$) subband has the baseband center frequency kB/K. At hop h and antenna m, the FH-MIMO radar-transmitted signal is

$$s_{hm}(t) = e^{j2\pi \frac{k_{hm}B}{K} t}, \quad 0 \le t - hT \le T, \tag{9.1}$$

where k_{hm} is the index of the subband selected for antenna m at hop h. To ensure waveform orthogonality, the following constraints are generally imposed on radar parameters [1–3]

$$k_{hm} \neq k_{hm'} \; \forall m \neq m', \quad BT/K = \mathrm{r}(\geq 1), \tag{9.2}$$

where r denotes a constant integer. Refer to Section 2.3.3 for the receiving processing on an FH-MIMO radar.

Denote the AoD of the LoS path between radar and Bob as ϕ and the complex channel gain of the path as β. Each hop of signals received at Bob is sampled into

L digital samples. Based on (9.1), the ith baseband signal sample received at Bob is given by

$$y_h(i) = \beta \sum_{m=0}^{M-1} e^{-jmu_\phi} e^{j2\pi r i k_{hm}/L} + \xi(i), \tag{9.3}$$

where $u_\phi = \frac{2\pi d \sin\phi}{\lambda}$ is referred to as *beamspace AoD* and $\xi(i)$ is additive white Gaussian noise (AWGN). Here, d is the antenna spacing of the radar transmitter array and λ is the radar wavelength. Taking the L-point discrete Fourier transform (DFT) of $y_h(i)$, the result at the lth discrete frequency is

$$Y_h(l) = \beta \sum_{m=0}^{M-1} e^{-jmu_\phi} \delta\left(l - rk_{hm}\right) + \Xi(l), \tag{9.4}$$

where $\delta(x)$ denotes the Kronecker delta function and $\Xi(l)$ is the DFT of $\xi(i)$, i.e. $\Xi(l) = \frac{1}{L}\sum_{i=0}^{i=L-1} \xi(i) e^{j\frac{2\pi l i}{L}}$ $\forall l$. Note that $\delta(x)$ takes one at $x = 0$ and zero elsewhere.

According to the waveform orthogonality imposed by (9.2), there are M different hopping frequencies per hop, which leads to M nonzero values of the Kronecker delta function in (9.4). Therefore, by detecting M peaks in $|Y_h(l)|$, the hopping frequenciesused at hop h can be identified. Let l_m^* $(m = 0, 1, \ldots, M-1)$ denote the index of the mth peak of $|Y_h(l)|$, satisfying

$$l_0^* < l_1^* < \cdots < l_{M-1}^*. \tag{9.5}$$

The M hopping frequenciescorresponding to the M peaks can be collected by the following set:

$$\mathcal{K}_h = \left\{l_0^*/r, l_1^*/r, \ldots, l_{M-1}^*/r\right\}. \tag{9.6}$$

Here, $\mathcal{K}_h$ is a combination set of taking M out of K hopping frequencies and is used as the hopping-frequency code selection (HFCS) subsymbol, as will be introduced in Section 9.1.3. From the above illustration, we notice that detecting $\mathcal{K}_h$ is independent of the specific entries in the set. This is because the waveform orthogonality of the FH-MIMO radar, as illustrated in (9.2), makes the signals orthogonal in the frequency domain, as manifested in (9.4). Hence, detecting the DFT peaks for identifying the entries in $\mathcal{K}_h$ (i.e. for demodulating an HFCS subsymbol) solely relies on the SNR in $Y_h(l)$ given in (9.4).

9.1.2 Signal Model of Eve

As FH-MIMO radar transmits signals in a wide angular range, Eve can receive and process radar signals as Bob does. Eve also takes an L-point DFT per hop. Let θ denote the AoD of Eve, α the channel gain and $X_h(l)$ the DFT result at the lth discrete frequency. With reference to (9.3) and (9.4), we can express $X_h(l)$ as

$$X_h(l) = \alpha \sum_{m=0}^{M-1} e^{-jmu_\theta} \delta\left(l - \mathrm{r}k_{hm}\right) + Z(l), \tag{9.7}$$

where $u_\theta = \frac{2\pi d \sin\theta}{\lambda}$ denotes the beamspace AoD of the LoS path between radar and Eve, and $Z(l)$ is AWGN at Eve.

Comparing (9.4) and (9.7), we see that the Kronecker delta function takes nonzero values at the same discrete frequencies, i.e. at $l_m^* \ \forall m$ given in (9.5). Without noise, the amplitudes of nonzero $|Y_h(l)|$ and $|X_h(l)|$ are $|\beta|$ and $|\alpha|$, respectively. Clearly, the powers of useful signals received at Bob and Eve are independent of their LoS AoDs with regards to radar. That means $\mathcal{K}_h$ can be readily identified at Eve, as described in Section 9.1.1 for Bob. This elaborates the low-physical layer security of solely using $\mathcal{K}_h$ for FH-MIMO DFRC.

9.1.3 Overall Description

To achieve a secure and high-speed FH-MIMO DFRC, we can employ the baseband waveform processing, as illustrated in Figure 9.1. There are four modules in the processing, including HFCS, hopping frequency permutation selection (HFPS), elementwise phase compensation, and random sign reversal. Below, we illustrate the first two in detail and then provide the motivations for developing the remaining two modules (which will be presented in Sections 9.2 and 9.3).

Besides combinations of hopping frequencies, we also use the permutation of hopping frequencies to convey information bits. Referring to Figure 9.1, we divide each communication symbol into two subsymbols. One is used for HFCS, which selects one out of C_K^M combinations of hopping frequencies for a radar hop. Given M hopping frequenciesthere are $M!$ permutations of the frequencies, each providing a pairing between hopping frequenciesand antennas. Thus, we use the second subsymbol to perform HFPS.

Remark 9.1 The joint use of HFCS and HFPS is, in essence, equivalent to permuting M out of K available hopping frequencies since $A_K^M = C_K^M \times M!$. Decoupling "A_K^M" into the "C_K^M" (HFCS) and "$M!$" (HFPS) is consistent with the way that the information is demodulated at Bob, as will be developed in Sections 9.2 and 9.3. Intuitively, this is because hopping frequencies (or equivalently HFCS) need to be identified (demodulated) first, either for a single modulation based on A_K^M or the adopted two-step modulation, i.e. first HFCS (C_K^M) and then HFPS ($M!$). As seen from (9.3), Bob only receives the superimposed single-tone sinusoidal signals. Therefore, hopping frequencies must be identified to separate signals from different radar transmitter antennas in the frequency domain. This then allows us to extract useful information for the demodulation of other communication subsymbols, e.g. HFPS, as will be developed in Section 9.2.2.

A straightforward benefit of introducing HFPS is the increased data rate given a large number of permutations. (This is to be illustrated shortly in Section 9.1.4.) However, the potential of HFPS in terms of rate enhancement is nontrivial to achieve in practice, due to the challenging AoD-dependent issue in demodulating HFPS, as will be illustrated in Section 9.2.1. To solve the issue, we devise an elementwise phase compensation processing, based on which an algorithm is designed for Bob to efficiently decode the HFPS subsymbol. This will be elaborated on in Section 9.2.2. Due to elementwise phase compensation, HFPS demodulation at Eve relies on not only her own AoD but also that of Bob. To this end, combining HFPS and elementwise phase compensation can enhance the physical layer security of FH-MIMO DFRC, particularly when Eve does not know the AoD of Bob. This will be illustrated in Section 9.2.3.

Considering that a powerful Eve can manage to acquire the AoD of Bob, we further present random sign reversal, introducing more randomness in communication symbols. Such randomness is unknown to either Bob or Eve, hence, increasing the difficulty for communication demodulation. Random sign reversal will be developed in Section 9.3.1. For Bob, owing to elementwise phase compensation, there is a sign rule available for removing random sign reversal with low complexity, as will be illustrated in Section 9.3.2. For Eve, however, even with maximum likelihood (ML) demodulation, her performance is no better than that of Bob under the same SNR and can only achieve the same performance as Bob at a limited number of angles. This will be analyzed in Section 9.3.3. Note that elementwise phase compensation and random sign reversal are only for HFPS (which is demodulated after HFCS demodulation). Therefore, elementwise phase compensation and random sign reversal do not change the demodulation performance of HFCS.

9.1.4 Maximum Achievable Rate (MAR)

Before developing the overall FH-MIMO DFRC scheme, we analyze its maximum achievable rate (MAR), which is defined as the upper limit of the achievable data rate and has been widely used in prior works [2–7]. These works are used as benchmark designs below. As illustrated in Section 9.1.3, the number of information bits, as conveyed by HFCS and HFPS per symbol, are up to $\lfloor \log_2 C_K^M \rfloor$ and $\lfloor \log_2(M!) \rfloor$, respectively. Then, the number of bits conveyed per radar pulse is given by $H(\lfloor \log_2 C_K^M \rfloor + \lfloor \log_2(M!) \rfloor)$, since there are H hops (i.e. communication symbols) per radar pulse. Thus, the MAR of the design in this chapter, as given in

Table 9.1 Maximum achievable rates (MARs) of different radar-based DFRC designs.

	HFCS [2]	$H\lfloor \log_2 C_K^M \rfloor f_r$
	Current	$H\left(\lfloor \log_2 C_K^M \rfloor + \lfloor \log_2 M! \rfloor\right) f_r$
DSSS	DSSS [4]	$\rho_{dc} BMJ_{PSK}/N_c$
NSS	quadrature amplitude modulation (QAM) [5]	$M\lfloor \log_2 J_{QAM} \rfloor f_r$
	Waveform permutation (WP) [6]	$\lfloor \log_2 M! \rfloor f_r$

Table 9.1, can be achieved, where f_r denotes the pulse repetition frequency (PRF) of the underlying FH-MIMO radar.

Similarly, we can obtain the MARs of FH-MIMO DFRC systems based on phase shift keying (PSK) [7] and HFCS [2], as given in Table 9.1. For the PSK-based FH-MIMO DFRC, a constant phase term $e^{j\varpi_{hm}}$, $\varpi_{hm} \in \{0, \frac{2\pi}{2^{J_{PSK}}}, \dots, \frac{2\pi(2^{J_{PSK}}-1)}{2^{J_{PSK}}}\}$, is multiplied onto the baseband radar signal to be transmitted by antenna m at hop h, i.e. $s_{hm}(t)$ given in (9.1). Here, J_{PSK} is the number of bits per PSK symbol.

Besides FH, direct sequence spread spectrum (DSSS) has also been considered for DFRC; see [4] for a comprehensive overview. For fair comparison, the MAR of the DSSS-based DFRC, as given in [4], is tailored according to our radar configuration. In the DSSS-based DFRC [4], each PSK symbol is spread by an N_c-length spreading code. Thus, for a single-antenna transmitter with bandwidth B, the MAR can be given by $\frac{J_{PSK}}{N_c/B}$. Note that the effective transmission time of a radar pulse is scaled by ρ_{dc} (the duty cycle) and also M orthogonal spreading codes can be used if M antennas are transmitting simultaneously. Accordingly, we obtain the MAR of the DSSS-based DFRC, as given in Table 9.1.

Table 9.1 also includes two nonspread spectrum (NSS) radar-based DFRC designs as benchmark DFRC systems, where the bandwidth of an NSS radar can be narrower than that of an FH radar, and a single information symbol is conveyed per radar pulse. In particular, NSS-QAM [5] transmits QAM symbols through the amplitudes and phases of the optimized sidelobes of the beam pattern of an antenna array. Several orthogonal signals, whose number is taken as M for a fair comparison, are transmitted simultaneously, each associated with their QAM symbol, thus resulting in the MAR in Table 9.1, where J_{QAM} denotes the number of bits per QAM symbol. NSS-WP [6] exploits the permutations of the orthogonal MIMO waveforms as communication symbols. Given M waveforms of NSS-WP, there will be $M!$ numbers of permutations of the waveforms, leading to the MAR given in Table 9.1.

9.2 Elementwise Phase Compensation

In this section, elementwise phase compensation is presented. Through elaborating on the AoD-dependent issue in HFPS demodulation, we develop elementwise phase compensation first. Then, the potential of using elementwise phase compensation to enhance physical layer security is unveiled.

9.2.1 AoD-Dependence Issue of Hopping Frequency Permutation Selection (HFPS) Demodulation

To show the AoD dependence of detecting HFPS, i.e. $\mathbf{k}_h$, we first formulate the detection problem. By multiplying $Y_h(l)$ given in (9.4) with β^*, the remaining exponential term e^{-jmu_ϕ} attached to the l_m^*th peak is the mth element of the following steering vector:

$$\mathbf{a}_\phi = [1, e^{-ju_\phi}, \ldots, e^{-j(M-1)u_\phi}]^{\mathrm{T}}. \tag{9.8}$$

Stacking the L peaks of $Y_h(l)$ into a vector, we have

$$\mathbf{y}_h = [Y_h(l_0^*), Y_h(l_1^*), \ldots, Y_h(l_{M-1}^*)]^{\mathrm{T}}.$$

Based on (9.4) and (9.5), we see that $\mathbf{y}_h$ is a permutation of the β-scaled $\mathbf{a}_\phi$, i.e.

$$\mathbf{y}_h = \beta \mathbf{P}_h \mathbf{a}_\phi + \mathbf{v}, \tag{9.9}$$

where $\mathbf{P}_h$ is a permutation matrix and $\mathbf{v}$ collects M independent noises $\Xi(l_m^*)\ \forall m$. Multiplying $\mathbf{P}_h^{\mathrm{T}}$ to both sides of (9.9) gives

$$\mathbf{P}_h^{\mathrm{T}} \mathbf{y}_h = \beta \mathbf{P}_h^{\mathrm{T}} \mathbf{P}_h \mathbf{a}_\phi + \mathbf{P}_h^{\mathrm{T}} \mathbf{v} = \beta \mathbf{a}_\phi + \mathbf{P}_h^{\mathrm{T}} \mathbf{v}, \tag{9.10}$$

where $\mathbf{P}_h^{\mathrm{T}} \mathbf{P}_h = \mathbf{I}$ is due to the orthogonal property of a permutation matrix. By comparing the two sides of (9.10) in a pointwise manner, the hopping frequenciesattached to the M antennas can be expressed as follows:

$$\mathbf{k}_h = \mathbf{P}_h^{\mathrm{T}} [l_0^*/\mathsf{r}, l_1^*/\mathsf{r}, \ldots, l_{M-1}^*/\mathsf{r}]^{\mathrm{T}}, \tag{9.11}$$

where l_m^*/r ($\forall m$), as given in (9.6), is the hopping frequency of the mth peak. Equation (9.11) indicates that the HFPS demodulation can be performed by identifying the permutation matrix $\mathbf{P}_h$.

According to (9.10), the following detector can be formulated to identify $\mathbf{P}_h$,

$$\mathbf{P}_h : \min_{\mathbf{P} \in \mathcal{P}} g(\mathbf{P}) \quad \text{s.t. } g(\mathbf{P}) = \|\beta \mathbf{a}_\phi - \mathbf{P}^{\mathrm{T}} \mathbf{y}_h\|_2^2, \tag{9.12}$$

where $\mathcal{P}$ is the set of possible permutation matrices. From (9.12), we see that the detecting performance of $\mathbf{P}_h$ varies with the AoD of Bob, i.e. ϕ. This leads to the AoD-dependent issue of HFPS demodulation. In particular, given a small ϕ, the distance between adjacent elements in $\mathbf{a}_\phi$ is also small, leading to a high-error probability of detecting $\mathbf{P}_h$ and a demodulating HFPS. This issue can be relieved with a large ϕ. However, a large ϕ can lead to an ambiguity issue. When $u_\phi > \frac{2\pi}{M}$, it can happen that the phases of e^{-jmu_ϕ} at two or more different values of m are identical.

9.2.2 Elementwise Phase Compensation and HFPS Demodulation at Bob

We see from (9.8) that the elements of $\mathbf{a}_\phi$ are similar to the constellation points of the M-PSK modulation. Thus, when e^{-jmu_ϕ}, $m = 0, 1, \ldots, M-1$, are uniformly distributed on the unit circle, the best detecting performance can be achieved [6]. In light of this, we introduce elementwise phase compensation to compensate the phase of each antenna-transmitted signal so that $\mathbf{a}_\phi$ given in (9.8) becomes the following AoD-independent steering vector

$$\mathbf{a} = [1, e^{-j2\pi/M}, \ldots, e^{-j2\pi(M-1)/M}]^{\mathrm{T}}. \tag{9.13}$$

Based on (9.1), (9.3), and (9.4), the following elementwise phase compensation is introduced for the mth antenna at hop h,

$$\tilde{s}_{hm} = s_{hm}(t)e^{-jm(2\pi/M-u_\phi)}. \tag{9.14}$$

We add a tilde sign above the relevant variables to reflect the impact of elementwise phase compensation. That is, $y_h(l)$, $Y_h(l)$, $\mathbf{y}_h$ and $g(\mathbf{P})$, as given in (9.3), (9.4), (9.9), and (9.12), respectively, are now denoted by $\tilde{y}_h(l)$, $\tilde{Y}_h(l)$, $\tilde{\mathbf{y}}_h$ and $\tilde{g}(\mathbf{P})$. Note that

- $\tilde{y}_h(l)$ has the same expression as $y_h(l)$ except that u_ϕ is replaced by $2\pi/M$. The same goes for $\tilde{Y}_h(l)$;
- $\tilde{\mathbf{y}}_h$ replaces $\mathbf{a}_\phi$ in (9.9) with $\mathbf{a}$ given in (9.13). The same permutation matrix is applicable for both $\tilde{\mathbf{y}}_h$ and $\mathbf{y}_h$, since elementwise phase compensation does not change the indexes of peaks in $\tilde{Y}_h(l)$.

Jointly considering the above changes caused by elementwise phase compensation, problem (9.12) can be equivalently turned into

$$\min_{\mathbf{P}\in\mathcal{P}} \ \tilde{g}(\mathbf{P}) = \|\beta\mathbf{a} - \mathbf{P}^{\mathrm{T}}\tilde{\mathbf{y}}_h\|_2^2, \tag{9.15a}$$

$$\Leftrightarrow \min_{\mathbf{P}\in\mathcal{P}} \ \|\beta\mathbf{a}\|_2^2 + \|\tilde{\mathbf{y}}_h\|_2^2 - 2\Re\left\{\beta^*\mathbf{a}^{\mathrm{H}}\mathbf{P}^{\mathrm{T}}\tilde{\mathbf{y}}_h\right\}, \tag{9.15b}$$

$$\Leftrightarrow \max_{\mathbf{P}\in\mathcal{P}} \ \Re\left\{\beta^*\mathbf{a}^{\mathrm{H}}\mathbf{P}^{\mathrm{T}}\tilde{\mathbf{y}}_h\right\}, \tag{9.15c}$$

where the ℓ_2-norm in (9.15a) is expanded to get (9.15b), and the first two terms in (9.15b) are suppressed to get (9.15c) since they are independent of the optimization variable $\mathbf{P}$. The maximization of (9.15c) is achieved when the pointwise phase difference between $\mathbf{a}$ and $\beta^*\mathbf{P}^{\mathrm{T}}\tilde{\mathbf{y}}_h$ is minimized. Due to elementwise phase compensation, the element phases of $\mathbf{a}$ are in descending order for sure. This implies that the correct $\mathbf{P}$ needs to sort the element phases of $\beta^*\mathbf{P}_h^{\mathrm{T}}\tilde{\mathbf{y}}_h$ in descending order as well.

Based on the above analysis, the optimal solution to (9.15c) is obtained as follows:

$$[\mathbf{P}_h]_{[\mathbf{i}]_m,m} = 1 \ (m = 0, 1, \dots, M-1), \tag{9.16}$$

where $\mathbf{P}_h$ is initialized as $\mathbf{0}_{M\times M}$, and the $M \times 1$ vector $\mathbf{i}$ collects the arrangements of the element phases of $\beta^*\tilde{\mathbf{y}}_h$ into the sorted version. The solution states that only entries at the $[\mathbf{i}]_m$th row and the mth column of $\mathbf{P}$ take ones. We notice that the performance of detecting $\mathbf{P}$, i.e. of HFPS demodulation, is independent of HFPS constellations (the set of $\mathbf{P}$ selected for DFRC). This is because elementwise phase compensation turns the HFPS demodulation into a simple sorting problem that is solely affected by the SNR in $\tilde{\mathbf{y}}_h$; see (9.15).

Based on the solution obtained in (9.16), Algorithm 9.1 is established to decode HFPS at Bob. In Step 2, directly taking the element angles makes the values in $\vec{\omega}$ fall in the region of $[-\pi, \pi]$. Thus, Step 3 recovers true angles by compensating -2π on elements larger than 0. This is because all the element angles of $\beta^*\tilde{\mathbf{y}}_h$ should be equal to those of $\mathbf{a}$ given in (9.13) in the absence of noise, and taking angle in Step 2 adds 2π onto any angle smaller than $-\pi$. Corrupted in noises, the phase of $[\beta^*\tilde{\mathbf{y}}_h]_0$ (which is zero without noise) can become a small positive value which will then be revised through Step 3 into a small negative value close to -2π. This phenomenon, known as zigzag [8], turns the largest element angle of $\beta^*\tilde{\mathbf{y}}_h$ into the smallest one. Step 5 is introduced to remove potential zigzag by comparing the Euclidean distances $\|e^{\mathrm{j}[\vec{\omega}]_{[\mathbf{i}]_{M-1}}} - 1\|_2$ and $\|e^{\mathrm{j}[\vec{\omega}]_{[\mathbf{i}]_0}} - 1\|_2$. Based on (9.11) and (9.16), Steps 6 and 7 produce $\mathbf{k}_h$ which is finally used for HFPS demodulation in Step 8.

Algorithm 9.1 has a low-computational complexity, since no computationally intensive operation (e.g. matrix inversion/decomposition) is involved. The major computations involved are the element sorting in Step 4, whose complexity is in

Algorithm 9.1 HFPS demodulation at Bob

1: **Input**: $\tilde{\mathbf{y}}_h$, β, $l_m^*(\forall m)$ and r (given in (9.2));
2: Take the element angles of $\beta^*\tilde{\mathbf{y}}_h$, and stack them $\vec{\omega}$;
3: At $\forall m$, if $[\vec{\omega}]_m > 0$, $[\vec{\omega}]_m = [\vec{\omega}]_m - 2\pi$;
4: Sorting the elements of $\vec{\omega}$ in descending order gives $\mathbf{i}$;
5: If $\|e^{\mathrm{j}[\vec{\omega}]_{[\mathbf{i}]_{M-1}}} - 1\|_2 < \|e^{\mathrm{j}[\vec{\omega}]_{[\mathbf{i}]_0}} - 1\|_2$, shift $\mathbf{i}$ circularly by a single element;
 ▷ Similar to "circshift()" in MATLAB [9];
6: Substitute $\mathbf{i}$ into (9.16) to construct $\mathbf{P}_h$;
7: Substituting $\mathbf{P}_h$, l_m^* and r into (9.11) leads to $\mathbf{k}_h$;
8: Look up $\mathbf{k}_h$ in the constellation set to decode HFPS.

the order of $\mathcal{O}(M \log M)$ on average [10]. Hence, the overall computational complexity of Algorithm 9.1 is of order $\mathcal{O}(M \log M)$. It is worth noting that the low complexity of Algorithm 9.1 largely owes to elementwise phase compensation, which turns the complexity-intensive searching into a simple sorting when solving the ML problem formulated in (9.12).

9.2.3 Enhancing Physical-Layer Security by Elementwise Phase Compensation

As the phase compensation of elementwise phase compensation is determined based on the AoD of Bob, the received signal at Eve is still AoD-dependent. To this end, we can conclude that elementwise phase compensation helps enhance the physical-layer security of FH-MIMO DFRC. Taking into account elementwise phase compensation in the frequency-domain signal received at Eve, i.e. $X_h(l)$ given in (9.7), the signal can be rewritten into

$$\tilde{X}_h(l) = \alpha \sum_{m=0}^{M-1} e^{-jm(\frac{2\pi}{M}+u_\theta-u_\phi)} \delta\left(l - rk_{hm}\right) + Z(l). \tag{9.17}$$

Again, we notice that detecting the set of hopping frequencies, i.e. $\mathcal{K}_h$, at Eve is not affected by AoD. The same set $\{l_m^* \ \forall m\}$ given in (9.6) can be identified at Eve as that obtained at Bob. Similar to $\mathbf{y}_h$ given in (9.9), the following vector can be obtained at Eve,

$$\tilde{\mathbf{x}}_h = [\tilde{X}_h(l_0^*), \tilde{X}_h(l_1^*), \dots, \tilde{X}_h(l_{M-1}^*)]^{\mathrm{T}}. \tag{9.18}$$

With reference to (9.10)–(9.12), the following detector is formulated at Eve to decode HFPS by identifying the permutation matrix $\mathbf{P}_h$,

$$\mathbf{P}_h : \min_{\mathbf{P}\in\mathcal{P}} f(\mathbf{P})$$
$$\text{s.t. } f(\mathbf{P}) = \begin{cases} \|\alpha \mathbf{a}_{\theta\phi} - \mathbf{P}^{\mathrm{T}}\tilde{\mathbf{x}}_h\|_2^2, & \text{if Eve knows } u_\phi, \\ \|\alpha \mathbf{a}_{\theta} - \mathbf{P}^{\mathrm{T}}\tilde{\mathbf{x}}_h\|_2^2, & \text{otherwise,} \end{cases} \tag{9.19}$$

where the two steering vectors are

$$\mathbf{a}_{\theta\phi} = [1, e^{-j(\frac{2\pi}{M}+u_\theta-u_\phi)}, \dots, e^{-j(M-1)(\frac{2\pi}{M}+u_\theta-u_\phi)}]^{\mathrm{T}};$$
$$\mathbf{a}_{\theta} = [1, e^{-j(\frac{2\pi}{M}+u_\theta)}, \dots, e^{-j(M-1)(\frac{2\pi}{M}+u_\theta)}]^{\mathrm{T}}. \tag{9.20}$$

Note that the same $\mathbf{P}_h$ is required for Eve and Bob to decode HFPS since $\mathbf{P}_h$ is added to the radar as part of the waveform.

From (9.19), we see that elementwise phase compensation makes demodulating HFPS at Eve rely on not only the AoD of Eve u_θ but also that of Bob u_ϕ. If u_ϕ is unknown to Eve, the demodulating performance can degrade drastically, since the actual steering vector contained in $\tilde{\mathbf{x}}_h$ is $\mathbf{a}_{\theta\phi}$. A powerful eavesdropper may manage to know u_ϕ, which can somewhat reduce the secrecy enhancement brought by elementwise phase compensation.

9.3 Random Sign Reversal

To further enhance the secrecy of FH-MIMO DFRC, we introduce random sign reversal to make communication demodulation more challenging at Eve. *In the following, we add a breve sign above relevant variables to indicate random sign reversal processing, e.g.* $\breve{s}_{hm}(t)$ *corresponding to the elementwise phase compensation-processed* $\tilde{s}_{hm}(t)$ *given in* (9.14) *and the original radar waveform* $s_{hm}(t)$ *given in* (9.3).

9.3.1 Random Sign Reversal and Maximum Likelihood (ML) Decoding

Random sign reversal is a waveform processing technique designed for the radar transmitter. Before the elementwise phase compensation-processed signals are transmitted, the radar randomly selects several antennas and reverses the signs of their signals. *The random selection, not known to either Bob or Eve, independently changes over hops.* Let b_{hm} denote a binary coefficient for the mth antenna at hop h. If antenna m is selected for random sign reversal, we have $b_{hm} = -1$; otherwise, $b_{hm} = 1$. Denoting the number of sign-reversed antennas as Q, we have

$$\mathbb{P}\{b_{hm} = 1\} = (M - Q)/M; \quad \mathbb{P}\{b_{hm} = -1\} = Q/M, \tag{9.21}$$

where $\mathbb{P}\{\}$ denotes taking probability. Multiplying b_{hm} onto $\tilde{s}_{hm}(t)$, as given in (9.14), leads to

$$\breve{s}_{hm}(t) = b_{hm}\tilde{s}_{hm}(t) = b_{hm}s_{hm}(t)e^{-jm(2\pi/M - u_\phi)}. \tag{9.22}$$

With $\breve{s}_{hm}(t)$ transmitted by the radar, the Bob-received signal, as given in (9.3), can be rewritten by replacing e^{-jmu_ϕ} with $b_{hm}e^{-jmu_\phi}$. The same applies to the frequency-domain signal given in (9.4). Accordingly, the vector $\mathbf{y}_h$, as given in (9.9) and comprised of the M peaks in the frequency-domain signal, becomes

$$\breve{\mathbf{y}}_h = \beta\mathbf{P}_h\breve{\mathbf{a}} + \mathbf{v}, \quad \text{s.t. } \breve{\mathbf{a}} = \left(\mathbf{b}_h \odot \mathbf{a}\right), \tag{9.23}$$

where $\mathbf{b}_h = [b_{h0}, b_{h1}, \dots, b_{h(M-1)}]^{\mathrm{T}}$ and $\mathbf{v}$ collects the AWGNs, i.e. $\Xi(l_m^*)\ \forall m$ given in (9.4). Note that $\mathbf{a}_\phi$ in (9.9) has been replaced with $\mathbf{a}$ given in (9.13) after elementwise phase compensation. Similarly, $\tilde{\mathbf{x}}_h$, as given in (9.18) and used by Eve for information decoding, can be rewritten into

$$\breve{\mathbf{x}}_h = \alpha\mathbf{P}_h\left(\mathbf{b}_h \odot \mathbf{a}_{\theta\phi}\right) + \mathbf{z}, \tag{9.24}$$

where $\mathbf{z}$ collects $Z(l_m^*)$ $(m = 0, 1, \dots, M - 1)$.

From (9.23) and (9.24), we see that random sign reversal introduces an extra unknown, i.e. $\mathbf{b}_h$, to both Bob and Eve. As a result, the communication demodulation at Bob and Eve requires the estimations of both $\mathbf{P}_h$ and $\mathbf{b}_h$. For Bob, its original ML detection, as given in (9.15), can be revised as follows:

$$\mathbf{P}_h : \min_{\mathbf{P}\in\mathcal{P},\mathbf{b}\in\mathcal{B}} \breve{g}(\mathbf{P},\mathbf{b}) = \|\beta\mathbf{a} - \mathbf{b}\odot\mathbf{P}^{\mathrm{T}}\breve{\mathbf{y}}_h\|_2^2, \tag{9.25}$$

where $\mathcal{B}$ denotes the set of possible values of $\mathbf{b}$ and $\breve{\mathbf{y}}_h$ is given in (9.23). Similarly, the ML detection at Eve, as originally formulated in (9.19), becomes

$$\mathbf{P}_h : \min_{\mathbf{P}\in\mathcal{P},\mathbf{b}\in\mathcal{B}} \breve{f}(\mathbf{P},\mathbf{b})$$
$$\text{s.t. } \breve{f}(\mathbf{P},\mathbf{b}) = \begin{cases} \|\alpha\mathbf{a}_{\theta\phi} - \mathbf{b}\odot\mathbf{P}^{\mathrm{T}}\breve{\mathbf{x}}_h\|_2^2, & \text{Eve knows } u_\phi, \\ \|\alpha\mathbf{a}_{\theta} - \mathbf{b}\odot\mathbf{P}^{\mathrm{T}}\breve{\mathbf{x}}_h\|_2^2, & \text{otherwise,} \end{cases} \tag{9.26}$$

where $\breve{\mathbf{x}}_h$ is given in (9.24).

As $\mathbf{b}$ introduces more unknowns to the ML demodulation, solving the revised ML problems (9.25) and (9.26) makes the final demodulation performance no greater than solving those in (9.15) and (9.19). Moreover, we will analyze in Section 9.3.3 that, given the same SNR, Bob, which solves (9.25), will always achieve a demodulation performance that is no worse than that of Eve which solves (9.26); only at a few discrete angles can Eve achieve the same performance as Bob. In addition, we notice that the searching space is now enlarged by $|\mathcal{B}|$ times, with $|\mathcal{B}|$ denoting the cardinality of $\mathcal{B}$. The demodulation complexity can be substantially reduced for Bob, as shown below.

9.3.2 Detecting Random Sign Reversal at Bob

Owing to elementwise phase compensation, there is a deterministic sign rule in $\breve{\mathbf{a}}$ – the random sign reversal-processed steering vector as given in (9.23). Specifically, given $b_{hm} = -1$, we have

$$[\breve{\mathbf{a}}]_m = b_{hm}[\mathbf{a}]_m = -e^{-\mathrm{j}\frac{2\pi m}{M}} = e^{\mathrm{j}\frac{2\pi M/2}{M}} e^{-\mathrm{j}\frac{2\pi m}{M}} = [\breve{\mathbf{a}}]_{m\pm\frac{M}{2}},$$

where $m \pm \frac{M}{2}$ depends on $m \lessgtr \frac{M}{2}$. Enabled by the sign rule, Bob can detect random sign reversal by identifying the most similar elements in $\breve{\mathbf{a}}$. There are two noteworthy issues, as illustrated below.

First, when $b_{hm} = -1$ $(m \leq \frac{M}{2} - 1)$ and $b_{h(m+\frac{M}{2})} = -1$ happen simultaneously, random sign reversal turns $[\breve{\mathbf{a}}]_m$ and $[\breve{\mathbf{a}}]_{m+\frac{M}{2}}$ into each other. In this case, two reversed antennas will be identified as a single one. To avoid this, we impose the constraint that $b_{hm} = -1$ and $b_{h(m+\frac{M}{2})} = -1$ cannot happen simultaneously:

$$b_{hm} + b_{h(m+M/2)} \neq -2 \quad \forall m < M/2. \tag{9.27}$$

Second, both $b_{hm} = -1$ and $b_{h(m+\frac{M}{2})} = -1$ lead to $[\breve{\mathbf{a}}]_m = [\breve{\mathbf{a}}]_{m+\frac{M}{2}}$. In turn, $[\breve{\mathbf{a}}]_m = [\breve{\mathbf{a}}]_{m+\frac{M}{2}}$ can be caused by either $b_{hm} = -1$ or $b_{h(m+\frac{M}{2})} = -1$, incurring ambiguity in random sign reversal detection. To remove the ambiguity, we need to enforce a protocol between radar and Bob that random sign reversal only happens on the antenna associated with smaller (or larger) hopping frequencies. This constraint can be expressed as follows:

$$b_{hm^*} = -1 \quad \text{s.t. } m^* : \min_{m,m+M/2} \{k_{hm}, k_{h(m+M/2)}\}. \tag{9.28}$$

Algorithm 9.2 is designed to remove random sign reversal for Bob based on $\breve{\mathbf{y}}_h$ given in (9.23). In Step 2, the power differences between each element in $\mathbf{y}_h$ and all the other elements are calculated. After Steps 3 and 4, the first several elements in $\mathbf{d}_1$ are related to the indexes of random sign reversal antennas. In Step 6, the indexes of the two antennas, whose received signals are most similar in power, are extracted. By comparing the associated hopping frequencies, random sign reversal is detected with the aid of constraint (9.28) and removed by reversing back the sign; see Step 7.

9.3.3 Random Sign Reversal Impact Analysis

Next, we analyze the impact of random sign reversal on Bob and Eve from a statistical perspective, which shows the secrecy enhancement brought by random sign reversal. Consider the case that Eve knows the AoD of Bob, i.e. u_ϕ. Based on (9.26), we can expand $\breve{f}(\mathbf{P}, \mathbf{b})$ as follows:

$$\begin{aligned} \breve{f}(\mathbf{P}, \mathbf{b}) &= \|\alpha \mathbf{a}_{\theta\phi} - \alpha \mathbf{b} \odot \tilde{\mathbf{a}} - \mathbf{P}^{\mathrm{T}}\mathbf{z}\|_2^2 = 2M|\alpha|^2 + \tilde{z} \\ &\quad - \underbrace{2|\alpha|^2 \Re\{\mathbf{a}_{\theta\phi}^{\mathrm{H}} \mathbf{b} \odot \tilde{\mathbf{a}}\}}_{f_r(\mathbf{P},\mathbf{b})}, \quad \text{s.t. } \tilde{\mathbf{a}} = \mathbf{P}^{\mathrm{T}}\mathbf{P}_h \left(\mathbf{b}_h \odot \mathbf{a}_{\theta\phi}\right), \end{aligned} \tag{9.29}$$

Algorithm 9.2 Removing random sign reversal for Bob

1: **Input**: Q, M and $\breve{\mathbf{y}}_h$;
2: Calculate $\mathbf{Y}_h = |\breve{\mathbf{y}}_h \mathbf{1}^{\mathrm{T}} - \mathbf{1}\breve{\mathbf{y}}_h^{\mathrm{T}}|^2$, where $|\cdot|^2$ takes elementwise absolute square;
3: Identify the minimum element (excluding diagonal element) and its index in each row of $\mathbf{Y}_h$, stacking them in $\mathbf{y}_{\min}$ and $\mathbf{d}$, respectively;
4: Sort $\mathbf{y}_{\min}$ in ascending order and denote the index vector from $\mathbf{y}_{\min}$ to the sorted version as $\mathbf{d}_1$;
5: **for** $q = 0 : Q-1$ **do** ▷ Index starts from 0.
6: Take $i = [\mathbf{d}_1]_q$ and $i' = [\mathbf{d}]_i$. Remove i' from $\mathbf{d}_1$;
7: $[\mathbf{y}_h]_{i^*} = -[\mathbf{y}_h]_{i^*}$, s.t. $i^* = \min_{i,i'} \left\{\frac{L-l_i^*}{r}, \frac{L-l_{i'}^*}{r}\right\}$;
8: **end for**
9: **Return** $\tilde{\mathbf{y}}_h = \breve{\mathbf{y}}_h$.

where $\|\mathbf{a}_{\theta\phi}\|_2^2 = M$ and $\|\mathbf{b} \odot \tilde{\mathbf{a}}\|_2^2 = M$ are plugged in, and $\check{z}$ is the sum of noise-related terms. Without noise, the minimization of $\check{f}(\mathbf{P}, \mathbf{b})$ is achieved at $(\mathbf{P}_h, \mathbf{b}_h)$, where $f_r(\mathbf{P}_h, \mathbf{b}_h) = M$ is maximized. Affected by the noise term $\check{z}$, however, $\check{f}(\mathbf{P}, \mathbf{b})$ may be minimized at $(\mathbf{P}, \mathbf{b}) \neq (\mathbf{P}_h, \mathbf{b}_h)$. Moreover, the pairs of $(\mathbf{P}, \mathbf{b})$'s in the following set are most likely to be selected as the solution to (9.26),

$$\Omega_{\text{Eve}} = \{(\mathbf{P}, \mathbf{b}) | f_r(\mathbf{P}, \mathbf{b}) \geq M - \epsilon\},$$

where ϵ is a small value. For Bob, there is a similar set of $(\mathbf{P}, \mathbf{b})$ candidates, as given below

$$\Omega_{\text{Bob}} = \{(\mathbf{P}, \mathbf{b}) | f_r(\mathbf{P}, \mathbf{b}) \geq M - \epsilon,\ u_\theta = u_\phi\},$$

where $\check{f}(\mathbf{P}, \mathbf{b})$ becomes equivalent to $\check{g}(\mathbf{P}, \mathbf{b})$ by taking $u_\theta = u_\phi$; see (9.25) and (9.26). Given the same ϵ, the numbers of entries in Ω_{Bob} and Ω_{Eve} determine the error probability of detecting $(\mathbf{P}_h, \mathbf{b}_h)$; the more entries, the higher the error probability. Below, we show that, given the same ϵ, the number of entries in Ω_{Eve} is no smaller than that in Ω_{Bob} across the whole angular region. The result will be validated by the simulation in Figure 9.6.

From (9.23), we see that $\tilde{\mathbf{a}}$ is a constant-modulus vector. Thus, we can express $f_r(\mathbf{P}, \mathbf{b})$ as

$$f_r(\mathbf{P}, \mathbf{b}) = \Re\left\{\sum_{m=0}^{M-1} b_m e^{j\Delta_{hm}}\right\},$$

where $b_m = [\mathbf{b}]_m$ and Δ_{hm} is the difference between the phases of $[\mathbf{a}_{\theta\phi}]_m$ and $[\tilde{\mathbf{a}}]_m$, i.e.

$$\Delta_{hm} = m(2\pi/M + u_{\theta\phi}) + \arg\left\{[\tilde{\mathbf{a}}]_m\right\}, \tag{9.30}$$

where $\mathbf{a}_{\theta\phi}$ is given in (9.19) and $u_{\theta\phi} = u_\theta - u_\phi$. As $b_0, b_1, \ldots, b_{M-1}$ are independent Bernoulli-like variables, a weighted sum of them with constant-modulus weights, i.e. $\sum_{m=0}^{M-1} b_m e^{j\Delta_{hm}}$, approaches to a normally distributed variable according to the central limit theorem (CLT) [11]. The real part of the sum, i.e. $f_r(\mathbf{P}, \mathbf{b})$, also conforms to a normal distribution, i.e. $f_r(\mathbf{P}, \mathbf{b}) \sim \mathcal{N}(\mu_f, \sigma_f^2)$.

As derived in Appendix A.10, the mean and variance of the normal distribution are

$$\mu_f = \frac{(M-2Q)^2}{M^3} S_M^2(u_{\theta\phi}); \quad \sigma_f^2 = \frac{\left(M - \left(\frac{M-2Q}{M}\right)^2 \mu_f\right)}{2}, \tag{9.31}$$

where $S_M(u_{\theta\phi})$ is defined as

$$S_M(u_{\theta\phi}) = \sin\left(\frac{M\left(u_{\theta\phi} + \frac{2\pi}{M}\right)}{2}\right) \Big/ \sin\left(\frac{\left(u_{\theta\phi} + \frac{2\pi}{M}\right)}{2}\right).$$

Refer to $f_{\mathrm{r}}(\mathbf{P}, \mathbf{b}) \geq M - \epsilon$ as the event $\mathcal{E}$. The probability of the event can be given by

$$\mathbb{P}\{\mathcal{E}\} = \frac{\text{erfc}\left\{h(\mu_f)\right\}}{2} \text{ with } h(\mu_f) = \frac{(M-\epsilon)-\mu_f}{\sqrt{M-\left(\frac{M-2Q}{M}\right)^2\mu_f}}, \tag{9.32}$$

where σ_f^2 given in (9.31) has been plugged in and erfc$\{\cdot\}$ denotes the complementary error function.

Remark 9.2 The first derivative of $h(\mu_f)$ w.r.t. μ_f is

$$h'(\mu_f) = -\frac{\left(2M - \left(\frac{M-2Q}{M}\right)^2(M+\mu_f) + \epsilon\left(\frac{M-2Q}{M}\right)^2\right)}{2(2\sigma_f^2)^{3/2}}.$$

Using (9.31), we can validate $\mu_f \leq M$. Applying this in the numerator of $h'(\mu_f)$, we have $M + \mu_f \leq 2M$. Since the coefficient of $(M + \mu_f)$ is constantly less than one, we see that the numerator of $h'(\mu_f)$ is always positive. This makes $h'(\mu_f) < 0$ hold constantly.

Note that $\mathbb{P}\{\mathcal{E}\}$ is minimized at $\mu_f = 0$ which, according to (9.31), is achieved at $u_\theta = u_\phi$ and hence achieved by Bob. As analyzed in Remark 9.2, $h(\mu_f)$ is a monotonically decreasing function of μ_f. This means $h(\mu_f)$ is maximized at $\mu_f = 0$. Since erfc$\{\cdot\}$ is a decreasing function, $\mathbb{P}\{\mathcal{E}\}$ is then minimized at $\mu_f = 0$. From (9.31), we see that $\mu_f = 0$ can also be achieved at $u_{\theta\phi} = 2\pi m/M,\ m = 1, \dots, M-1$. However, these angles can lead to identical entries in the steering vector of Eve and hence, the ambiguity in detecting $\mathbf{P}_h$ for demodulation. Take $u_{\theta\phi} = 2\pi(M-1)/M$ for an example. Substituting the angle into (9.20), $\mathbf{a}_{\theta\phi}$ becomes an all-one vector. This invalidates the ML detection of $\mathbf{P}_h$ at Eve; see (9.26). Since we need the phase differences among the entries in the steering vector for $\mathbf{P}_h$ detection, these angels will yield the worst demodulation performance for Eve.

Since the overall sets of **P**'s and **b**'s are the same for Bob and Eve, the minimization of $\mathbb{P}\{\mathcal{E}\}$ at $u_\theta = u_\phi$ also means the number of entries in Ω_{Bob} is no greater than that in Ω_{Eve} across the whole angular region. As mentioned earlier, the number of entries in the two sets determines the error probability of demodulating HFPS symbols. Thus, we can translate the above analysis into the following result:

Proposition 9.1 *Given the same SNR, random sign reversal plus element-wise phase compensation makes the demodulation performance of Eve upper bounded by that of Bob in the whole angular region, with the bound attained only at $u_\theta = u_\phi$, where u_θ is the AoD of Eve and u_ϕ the AoD of Bob.*

Proposition 9.1 provides a theoretical guarantee of the secrecy enhancement brought by random sign reversal and elementwise phase compensation. It also points out that the design is not able to combat eavesdropping when Eve has the same AoD as Bob, i.e. $u_\theta = u_\phi$. However, it is worth remarking that $u_\theta \neq u_\phi$ is typically assumed in many existing works on physical-layer security [11–13]. The case of $u_\theta \approxeq u_\phi$ is known as the "mainlobe eavesdropping," which is generally treated as individual research [14]. The countermeasure of the mainlobe eavesdropping in FH-MIMO DFRC is left for future work.

Moreover, we remark that it is generally challenging for Eve to know the AoD of Bob in the considered system with the presented signaling, when, e.g. the radar is mechanically steerable too. In this case, even when the physical positions of Bob and the radar is known to Eve, u_ϕ (the AoD of Bob) changes with the physical orientation of the antenna array. It is very hard for Eve to fast estimate and track u_ϕ as the received signals are in the form of (26) with unknown u_ϕ. *For Bob, however, the change of radar antenna orientation does not affect his communication demodulation, as he does not need his own AoD for the demodulation.* This owes to elementwise phase compensation, which can take into account any orientation change and make the steering vector, that underlies Bob-received signals, always independent of his actual AoD; see Section 9.2.2.

9.3.4 Impact of Presented Design on Radar Performance

Subsequently, we illustrate the impact of each module in the presented baseband waveform processing on radar performance using the range ambiguity function. Consider an FH-MIMO radar with M antennas and H hops per pulse. Let τ denote time delay. Based on [1, Eq. (27)], we can express the range ambiguity function of the radar as follows:

$$R(\tau) = \left| \sum_{m=0}^{M-1} \sum_{m'=0}^{M-1} \sum_{h,h'=0}^{H-1} \underbrace{\chi(\tilde{\tau}, \nu) e^{j2\pi\nu hT}}_{B} \underbrace{e^{j2\pi f_{h'm'}\tau}}_{D} \right|, \tag{9.33}$$

where $\tilde{\tau} = \tau - T(h' - h)$, $\nu = f_{hm} - f_{h'm'}$ and $\chi(x, y)$ is the ambiguity function of a standard rectangular pulse with x and y spanning range and Doppler domains, respectively. According to [1, Eq. (26)], we have

$$\chi(x, y) = (T - |x|)\, S\,(y\,(T - |x|))\, e^{j\pi y(x+T)}, \quad \text{if } |x| < T;$$

and otherwise $\chi(x, y) = 0$, where $S(\alpha) = \frac{\sin(\pi\alpha)}{\pi\alpha}$. The impact of presented processing on $R(\tau)$ is analyzed below.

9.3.4.1 Impact of HFCS on $R(\tau)$

HFCS selects M out of K *different* hopping frequency per hop based on varying information bits to be transmitted. The waveform orthogonality condition given in (9.2) is hence always satisfied under HFCS processing. As conventional FH-MIMO radars randomly selects hopping frequencies [1], HFCS, resembling the random selection, incurs negligible changes to the key features of $R(\tau)$, e.g. main lobe width and mainlobe-to-sidelobe ratio (MSR) etc.

9.3.4.2 Impact of HFPS on $R(\tau)$

We see from (9.33) that $R(\tau)$ is determined by the combinations of $(\mathcal{B}, \mathcal{D})$ which is in essence relied on the combinations of $(\nu, f_{h'm'})$. By fixing $f_{h'm'}$, the combinations of $(\nu, f_{h'm'})$ remain the same despite the ordering of the hopping frequencies at hop h. The same conclusion holds by fixing f_{hm} and randomly changing the ordering of the hopping frequencies at hop h'. This is validated by the example given in Tables 9.2 and 9.3, where, C1 and C2 in Table 9.2 give two orderings of the same hopping frequencies, and, clearly, the overall combination set of $\left(\nu, f_{h'm'}\right)$ obtained under C1 is identical to that of $\left(\tilde{\nu}, \tilde{f}_{h'm'}\right)$ under C2. Therefore, we can claim that HFPS does not incur any change to $R(\tau)$ after hopping frequenciesare selected by HFCS.

9.3.4.3 Impact of Elementwise Phase Compensation and Random Sign Reversal on $R(\tau)$

According to (9.14) and (9.23), the joint impact of elementwise phase compensation and random sign reversal is that the phases of radar-transmitted signals are randomly modulated across antennas and hops. As analyzed in [3], PSK modulations can prevent periodic coherent accumulation (which occurs whenever τ is integer times of a hop duration), hence, suppressing periodic sidelobe spikes of R_τ. Given the equivalence between the impact of elementwise phase compensation and random sign reversal on radar signals and that of PSK [3], we conclude that

Table 9.2 Different FH sequences.

	(h, m)	$(0,0)$	$(0,1)$	$(1,0)$	$(1,1)$
C1	f_{hm} (MHz)	20	10	45	30
C2	$\tilde{f}_{hm}$ (MHz)	10	20	30	45

Table 9.3 Combinations of $\left(\nu, f_{h'm'}\right)$, where the frequency is in MHz.

(m, m')	$(0,0)$	$(0,1)$	$(1,0)$	$(1,1)$
(h, h')		$(0,1)$		
$\left(\nu, f_{h'm'}\right)$, C1	$(-20, 30)$	$(-35, 45)$	$(-10, 30)$	$(-25, 45)$
$\left(\tilde{\nu}, \tilde{f}_{h'm'}\right)$, C2	$(-25, 45)$	$(-10, 30)$	$(-35, 45)$	$(-20, 30)$

elementwise phase compensation and random sign reversal can suppress periodic sidelobe spikes of $R(\tau)$. A benefit of the suppression is the reduced mutual interference among radar targets. This will be validated in Section 9.4.

9.3.4.4 Limitations of Presented Design for Radar Applications

The price paid by the underlying FH-MIMO radar to achieve secure data communications is the loss of the degrees of freedom (DoF) in the radar waveform design. In conventional FH-MIMO radars, the hopping frequency sequences and signal phases can be specifically designed for different application scenarios [1, 15]. However, in FH-MIMO DFRC, the hopping sequence is randomly determined by the information bits to be transmitted, and the signal phases are subject to the manipulations designed for data communications. Specifically, the demodulation of HFPS, as developed in Algorithm 9.1, requires that the phases of signals from different radar antennas conform to the steering vector given in (9.13). Moreover, random sign reversal adds negative signs to randomly selected antennas per hop; see (9.23). It is worth noting that, as a by-product, the processing can reduce the sidelobe levels of the conventional FH-MIMO radar, as illustrated in Figure 9.7.

9.3.5 Extension to Multipath and Multiuser Scenarios

The presented design has the potential to be extended to multipath and multiuser scenarios, which is illustrated below.

9.3.5.1 Multipath Scenario

Our presented design can be extended to multipath scenarios through performing elementwise phase compensation based on the $M \times 1$ channel response vector $\mathbf{h}$ instead of the scaled steering vector in the LoS case, where $\mathbf{h}$ is given in (9.3). The mth element of $\mathbf{h}$, denoted by $[\mathbf{h}]_m$, is the channel coefficient between the mth radar antenna and the receiver at Bob. To perform elementwise phase compensation, we can replace the amount of phase compensation, i.e. mu_ϕ given in (9.14), with $\angle[\mathbf{h}]_m$ which is the angle of $[\mathbf{h}]_m$. Enabled by the modified elementwise phase compensation, the optimal solution to the ML detection of the permutation matrix, as given in (9.16), still holds for the multipath scenarios. This can be readily validated by replacing $\beta\mathbf{a}$ with $|\mathbf{h}| \odot \mathbf{a}$ in (9.15), where $|\cdot|$ takes pointwise magnitude and $\odot$ denotes elementwise product.

Moreover, since random sign reversal and its removal solely rely on the phases of the elementwise phase, compensation-processed channel vector, Algorithm 9.2 can still be directly applied to Bob for random sign reversal removal. Note that there is a common problem for FH-MIMO DFRC in multipath scenarios – signals of different hopping frequencies can experience various attenuation levels due to multipath fading [16, Fig. 8]. In the worst case, an information-bearing signal can be fully buried in noises, leading to the incorrect detection of hopping frequencies and hence an error in demodulating an HFPS symbol. Antenna selection may be performed at the radar when embedding information bits to address this issue. The detailed design is left for future work.

9.3.5.2 Multiuser Scenario

In this scenario, a key challenge to be addressed is selecting the multiple access (MA) methods, which is still an open issue in FH-MIMO DFRC. Once an MA method is selected, the presented design can be adapted for serving multiple users. For example, if we consider time-division multiple access (TDMA) based on radar pulses, e.g. each radar pulse is assigned for a user, the presented methods can be directly performed based on the channel parameters of the user currently being served. When frequency division multiple access (FDMA) and space division multiple access (SDMA) are applied, the constellation size for each user can be smaller than that for a single user. The detailed implementation and comparison of these MA methods need further investigation for FH-MIMO DFRC.

9.4 Simulation Results

In this section, simulation results are presented to validate the presented design. Unless otherwise specified, the FH-MIMO radar is configured as[1]: $M = 4$, $Q = \frac{M}{2}$, $K = 20$, $H = 15$, $B = 100\,\text{MHz}$, $T = 1\,\mu\text{s}$, and $L = 200$ (based on the sampling frequency of $2B$), and the communication parameters are the following: $\phi \sim \mathcal{U}_{[-90^\circ,90^\circ]}$, $\theta \sim \mathcal{U}_{[-90^\circ,90^\circ]}$, $\alpha = e^{jx}$ ($x \sim \mathcal{U}_{[0,2\pi]}$), and $\beta = e^{jy}$ ($y \sim \mathcal{U}_{[0,2\pi]}$). Here, $\mathcal{U}_{[\cdot,\cdot]}$ stands for the uniform distribution in the subscript region. Throughout the simulation, Eve knows the AoD of Bob, if not otherwise specified. The time-domain SNR at Bob is defined based on (9.3), as given by $\gamma_B = \frac{M|\beta|^2}{\sigma_\xi^2}$, where σ_ξ^2 is the noise power of $\xi(i)$. Based on (9.9), the demodulation SNR at Bob is $L\gamma_B$, where the L times improvement is brought by DFT; see (9.4). Likewise, the time-domain and demodulation SNRs at Eve are given by $\gamma_E = \frac{M|\alpha|^2}{\sigma_z^2}$ and $L\gamma_E$, respectively, where σ_z^2/L is the noise power of $Z(l)$ given in (9.7). When presenting demodulation performance, we also use E_b/N_0, defined as energy per bit to noise power density ratio, specifically

$$E_b/N_0 = L\gamma_B BT/E, \tag{9.34}$$

1 Note that the presented methods do not rely on any specific parameter settings. The parameters set here are solely for illustration convenience and clarity. In all the simulation results given below, different values/regions of critical parameters are taken to demonstrate the scalability of the presented methods.

where E is the number of bits conveyed per radar hop and γ_B can be replaced with γ_E to obtain E_b/N_0 for Eve.

The labels used in the figures are interpreted as follows:

- *Bob-proposed*: Element-wise phase compensation and random sign reversal are performed at the radar, and Algorithms 9.1 and 9.2 at Bob;
- *Bob-ML*: Same as above except that the ML problem (9.25) is solved at Bob. This is a lower bound of "Bob-proposed";
- *Eve-ML*: Elementwise phase compensation and random sign reversal are performed at the radar, and the ML problem (9.26) is solved for demodulation at Eve, who knows the AoD of Bob;
- *Eve-without Bob's AoD*: Same as above except that the AoD of Bob is unavailable at Eve;
- *Eve-without random sign reversal*: Elementwise phase compensation is performed at the radar, but random sign reversal is not, and (9.19) is solved by Eve. This acts as a performance indicator of a general HFPS demodulation without conducting elementwise phase compensation and random sign reversal at radar.

Figure 9.2 compares the symbol error rates (SERs) achieved by Bob and Eve as E_b/N_0 increases. We see that the presented scheme outperforms the previous FH-MIMO DFRC designs based on BPSK [7] and HFCS [2], in terms of SER per E_b/N_0. In particular, for the SER of 10^{-4}, the presented design reduces E_b/N_0 by 6 dB and 1.5 dB, compared with BPSK and FHCS, respectively. This improvement owes to (i) the use of HFPS, which increases the number of bits conveyed per radar

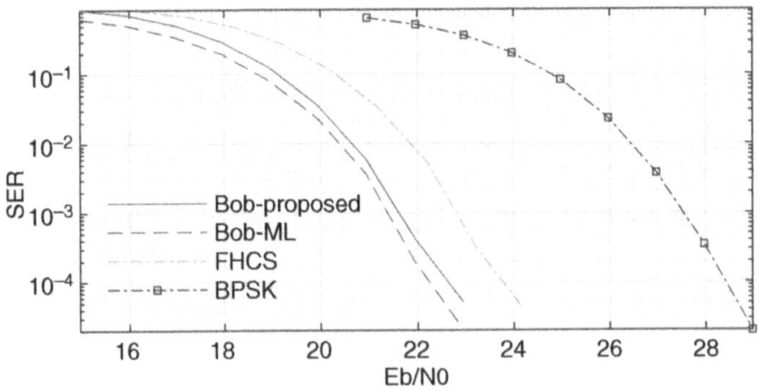

Figure 9.2 SER against E_b/N_0, where γ_B and γ_E are in the region of $[-10, -2]$ dB. The radar configuration leads to $\lfloor \log_2 C_K^M \rfloor = 12$ bits conveyed by HFCS subsymbol and $\lfloor \log_2 M! \rfloor = 4$ bits by HFPS. Substituting $E = 12 + 4$ into (9.34) gives the E_b/N_0 region in the figure. The FH-MIMO DFRC systems based on binary PSK (BPSK) [7] and FHCS [2] are provided as benchmark methods.

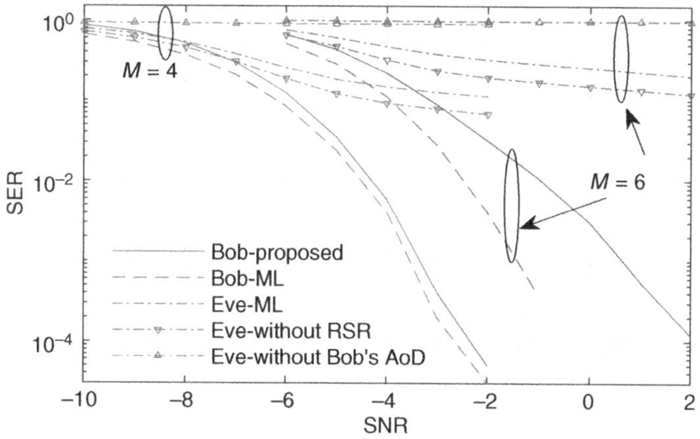

Figure 9.3 SER against SNR under different values of M.

hop; (ii) elementwise phase compensation, which solves the AoD-dependent issue of HFPS demodulation.

Figure 9.3 compares the SER performance achieved under different values of M. We see that Bob achieves a much-improved SER performance than Eve. This demonstrates the secrecy enhancement of the presented FH-MIMO DFRC scheme. Despite that the gap between "Bob-proposed" and "Bob-ML" increases slightly with M, the SERs of Bob can still reduce fast to a considerably small value as SNR increases. For Eve, we see that, even if she knows the AoD of Bob, the AoD-dependent issue, as illustrated in Section 9.2.3, degrades her SER performance substantially; moreover, random sign reversal makes her performance further degraded. In addition, if Eve does not know the AoD of Bob, her eavesdropping would fully fail to work, as indicated by the curves "Eve-without Bob's AoD."

Figure 9.4 compares the error probability of detecting random sign reversal at Bob and Eve. We see that Bob achieves a much smaller error probability than Eve. We also see that the presented low-complexity algorithm for random sign reversal removal, i.e. Algorithm 9.2, achieves a near-ML performance, particularly for $M = 4$. Jointly observing Figures 9.3 and 9.4, we see that the SER performance closely complies with the error probability of random sign reversal detection. This manifests the critical role of random sign reversal in enhancing the communication secrecy for the presented FH-MIMO DFRC.

Figure 9.5 illustrates the impact of Q on SER performance, where Q is the number of sign-reversed antennas in random sign reversal. We see that increasing Q has a more prominent impact on Eve. For Bob, Algorithm 9.2 is applied for random sign reversal removal. Thus, the results in the figure also validate the effectiveness

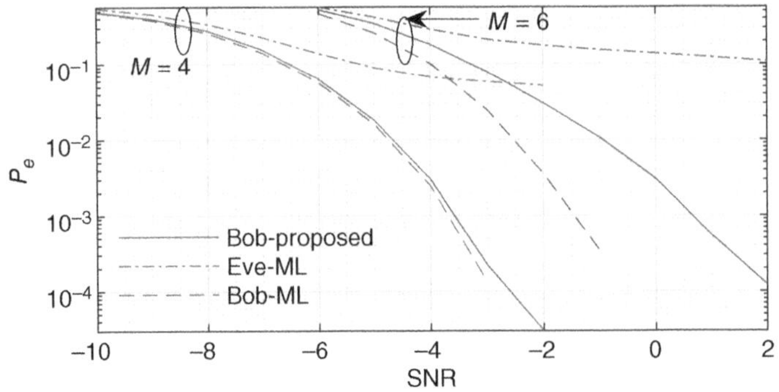

Figure 9.4 Illustration of the error probability of detecting random sign reversal, as denoted by P_e in the figure.

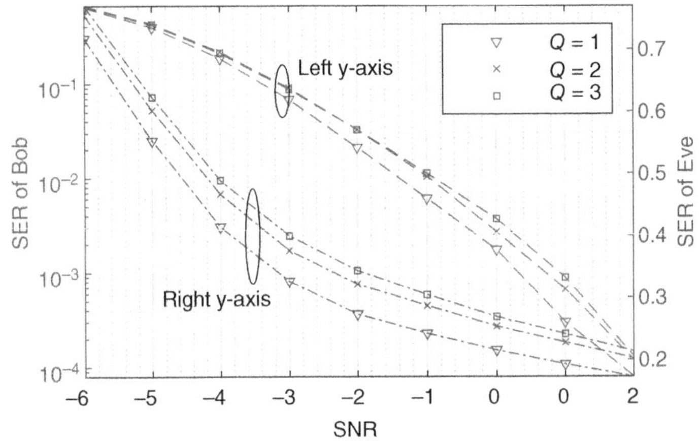

Figure 9.5 SER against SNR under different values of Q, where $M = 6$. Note that "Bob-proposed" and "Eve-ML" are presented here.

of Algorithm 9.2 in handling different values of Q. It is worth mentioning that when Q increases, the complexity of solving the ML problem (9.26) for Eve also increases substantially. For this simulation, the number of candidates under test is $C_M^Q \times 512 = 3072,\ 7680$, and 10 240 for $Q = 1,\ 2$, and 3, respectively. Given $M = 6$, there are 720 numbers of permutations of hopping frequencies. Among them, $512 = 2^9$ numbers of permutations are used for information embedding.

Figure 9.6 observes the SER performance against the spatial angle, where 2×10^3 independent trials are performed for most angle grids and 2×10^5 trials for the three angles around 60°. We see that "Bob-ML," obtained by solving (9.25), can achieve an SER which is lower than that of "Eve-ML" by orders of magnitude.

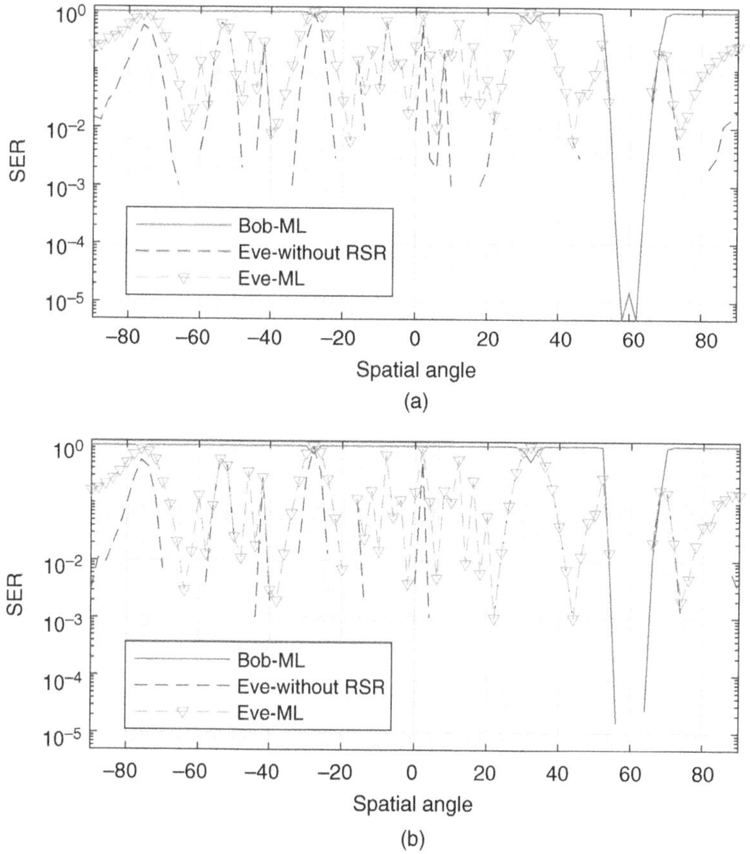

Figure 9.6 SER against spatial angle from $-90°$ to $90°$ with a grid of $2°$, where $\phi = 60°$, $M = 6$, $Q = 2$, $\gamma_B = 0$ dB for Figure 9.6a and 1.5 dB for Figure 9.6b.

The results under "Eve-ML" are obtained by solving (9.26). This shows the high communication secrecy of the presented FH-MIMO DFRC. This also validates the analysis in Section 9.3.3. For the curves labeled "Bob-ML," the ML demodulation is performed in the whole angular region. Thus, the flat close-to-one SER indicates the poor demodulation performance of Eve when she does not know the AoD of Bob. Moreover, some discontinuities are seen at some angles in the curves labeled "Eve-without random sign reversal," which means the orders of the SERs at the angles can be lower than 10^{-3}. Nevertheless, we see that when random sign reversal is performed, the SER performance under those angles can be substantially degraded.

Last but not least, we illustrate the impact of the presented waveform processing on FH-MIMO radar performance. Figure 9.7a is provided to illustrate the impact of

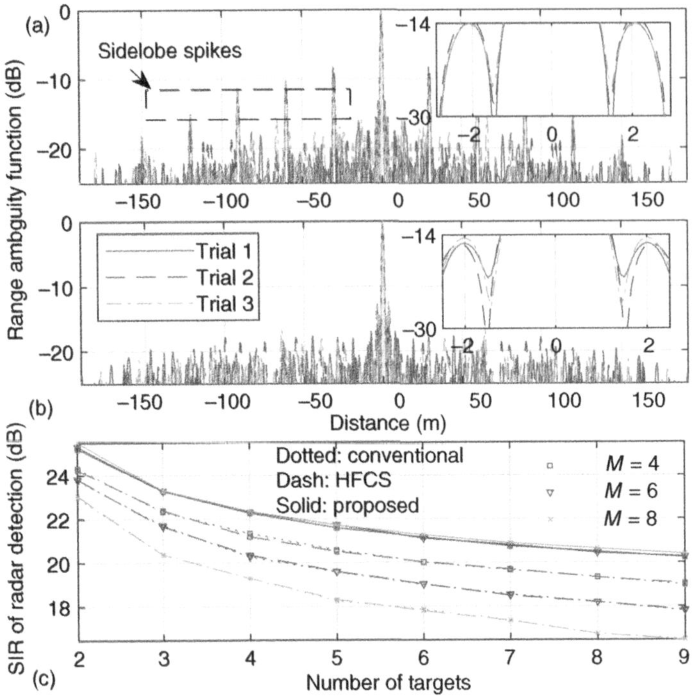

Figure 9.7 Impact of the presented baseband waveform processing on radar detection, where (a) range ambiguity functions (RAFs) under three groups of independently and randomly generated hopping frequencies; (b) RAFs after performing HFPS, elementwise phase compensation and random sign reversal on the waveform generated for Figure 9.7a; and (c) SIR against the number of targets. In Figure 9.7a,b, $M = 8$ is taken. In Figure 9.7c, 5×10^3 independent trials are performed for each number of targets.

HFCS on radar-ranging performance. We see that the mainlobes and MSRs under three realizations are almost identical. We also see that the periodicity of sidelobe spikes is the same for different random sets of hopping frequencies. From Figure 9.7b, we see that the presented waveform processing suppresses the sidelobe spikes, improving the minimum MSR by more than 10 dB. The spike suppression is achieved due to the randomness introduced by the presented random sign reversal; refer to [7] for a detailed analysis. Since the shape of the signal has not changed the randomness, the mainlobe width remains the same as that of the conventional radar. This is validated by comparing the zoomed-in subfigures in Figure 9.7a,b.

Note that the improved MSR can help reduce mutual interference among radar targets. The following metric is defined based on the range ambiguity function $R(\tau)$, as given in (9.33), to measure signal-to-interference ratio (SIR),

$$\text{SIR} = |R(0)|^2 \bigg/ \left(\max_{p=1,\ldots,P} |R(\tau_p)|^2 \right),$$

where τ_p is the relative delay between the pth target and the target of interest. Note that $\max_p |R(\tau_p)|^2$ represents the strongest interference incurred by one of P unit-power targets whose distances are randomly distributed in the range sidelobes of $R(\tau)$. From Figure 9.7c, we see that the improvement of MSR earned by the presented design increases the SIR of radar detection by up to 4 dB at $M = 8$. We also see that, despite the values of M, the presented processing leads to similar SIRs. We further see that the HFCS waveform achieves the same SIR as the conventional FH-MIMO radar waveform. These observations of Figure 9.7 validate our analysis in Section 9.3.4.

9.5 Conclusions

In this chapter, a secure and high date rate FH-MIMO DFRC system is developed. This is achieved by introducing HFPS constellations to fully exploit the information embedding capability embodied in hopping frequencies. This is also accomplished by a new elementwise phase compensation processing which addresses the AoD-dependent issue for Bob and enhances the physical layer security if the AoD of Bob is unknown to Eve. This is further fulfilled by the proposed random sign reversal which, in combination with elementwise phase compensation, enables Bob to outperform Eve in HFPS demodulation over the most angular region. Validated by simulations, the design presented in this chapter achieves high communication secrecy, increases data rate over previous designs of FH-MIMO DFRC, and improves SIR of radar detection.

References

1 C. Chen and P. P. Vaidyanathan. MIMO radar ambiguity properties and optimization using frequency-hopping waveforms. *IEEE Transactions on Signal Processing*, 56(12):5926–5936, December 2008. ISSN 1941-0476. doi: 10.1109/TSP.2008.929658.

2 W. Baxter, E. Aboutanios, and A. Hassanien. Dual-function MIMO radar-communications via frequency-hopping code selection. In *2018 52nd*

Asilomar Conference on Signals, Systems, and Computers, pages 1126–1130, October 2018. doi: 10.1109/ACSSC.2018.8645212.

3 I. P. Eedara, A. Hassanien, M. G. Amin, and B. D. Rigling. Ambiguity function analysis for dual-function radar communications using PSK signaling. In *2018 52nd Asilomar Conference on Signals, Systems, and Computers*, pages 900–904, October 2018. doi: 10.1109/ACSSC.2018.8645328.

4 C. Sturm and W. Wiesbeck. Waveform design and signal processing aspects for fusion of wireless communications and radar sensing. *Proceedings of the IEEE*, 99(7):1236–1259, 2011.

5 A. Ahmed, Y. D. Zhang, and Y. Gu. Dual-function radar-communications using QAM-based sidelobe modulation. *Digital Signal Processing*, 82:166–174, 2018.

6 A. Hassanien, E. Aboutanios, M. G. Amin, and G. A. Fabrizio. A dual-function MIMO radar-communication system via waveform permutation. *Digital Signal Processing*, 83:118–128, 2018.

7 I. P. Eedara, M. G. Amin, and A. Hassanien. Analysis of communication symbol embedding in FH MIMO radar platforms. In *2019 IEEE Radar Conference (RadarConf)*, pages 1–6, April 2019. doi: 10.1109/RADAR.2019.8835532.

8 K. Wu, W. Ni, T. Su, R. P. Liu, and Y. J. Guo. Robust unambiguous estimation of angle-of-arrival in hybrid array with localized analog subarrays. *IEEE Transactions on Wireless Communications*, 17(5):2987–3002, May 2018. ISSN 1536-1276. doi: 10.1109/TWC.2018.2805713.

9 MathWorks. Circshift. https://www.mathworks.com/help/matlab/ref/circshift.html. Accessed: 2020-10-12.

10 F. Gini, A. De Maio, and L. Patton. *Waveform Design and Diversity for Advanced Radar Systems*. Institution of Engineering and Technology London, 2012.

11 M. E. Eltayeb, J. Choi, T. Y. Al-Naffouri, and R. W. Heath. Enhancing secrecy with multiantenna transmission in millimeter wave vehicular communication systems. *IEEE Transactions on Vehicular Technology*, 66(9):8139–8151, 2017.

12 Y. Hong, X. Jing, and H. Gao. Programmable weight phased-array transmission for secure millimeter-wave wireless communications. *IEEE Journal on Selected Topics in Signal Processing*, 12(2):399–413, 2018.

13 K. Wu, W. Ni, J. A. Zhang, R. P. Liu, and J. Guo. Secrecy rate analysis for millimeter-wave lens antenna array transmission. *IEEE Communications Letters*, 24(2):272–276, 2020.

14 X. Chen, D. W. K. Ng, W. H. Gerstacker, and H.-H. Chen. A survey on multiple-antenna techniques for physical layer security. *IEEE Communication Surveys and Tutorials*, 19(2):1027–1053, 2016.

15 K. Han and A. Nehorai. Jointly optimal design for MIMO radar frequency-hopping waveforms using game theory. *IEEE Transactions on Aerospace and Electronic Systems*, 52(2):809–820, 2016.
16 K. Wu, J. A. Zhang, X. Huang, and Y. J. Guo. Frequency-hopping MIMO radar-based communications: An overview. *IEEE Aerospace and Electronic Systems Magazine*, 37(4):42–54, 2021.

A

Proofs, Analyses, and Derivations

A.1 Proof of Lemma 5.1

As illustrated in Section 5.1.1, $s_n[l] \sim \mathcal{CN}(0, \sigma_d^2)$ and $s_n[l]$ ($\forall n$) is i.i.d. over l. As the unitary DFT of $s_n[l]$, $S_n[m]$ ($\forall n$) conforms to the same distribution and is i.i.d. over m. Likewise, the distribution of $W_n[m]$, which is the unitary DFT of $w_n[l]$ given in (5.13), conforms to a complex Gaussian distribution and is i.i.d. over m. Since $w_n[l]$ ($\forall n$) has the nonidentical variance over l, the variance of $W_n[m]$ is not equal to that of $w_n[l]$ and instead can be calculated, based on (5.13) and (5.14); specifically

$$\left(2\sigma_w^2 \tilde{Q} + \sigma_w^2(\tilde{M} - \tilde{Q})\right) / \tilde{M} = \left(1 + \tilde{Q}/\tilde{M}\right) \sigma_w^2.$$

Next, we analyze the distribution of $Z_n[m]$. From (5.14), we see that the core of $Z_n[m]$ is the DFT (w.r.t. l) of $z_n^{(p)}[l] g_{\tilde{Q}}[l]$ ($\forall p$). From Figure 5.1, we see that for any p, $z_n^{(p)}[l] g_{\tilde{Q}}[l]$ consists of two parts, one from the essential signal of the previous subblock and the other from the sequential. Thus, $z_n^{(p)}[l] g_{\tilde{Q}}[l]$ ($\forall p$, $\forall n$) satisfies $z_n^{(p)}[l] g_{\tilde{Q}}[l] \sim \mathcal{CN}(0, \sigma_d^2)$, is i.i.d. over l, and is independent from $s_n[l]$. As a result, the DFT of $z_n^{(p)}[l] g_{\tilde{Q}}[l]$ conforms to

$$\sum_{l=0}^{\tilde{M}-1} z_n^{(p)}[l] g_{\tilde{Q}}[l] \mathcal{Z}_{\tilde{M}}^{lm} \sim \mathcal{CN}(0, \tilde{Q}\sigma_d^2/\tilde{M}).$$

With the assumption that α_p is independent over p, the p-related summation in (5.14) leads to the final distribution of $Z_n[m]$ given in the statement of the lemma.

A.2 Proof of Lemma 5.2

The first statement of the lemma arises from the fact that $W_n[m]$ is only related to the receiver noise while the other two components to the essential signals. Next, we illustrate the independence of $S_n[m]$ over n.

Joint Communications and Sensing: From Fundamentals to Advanced Techniques, First Edition.
Kai Wu, J. Andrew Zhang, and Y. Jay Guo.

Using the expression of $S_n[m]$ given in (5.14), we can have

$$\begin{aligned}\mathbb{E}\left\{S_n[m]S^*_{n+1}[m]\right\} &= \mathbb{E}\left\{\left(\sum_{l_1=0}^{\tilde{M}-1} s_n[l_1]\mathcal{Z}_{\tilde{M}}^{l_1 m}\right)\times\left(\sum_{l_2=0}^{\tilde{M}-1} s^*_{n+1}[l_2]\mathcal{Z}_{\tilde{M}}^{-l_2 m}\right)\right\}\\ &= \bar{Q}\sigma_d^2 e^{-\mathrm{j}\frac{2\pi(\tilde{M}-\bar{Q})}{\tilde{M}}}/\tilde{M}.\end{aligned} \tag{A.1}$$

The last result is based on two facts: *First*, only at the indexes given in (A.2), we can have nonzero expectation; otherwise, the summands involved are uncorrelated.

$$l_1 = \tilde{M} - \bar{Q}, \cdots, \tilde{M} - 1 \text{ and } l_2 = l_1 - (\tilde{M} - \bar{Q}) \tag{A.2}$$

Second, from Fig. 5.1, we can see that at the indexes given in (A.2), $s_n[l_1] = s_{n+1}[l_2]$. Based on (A.1), we can validate $\mathbb{E}\left\{|S_n[m]|^2\right\} = \sigma_d^2$ and $\mathbb{E}\left\{|S_{n+1}[m]|^2\right\} = \sigma_d^2$. Combining these expectations with (A.1), the correlation coefficient between $S_n[m]$ and $S_{n+1}[m]$ can be obtained, as given in the statement of Lemma 5.2. Following the above analysis procedure for $S_n[m]$, we can similarly calculate the correlation coefficient between $W_n[m]$ and $W_{n+1}[m]$. For brevity, we suppress the details here.

A.3 Proof of Lemma 5.3

As said in Appendix A.1, $S_n[m]$ ($\forall n$) is the unitary DFT (w.r.t. l) of $s_n[l]$ that is i.i.d. over l. Thus, $S_n[m]$ ($\forall n$) is i.i.d. over m. For the same reason, $W_n[m]$ ($\forall n$) is i.i.d. over m. However, $Z_n[m]$ is not i.i.d. over m, since the length of $z_n^{(p)}[l]g_{\bar{Q}}[l]$ ($\forall p$) is smaller than the DFT dimension. Using the expression of $Z_n[m]$ given in (5.14), we can calculate the correlation coefficient between $Z_n[m_1]$ and $Z_n[m_2]$. Specifically, we have

$$\begin{aligned}\mathbb{E}\left\{Z_n[m_1]Z_n^*[m_2]\right\} &\overset{(a)}{=} \sigma_P^2\mathbb{E}\left\{\left(\sum_{l_1=0}^{\tilde{M}-1} z_n^{(p)}[l_1]g_{\bar{Q}}[l_1]\mathcal{Z}_{\tilde{M}}^{l_1 m_1}\right)\right.\\ &\quad\left.\times\left(\sum_{l_2=0}^{\tilde{M}-1} z_n^{(p)}[l_2]g_{\bar{Q}}[l_2]\mathcal{Z}_{\tilde{M}}^{l_2 m_2}\right)^*\right\}\\ &\overset{(b)}{=} \sigma_P^2\sigma_d^2\frac{\sum_{l=0}^{\bar{Q}-1}\mathcal{Z}_{\tilde{M}}^{l(m_1-m_2)}}{\sqrt{\tilde{M}}},\end{aligned} \tag{A.3}$$

where $\sigma_P^2 = \sum_{p=0}^{P-1}\sigma_p^2$, $\overset{(a)}{=}$ is obtained based on the uncorrelated α_p ($\forall p$), and $\overset{(b)}{=}$ is due to the independence of $z_n^{(p)}[l]g_{\bar{Q}}[l]$ over l. Calculating the l-related summation on the RHS of $\overset{(b)}{=}$ by plugging in the definition of the DFT basis given in (5.3), we can obtain the following correlation coefficient:

$$\frac{|\mathbb{E}\left\{Z_n[m_1]Z_n^*[m_2]\right\}|}{\sqrt{\mathbb{E}\left\{|Z_n[m_1]|^2\right\}\mathbb{E}\left\{|Z_n[m_2]|^2\right\}}} = |f(m_1 - m_2)|,$$

$$\text{s.t. } f(m_1 - m_2) = \sin\left(\frac{2\pi}{\tilde{M}}\frac{\tilde{Q}(m_1 - m_2)}{2}\right) \Big/ \left(\tilde{Q}\sin\left(\frac{2\pi}{\tilde{M}}\frac{(m_1 - m_2)}{2}\right)\right)$$

where $\mathbb{E}\left\{|Z_n[m_1]|^2\right\}$ can be readily attained by setting $m_2 = m_1$ in (A.3); likewise, for $\mathbb{E}\left\{|Z_n[m_2]|^2\right\}$.

A.4 Proof of Proposition 5.1

As said above (5.27), with a sufficiently large α introduced, $|\alpha S_n[m]| < 1$ can barely happen. Moreover, the operator $\mathbb{I}_{\mathcal{E}}\{\cdot\}$ fully removes the cases of $|\alpha S_n[m]| < 1$. Therefore, we have

$$\left|D_{n,m}^{k,l}/\alpha S_n[m]\right| \leq \max\left\{\left|\alpha D_{n,m}^{k,l}S_n^*[m]\right|\right\}, \tag{A.4}$$

where $D_{n,m}^{k,l}$ is defined in (5.24). Since $D_{n,m}^{k,l}/\alpha S_n[m]$ conforms to a truncated Cauchy distribution, we are now able to invoke the CLT on deriving the distribution of $\tilde{W}_k^{\mathrm{r}}[l] + \tilde{Z}_k^{\mathrm{r}}[l]$. Based on Lemma 5.2, we know that $Z_n[m]$ is independent over n and, under the condition $\tilde{M} \gg (\tilde{Q} + \bar{Q})$, such independence is also owned by $S_n[m]$ and $W_n[m]$. Accordingly, the ratio $D_{n,m}^{k,l}/\alpha S_n[m]$ $(\forall m)$ is independent over n. Invoking the CLT under large $\tilde{N}$, we attain

$$\begin{aligned} &\sum_{n=0}^{\tilde{N}-1}\frac{D_{n,m}^{k,l}}{\alpha S_n[m]} \sim \mathcal{CN}\left(0, \tilde{N}\left(\rho\beta(\epsilon)\right)\right), \text{ s.t. } \rho = \frac{(\sigma_Z^2 + \sigma_W^2)}{\tilde{M}\tilde{N}\alpha^2\sigma_d^2}, \\ &b(\epsilon) = 2\left(\ln\left(\frac{2(1-\epsilon)}{\sqrt{\epsilon(2-\epsilon)}}\right) - 1\right), \end{aligned} \tag{A.5}$$

where ϵ is a sufficiently small value and $\rho\beta(\epsilon)$ is the variance of each summand according to [1, Proposition 1]. Note that ρ is the ratio between the variance of $D_{2\tilde{n},m}^{k,l}$, as given in (5.25), and that of $\alpha S_{2\tilde{n}}[m]$, as easily deduced from Lemma 5.1.

With (A.5) attained, we know that $\tilde{W}_k^{\mathrm{r}}[l] + \tilde{Z}_k^{\mathrm{r}}[l]$ also conforms to a Gaussian distribution, as it is the summation of $\sum_{n=0}^{\tilde{N}-1}\frac{D_{n,m}^{k,l}}{\alpha S_n[m]}$ over m. Moreover, as shown below, $\sum_{n=0}^{\tilde{N}-1}\frac{D_{n,m}^{k,l}}{\alpha S_n[m]}$ is approximately independent over m. This leads to the final distribution of $\tilde{W}_k^{\mathrm{r}}[l] + \tilde{Z}_k^{\mathrm{r}}[l]$, as given in the statement of Proposition 5.1.

Independence of $\sum_{n=0}^{\tilde{N}-1} \frac{D_{n,m}^{k,l}}{\alpha S_n[m]}$ *over* m: As Gaussian-distributed, $\sum_{n=0}^{\tilde{N}-1} \frac{D_{n,m_1}^{k,l}}{\alpha S_n[m_1]}$ and $\sum_{n=0}^{\tilde{N}-1} \frac{D_{n,m_2}^{k,l}}{\alpha S_n[m_2]}$ are independent if they are uncorrelated. Replacing $D_{n,m}^{k,l}$ with its full expression given in (5.24), we can have

$$
\begin{aligned}
&\mathbb{E}\left\{\begin{matrix}\left(\sum_{n_1=0}^{\tilde{N}-1} \frac{(Z_{n_1}[m_1]+W_{n_1}[m_1])\mathcal{Z}_{\tilde{M}}^{-m_1 l}\mathcal{Z}_{\tilde{N}}^{n_1 k}}{\alpha S_{n_1}[m_1]}\right) \times \\ \left(\sum_{n_2=0}^{\tilde{N}-1} \frac{\left(Z_{n_2}[m_2]+W_{n_2}[m_2]\right)\mathcal{Z}_{\tilde{M}}^{-m_2 l}\mathcal{Z}_{\tilde{N}}^{n_2 k}}{\alpha S_{n_2}[m_2]}\right)^*\end{matrix}\right\} \\
&\overset{(a)}{=} \mathbb{E}\left\{\sum_{n=0}^{\tilde{N}-1} \frac{\left(Z_n[m_1]+W_n[m_1]\right)\left(Z_n^*[m_2]+W_n^*[m_2]\right)e^{\mathrm{j}\frac{2\pi l(m_1-m_2)}{\tilde{M}}}}{\alpha^2 S_n[m_1]S_n^*[m_2]\tilde{N}\tilde{M}}\right\} \\
&= \sum_{n=0}^{\tilde{N}-1} \mathbb{E}\left\{f/g\right\} \overset{(b)}{\approx} \sum_{n=0}^{\tilde{N}-1} \mu_f/\mu_g = 0, \\
&\text{s.t. } f = \left(Z_n[m_1]Z_n^*[m_2] + Z_n[m_1]W_n^*[m_2] + W_n[m_1]Z_n^*[m_2]\right. \\
&\qquad \left. + W_n[m_1]W_n^*[m_2]\right) S_n^*[m_1]S_n[m_2]e^{\mathrm{j}\frac{2\pi l(m_1-m_2)}{\tilde{M}}} \\
&\quad g = \alpha^2 |S_n[m_1]S_n[m_2]|^2 \tilde{N}\tilde{M},
\end{aligned} \tag{A.6}
$$

where $\overset{(a)}{=}$ is obtained by suppressing all cross-terms at $n_1 \neq n_2$ since they are uncorrelated, and $\overset{(b)}{\approx}$ is attained by applying the first-order approximation of the mean of the ratio f/g [2]. The last result is based on $\mu_f = 0$ which can be readily obtained by applying Lemmas 5.2 and 5.3.

A.5 Proof of Proposition 5.2

Consider $S_k^c[l]$ first. According to Lemma 5.1, we have $S_n[m] \sim \mathcal{CN}(0, \sigma_d^2)$. This then yields $\frac{|S_n[m]|^2}{\sigma_d^2/2} \sim \chi_2^2$ and

$$
\mathbb{E}\left\{\frac{|S_n[m]|^2}{\sigma_d^2/2}\right\} = 2, \ \mathbb{V}\left\{\frac{|S_n[m]|^2}{\sigma_d^2/2}\right\} = 4, \tag{A.7}
$$

where χ_2^2 denotes the chi-square distribution with two degrees of freedom (DoF). Given that $S_n[m]$ ($\forall n$) is statistically independent of m, we first consider the m-related summation in calculating $S_k^c[l]$; see (5.16). Invoking the CLT, we know that the summation leads to a Gaussian distribution whose variance is given by

$$\mathbb{V}\left\{\frac{\sigma_d^2}{2}\sum_{m=0}^{\tilde{M}-1}\frac{|S_n[m]|^2}{\sigma_d^2/2}e^{-j\frac{2\pi m l_p}{\tilde{M}}}\mathcal{Z}_{\tilde{M}}^{-ml}\right\}=\sigma_d^4.$$

The mean of the Gaussian distribution is more complicated, as we need to consider the cases of $l = l_p$ and $l \neq l_p$. For the first case, we have

$$\mathbb{E}\left\{\frac{\sigma_d^2}{2}\sum_{m=0}^{\tilde{M}-1}\frac{|S_n[m]|^2}{\sigma_d^2/2}e^{-j\frac{2\pi m l_p}{\tilde{M}}}\mathcal{Z}_{\tilde{M}}^{-ml_p}\right\}=\sigma_d^2\sqrt{\tilde{M}},$$

where $e^{-j\frac{2\pi m l_p}{\tilde{M}}}\mathcal{Z}_{\tilde{M}}^{-ml_p}=1/\sqrt{\tilde{M}}$ is applied. For the case of $l \neq l_p$, we have

$$\begin{aligned}\mathbb{E}\left\{\frac{\sigma_d^2}{2}\sum_{m=0}^{\tilde{M}-1}\frac{|S_n[m]|^2}{\sigma_d^2/2}e^{-j\frac{2\pi m l_p}{\tilde{M}}}\mathcal{Z}_{\tilde{M}}^{-ml}\right\}&=\sigma_d^2\sum_{m=0}^{\tilde{M}-1}e^{j\frac{2\pi m(l-l_p)}{\tilde{M}}}\\&=0,\end{aligned}\tag{A.8}$$

where we have applied the fact that summing a discrete single-tone exponential signal (with a nonzero frequency) over integer cycles leads to zero. Combining the above discussion, we conclude

$$\sum_{m=0}^{\tilde{M}-1}|S_n[m]|^2 e^{-j\frac{2\pi m l_p}{\tilde{M}}}\mathcal{Z}_{\tilde{M}}^{-ml}\sim\begin{cases}\mathcal{N}\left(\sigma_d^2\sqrt{\tilde{M}},\sigma_d^4\right) & l=l_p\\ \mathcal{N}\left(0,\sigma_d^4\right) & l\neq l_p\end{cases}.$$

Based on Lemma 5.2, $\sum_{m=0}^{\tilde{M}-1}|S_n[m]|^2 e^{-j\frac{2\pi m l_p}{\tilde{M}}}\mathcal{Z}_{\tilde{M}}^{-ml}$ is independent over the set of either odd n or even n, but shows dependence between adjacent pair. However, under the assumption of $\tilde{M} \gg (\tilde{Q}+\bar{Q})$, the dependence is weak as the correlation coefficient approaches zero. Thus, summing the LHS of the above equation over n converges in distribution to another Gaussian with its statistical properties depicted in the statement of Proposition 5.2. The distribution of $\mathsf{X}_k^{\mathrm{r}}[l]$, $\mathsf{X} \in \{W, Z\}$ can be similarly derived with reference to Appendix A.4. Thus, we suppress the details for brevity.

A.6 Proof of Proposition 6.1

By definition, we have

$$z=\frac{x}{y}=\frac{xy^*}{|y|}=\underbrace{\frac{x_r y_r + x_i y_i}{|y|}}_{z_r}-j\underbrace{\frac{(x_i y_r - x_r y_i)}{|y|}}_{z_i},\tag{A.9}$$

where adding the subscripts, $(\cdot)_r$ and $(\cdot)_i$, to a complex variable denotes the real and imaginary parts of the variable, respectively. The variance of z, denoted by σ_z^2, can be calculated as follows: $\sigma_z^2 = \sigma_{z_r}^2 + \sigma_{z_i}^2$, where $\sigma_{z_r}^2$ (or $\sigma_{z_i}^2$) denotes the variance of z_r (or z_i). Since z_r and z_i have the same PDF [3] and occupy the same region, they have the same variance as well. Thus, we only illustrate the calculation of $\sigma_{z_r}^2$ below.

With reference to [3, Eq. (18)], the cumulative density function of z_r can be expressed as follows: $F_{z_r}(u) = (\lambda(u)+1)/2$; where $\lambda(u) = \frac{u}{\sqrt{\rho+u^2}}$ and $\rho = \sigma_x^2/\sigma_y^2$. Taking the derivative of $F_{z_r}(u)$ w.r.t. u, we obtain the PDF of z_r, as given by

$$f_{z_r}(u) = \frac{\rho}{2(\rho+u^2)^{\frac{3}{2}}}.$$

The PDF is an even function of u and, hence, $\mathbb{E}\{z_r\} = 0$. Directly calculating the variance of z_r with an infinite region of u will result in an infinite variance. Thus, we put a finite upper limit on $|z_r|$ and approximate its variance as follows:

$$\sigma_{z_r}^2 = 2\int_0^{\bar{u}} u^2 f_{z_r}(u)\mathrm{d}u = \rho^2\left(\operatorname{asinh}\left(\frac{\bar{u}}{\sqrt{\rho}}\right) - \frac{\bar{u}}{\sqrt{\rho+\bar{u}^2}}\right)$$

$$\stackrel{\bar{u}=\bar{k}\sqrt{\rho}}{=} \rho\left(\operatorname{asinh}\left(\bar{k}\right) - \frac{\bar{k}}{\sqrt{1+\bar{k}^2}}\right) \stackrel{\bar{k}\gg 1}{\approx} \rho\left(\ln(2\bar{k})-1\right),$$

where under $\bar{k} \gg 1$, two approximations are used, i.e. $\operatorname{asinh}\left(\bar{k}\right) \approx \ln\left(2\bar{k}\right)$ and $\frac{\bar{k}}{\sqrt{1+\bar{k}^2}} \approx 1$. Based on the PDF $f_{z_r}(u)$, the probability that $|z_r| \le \bar{u}$ can be calculated as follows:

$$\mathsf{P}\{|z_r| \le \bar{u}\} = \int_{-\bar{u}}^{\bar{u}} f_{z_r}(u)\mathrm{d}u = \frac{\bar{u}}{\sqrt{\bar{u}^2+\rho}} = \frac{\bar{k}}{\sqrt{\bar{k}^2+1}},$$

where the substitution $\bar{u} = \bar{k}\sqrt{\rho}$ is performed. Assume that $\mathsf{P}\{|z_r| \le \bar{u}\} = 1-\epsilon$, where ϵ is a sufficiently small probability claimed in the condition of Proposition 6.1. Then we can solve that $\bar{k} = \sqrt{\frac{\bar{\epsilon}}{1-\bar{\epsilon}^2}}$ and, moreover,

$$\sigma_{z_r}^2 \approx \rho\left(\ln\left(\sqrt{\frac{4\bar{\epsilon}}{1-\bar{\epsilon}^2}}\right)-1\right), \quad \text{s.t. } \bar{\epsilon} = 1-\epsilon. \tag{A.10}$$

As mentioned earlier, z_i has the same variance as z_r. Thus, the final variance of z becomes the one given in (6.35).

A.7 Deriving the Powers of the Four Terms of $\tilde{X}_n[l]$ Given in (6.33)

Power of $\tilde{X}_n^{\mathrm{S}}[l]$: Applying Lemma 6.4, the variance of $\tilde{X}_n^{\mathrm{S}}[l]$ given in (6.33), as denoted by σ_{S}^2, can be calculated as follows:

$$\sigma_{\mathrm{S}}^2 = \frac{1}{\mathrm{k}^2\bar{M}^2} \underbrace{\sum_{p=0}^{P-1} \sigma_p^2 \mathbb{E}\left\{\begin{array}{l} \left(\sum_{m_1=0}^{\bar{M}-1} e^{\mathrm{j}2\pi\nu_p m_1 T_{\mathrm{s}}}\right) \\ \times \left(\sum_{m_2=0}^{\bar{M}-1} e^{-\mathrm{j}2\pi\nu_p m_2 T_{\mathrm{s}}}\right) \end{array}\right\}}_{\Pi} \approx \frac{1}{\mathrm{k}^2}\left(\sigma_P^2 + \alpha - \alpha\bar{M}^2\right),$$

$$\text{s.t. } \alpha = \frac{\pi^2 T_{\mathrm{s}}^2}{3} \sum_{p=0}^{P-1} \sigma_p^2 \nu_p^2,\ \sigma_P^2 = \sum_{p=0}^{P-1} \sigma_p^2, \tag{A.11}$$

where the approximation is due to the Taylor series used for calculating Π, as illustrated below.

Based on the expression of Π given in (A.11), we can have

$$\begin{aligned} \Pi &\overset{(a)}{=} \sum_{p=0}^{P-1} \sigma_p^2 \mathbb{E}\left\{ \sum_{\substack{m_1=0 \\ m_2=m_1}}^{\bar{M}-1} + \sum_{\substack{m=|m_1-m_2| \\ =1}}^{\bar{M}-1} \begin{array}{c} (\bar{M}-m)\times \\ (e^{\mathrm{j}2\pi\nu_p m T_{\mathrm{s}}} + e^{-\mathrm{j}2\pi\nu_p m T_{\mathrm{s}}}) \end{array} \right\} \\ &= \sum_{p=0}^{P-1} \sigma_p^2 \left(\bar{M} + 2\sum_{m=1}^{\bar{M}-1} (\bar{M}-m)\cos(2\pi\nu_p m T_{\mathrm{s}}) \right) \\ &\overset{\omega_p=2\pi\nu_p T_{\mathrm{s}}}{\approx} \sum_{p=0}^{P-1} \sigma_p^2 \left(\bar{M} + 2\sum_{m=1}^{\bar{M}-1} (\bar{M}-m)\left(1 - \frac{m^2\omega_p^2}{2}\right) \right) \\ &\overset{(b)}{=} \bar{M}^2\left(\sigma_P^2 + \alpha - \alpha\bar{M}^2\right), \end{aligned} \tag{A.12}$$

where $\alpha = \sum_{p=0}^{P-1} \sigma_p^2 \frac{\omega_p^2}{12} = \frac{\pi^2 T_{\mathrm{s}}^2}{3} \sum_{p=0}^{P-1} \sigma_p^2 \nu_p^2$ and $\sigma_P^2 = \sum_{p=0}^{P-1} \sigma_p^2$. Note that the result $\overset{(a)}{=}$ is obtained by separately calculating the cases of $m_1 = m_2$ and $m_1 \neq m_2$; the approximation is obtained based on the second-degree Taylor expansion of the cosine function w.r.t. m, and the result $\overset{(b)}{=}$ is obtained based on the following formula of summations: $\sum_{m=1}^{\bar{M}-1} m = \bar{M}^2/2 - \bar{M}/2$, $\sum_{m=1}^{\bar{M}-1} m^2 = \bar{M}^3/3 - \bar{M}^2/2 + \bar{M}/6$ and $\sum_{m=1}^{\bar{M}-1} m^3 = \bar{M}^4/4 - \bar{M}^3/2 + \bar{M}^2/4$ [4].

Power of $\tilde{X}_n^{\mathrm{I}}[l]$: Applying Lemma 6.4, the variance of $\tilde{X}_n^{\mathrm{I}}[l]$ given in (6.33), as denoted by σ_{I}^2, can be calculated as follows:

$$\sigma_{\mathrm{I}}^2 = \mathbb{E}\left\{\left|\frac{\tilde{S}_n[l']}{\mathsf{k}\tilde{S}_n[l]}\right|^2\right\} \times \underbrace{\sum_{p=0}^{P-1}\sigma_p^2 \mathbb{E}\left\{\begin{array}{c}\sum_{l'=0,l'\neq l}^{\bar{M}-1}\left(\sum_{m_1=0}^{\bar{M}-1} e^{\mathrm{j}2\pi\nu_p m_1 T_\mathrm{s}}\frac{\mathcal{Z}_{\bar{M}}^{(l-l')m_1}}{\sqrt{\bar{M}}}\right)\\ \times\left(\sum_{m_2=0}^{\bar{M}-1} e^{-\mathrm{j}2\pi\nu_p m_2 T_\mathrm{s}}\frac{\mathcal{Z}_{\bar{M}}^{-(l-l')m_2}}{\sqrt{\bar{M}}}\right)\end{array}\right\}}_{\Gamma},$$

$$= \beta(\epsilon)\mathsf{k}^{-2}(\alpha\bar{M}^2 - \alpha), \text{ s.t. } \alpha = \frac{\pi^2 T_\mathrm{s}^2}{3}\sum_{p=0}^{P-1}\sigma_p^2\nu_p^2 \tag{A.13}$$

where $\mathbb{E}\left\{\left|\frac{\tilde{S}_n[l']}{\mathsf{k}\tilde{S}_n[l]}\right|^2\right\} = \beta(\epsilon)\mathsf{k}^{-2}$ according to Lemma 6.2 and Proposition 6.1, and Γ is calculated below

$$\Gamma = \sum_{p=0}^{P-1}\sigma_p^2 \sum_{l'=0,l'\neq l}^{\bar{M}-1}\sum_{m_1=0}^{\bar{M}-1}\sum_{m_2=0}^{\bar{M}-1} e^{\mathrm{j}2\pi\nu_p(m_1-m_2)T_\mathrm{s}}\frac{\mathcal{Z}_{\bar{M}}^{(l-l')(m_1-m_2)}}{\bar{M}^{\frac{3}{2}}}$$

$$\overset{(a)}{=} \underbrace{\sum_{p=0}^{P-1}\sigma_p^2\left(\begin{array}{c}\sum_{l'=0}^{\bar{M}-1}\sum_{m_1=0}^{\bar{M}-1}\sum_{m_2=0}^{\bar{M}-1}\\ e^{\mathrm{j}2\pi\nu_p(m_1-m_2)T_\mathrm{s}}\frac{\mathcal{Z}_{\bar{M}}^{(l-l')(m_1-m_2)}}{\bar{M}^{\frac{3}{2}}}\end{array}\right)}_{\Gamma_1}$$

$$-\underbrace{\sum_{p=0}^{P-1}\sigma_p^2\left(\begin{array}{c}\sum_{l'=l}\sum_{m_1=0}^{\bar{M}-1}\sum_{m_2=0}^{\bar{M}-1}\\ e^{\mathrm{j}2\pi\nu_p(m_1-m_2)T_\mathrm{s}}\frac{\mathcal{Z}_{\bar{M}}^{0\times(m_1-m_2)}}{\bar{M}^{\frac{3}{2}}}\end{array}\right)}_{\Pi/\bar{M}^2} \overset{(b)}{=} \sigma_p^2 - \left(\sigma_p^2 + \alpha - \alpha\bar{M}^2\right) = \alpha\bar{M}^2 - \alpha. \tag{A.14}$$

In the calculation, the result $\overset{(a)}{=}$ is obtained by separately considering the summation over $l' = 0, 1, \cdots, \bar{M}-1$ (leading to Γ_1), and the special case of $l' = l$ (which happens to become $\Pi/\bar{M}^2$). As calculated in Remark A.1, $\Gamma_1 = \sigma_p^2$. Then, combining the result of Π calculated in (A.12), we obtain the result $\overset{(b)}{=}$ in (A.14).

Remark A.1 For convenience, we provide below the expression of Γ_1,

$$\sum_{p=0}^{P-1}\sigma_p^2\left(\sum_{l'=0}^{\bar{M}-1}\sum_{m_1=0}^{\bar{M}-1}\sum_{m_2=0}^{\bar{M}-1} e^{\mathrm{j}2\pi\nu_p(m_1-m_2)T_\mathrm{s}}\frac{\mathcal{Z}_{\bar{M}}^{(l-l')(m_1-m_2)}}{\bar{M}^{\frac{3}{2}}}\right),$$

where we notice that the summations enclosed in the round brackets can be calculated first. Similar to the way Π is calculated, see Section 6.3.2, we calculate Γ_1 by considering $m_1 = m_2$ and $m_1 \neq m_2$. For the first case, we have

$$\sum_{l'=0}^{\bar{M}-1} \sum_{\substack{m_1=0 \\ m_2=m_1}}^{\bar{M}-1} e^{j2\pi\nu_p(m_1-m_2)T_s} \frac{\mathcal{Z}_{\bar{M}}^{(l-l')(m_1-m_2)}}{\bar{M}^{\frac{3}{2}}} = 1,$$

where we have used $\mathcal{Z}_{\bar{M}}^{(l-l')\times 0} = \frac{1}{\sqrt{\bar{M}}}$; see (6.11). For the case of $m_1 \neq m_2$, the summation w.r.t. l' is always zero since the summands, given $\forall m_1, m_2$, are the samples of a complex exponential signal within an integer multiple of periods. Combining the two cases, we have $\Gamma_1 = \sum_{p=0}^{P-1} \sigma_p^2 \times 1 = \sigma_P^2$.

Power of $\tilde{X}_n^{\mathrm{Z}}[l]$: According to Lemma 6.2, $Z_n^{\mathrm{A}}[l]$, $Z_n^{\mathrm{B}}[l]$ and $\tilde{S}_n[l]$ are zero-mean complex Gaussian variables. Thus, applying Lemma 6.1, we have

$$\mathbb{V}\left\{\frac{Z_n^{\mathrm{A}}[l]}{\mathrm{k}\tilde{S}_n[l]}\right\} = \mathbb{E}\left\{\left|\frac{Z_n^{\mathrm{A}}[l]}{\mathrm{k}\tilde{S}_n[l]}\right|^2\right\} = \frac{\beta(\epsilon)\sigma_{Z^{\mathrm{A}}}^2}{\mathrm{k}^2\sigma_d^2} = \frac{\beta(\epsilon)\mathrm{k}^{-2}\mathrm{i}_{\max}\sum_{p=0}^{P-1}(p+1)\sigma_p^2}{P\bar{M}},$$

where $\sigma_{Z^{\mathrm{A}}}^2$ is given in (6.13). Likewise, we obtain

$$\mathbb{V}\left\{\frac{Z_n^{\mathrm{B}}[l]}{\mathrm{k}\tilde{S}_n[l]}\right\} = \frac{\beta(\epsilon)\mathrm{k}^{-2}(Q-\mathrm{i}_{\min})\sum_{p=0}^{P-1}(P-p)\sigma_p^2}{P\bar{M}}.$$

Using the above two results, the variance of $\tilde{X}_n^{\mathrm{Z}}[l]$ given in (6.33), as denoted by σ_{Z}^2, can be calculated as follows:

$$\begin{aligned}
\sigma_{\mathrm{Z}}^2 &= \mathbb{V}\left\{\frac{Z_n^{\mathrm{A}}[l]}{\mathrm{k}\tilde{S}_n[l]}\right\} + \mathbb{V}\left\{\frac{Z_n^{\mathrm{B}}[l]}{\mathrm{k}\tilde{S}_n[l]}\right\} + \mathbb{E}\left\{\left(\frac{Z_n^{\mathrm{A}}[l]}{\mathrm{k}\tilde{S}_n[l]}\right)^* \frac{Z_n^{\mathrm{B}}[l]}{\mathrm{k}\tilde{S}_n[l]}\right\} \\
&\quad + \mathbb{E}\left\{\left(\frac{Z_n^{\mathrm{B}}[l]}{\mathrm{k}\tilde{S}_n[l]}\right)^* \frac{Z_n^{\mathrm{A}}[l]}{\mathrm{k}\tilde{S}_n[l]}\right\} \\
&\approx \frac{\beta(\epsilon)\begin{pmatrix} \mathrm{i}_{\max}\sum_{p=0}^{P-1}(p+1)\sigma_p^2 \\ +(Q-\mathrm{i}_{\min})\sum_{p=0}^{P-1}(P-p)\sigma_p^2 \end{pmatrix}}{\mathrm{k}^2 P\bar{M}} \overset{\substack{\mathrm{i}_{\min}=0 \\ \mathrm{i}_{\max}=Q}}{\leq} \frac{\beta(\epsilon)Q(P+1)\sigma_P^2}{\mathrm{k}^2 P\bar{M}}, \\
&\text{s.t. } \sigma_P^2 = \sum_{p=0}^{P-1}\sigma_p^2,
\end{aligned} \tag{A.15}$$

where the two expectations are approximately zeros, as proved in Remark A.2. The equality in the last line can be achieved when the two conditions are satisfied, $\mathrm{i}_{\min} = 0$ and $\mathrm{i}_{\max} = Q$. These conditions are likely to hold for a large P, since the larger P is, the more variety can present among targets.

Remark A.2 For convenience, we define some shorthand expressions: $Z_n^{\mathrm{A}}[l] = a + \mathrm{j}b$, $Z_n^{\mathrm{B}}[l] = c + \mathrm{j}d$ and $\tilde{S}_n[l] = e + \mathrm{j}f$. Based on (6.18), we have

$$\frac{Z_n^{\mathrm{A}}[l](Z_n^{\mathrm{B}}[l])^*}{\mathrm{k}^2|\tilde{S}_n[l]|^2} = \frac{\overbrace{(ac+bd)+\mathrm{j}(bc-ad)}^{R}}{\underbrace{\mathrm{k}^2(e^2+f^2)}_{S}},$$

The expectation of R/S can be calculated as follows: [2]

$$\mathbb{E}\{R/S\} \approx \mu_R/\mu_S - \sigma_{RS}/\mu_S^2 + \sigma_S^2\mu_R/\mu_S^3, \tag{A.16}$$

where μ_R and μ_S are means, σ_{RS} denotes covariance and σ_S^2 denotes variance. Based on Lemma 6.1, a, b, c, d, e, and f are independent centered Gaussian variables. Therefore, we have $\mu_R = 0$ and, moreover,

$$\begin{aligned}\sigma_{RS} &= \mathbb{E}\left\{(R-\mu_R)^*(S-\mu_S)\right\} = \mathbb{E}\{R^*S\} \\ &= \mathbb{E}\left\{\left((ad-bc)\left(e^2+f^2\right)\right)\mathrm{j} + (ac+bd)\left(e^2+f^2\right)\right\} = 0.\end{aligned}$$

These further lead to $\mathbb{E}\left\{\left(\frac{Z_n^{\mathrm{A}}[l]}{\mathrm{k}\tilde{S}_n[l]}\right)^* \frac{Z_n^{\mathrm{B}}[l]}{\mathrm{k}\tilde{S}_n[l]}\right\} \approx 0$. Likewise, we can show $\mathbb{E}\left\{\left(\frac{Z_n^{\mathrm{B}}[l]}{\mathrm{k}\tilde{S}_n[l]}\right)^* \frac{Z_n^{\mathrm{A}}[l]}{\mathrm{k}\tilde{S}_n[l]}\right\} \approx 0$.

Power of $\tilde{\boldsymbol{W}}_{\boldsymbol{n}}[l]$: According to (6.17), we have $\tilde{W}_n[l] = W_n[l]/(\mathrm{k}\tilde{S}_n[l])$ which is the ratio of two complex Gaussian. Thus, directly applying Proposition 6.1 gives us

$$\sigma_W^2 = \mathbb{V}\left\{\tilde{W}_n[l]\right\} = \frac{\beta(\epsilon)\sigma_w^2}{\mathrm{k}^2\sigma_d^2}, \tag{A.17}$$

where $\mathbb{V}\left\{W_n[l]\right\} = \sigma_w^2$, as proved in Lemma 6.2.

A.8 Proof of Proposition 6.2

Based on (A.11), we have

$$f(\bar{M}) \approx \frac{\bar{M}I}{(\bar{M}+Q)}\frac{1}{\mathrm{k}^2}\left(\sigma_P^2 + \alpha - \alpha\bar{M}^2\right)$$

where the approximation is due to $\tilde{N} = \left\lfloor \frac{I}{\tilde{M}} \right\rfloor \approx \frac{I}{\tilde{M}} = \frac{I}{\bar{M}+Q}$. The first derivative of $f(\bar{M})$ w.r.t. $\bar{M}$ is

$$f'(\bar{M}) = \frac{-\mathrm{C}(2\bar{M}^3\alpha + 3Q\bar{M}^2\alpha - Q\alpha - Q\sigma_P^2)}{\left(\bar{M}+Q\right)^2}, \tag{A.18}$$

and the second derivative is

$$f''(\bar{M}) = -\frac{2\mathsf{C}\left(\bar{M}^3\alpha + 3\bar{M}^2 Q\alpha + 3\bar{M}Q^2\alpha + Q\alpha + Q\sigma_P^2\right)}{\left(\bar{M}+Q\right)^3} \tag{A.19}$$

where $\mathsf{C}(>0)$ absorbs the $\bar{M}$-independent coefficients. Since $f''(\bar{M}) < 0$ always holds, $f'(\bar{M})$ monotonically decreases. We notice that $f'(0) > 0$. Thus, there exists $\bar{M}_f^*$ such that $f'(\bar{M}) > 0$ for $\bar{M} < \bar{M}_f^*$, $f'(\bar{M}) = 0$ for $\bar{M} = \bar{M}_f^*$, and $f'(\bar{M}) < 0$ for $\bar{M} > \bar{M}_f^*$. The value of $\bar{M}_f^*$ can be determined by solving the equation $f'(\bar{M}) = 0$ which is essentially:

$$2\bar{M}^3\alpha + 3Q\bar{M}^2\alpha - Q\alpha - Q\sigma_P^2 = 0.$$

Directly using the cubic formula leads to a solution with a complex structure and does not provide much insight. To this end, we consider the case of $\bar{M} \gg 3Q/2$ and simplify the equation by dropping the quadratic term, obtaining $2\bar{M}^3 a - Qa - Q\sigma_P^2 = 0$. Thus, an approximation of $\bar{M}_f^*$ is achieved as follows:

$$\bar{M}_f^* \overset{\bar{M}\gg 3Q/2}{\approx} \sqrt[3]{\frac{Q\alpha + Q\sigma_P^2}{2\alpha}}.$$

A.9 Proof of Proposition 6.3

Based on (A.13), (A.15), and (A.17), we have

$$\begin{aligned} \mathrm{g}(\bar{M}) &= \frac{\beta(\epsilon)}{\mathsf{k}^2}\left((\alpha\bar{M}^2 - \alpha) + \frac{Q(P+1)\sigma_P^2}{\bar{M}P} + \frac{\sigma_w^2}{\sigma_d^2}\right) \\ &= \frac{\beta(\epsilon)}{\mathsf{k}^2}\,\frac{\alpha\bar{M}^3 + \left(\frac{\sigma_w^2}{\sigma_d^2} - \alpha\right)\bar{M} + \frac{Q(P+1)\sigma_P^2}{P}}{\bar{M}}. \end{aligned}$$

The first derivative of $\mathrm{g}(\bar{M})$ is

$$\mathrm{g}'(\bar{M}) = \frac{\frac{\beta(\epsilon)}{\mathsf{k}^2}\left(2\bar{M}^3\alpha - \frac{Q(P+1)\sigma_P^2}{P}\right)}{\bar{M}^2}.$$

Moreover, we can readily validate that the second derivative of $\mathrm{g}(\bar{M})$ is constantly positive, which is suppressed here for brevity. Therefore, there exists $\bar{M}_g^*$ such that $\mathrm{g}'(\bar{M}) < 0$ for $\bar{M} < \bar{M}_g^*$, $\mathrm{g}'(\bar{M}) = 0$ for $\bar{M} = \bar{M}_g^*$, and $\mathrm{g}'(\bar{M}) > 0$ for $\bar{M} > \bar{M}_g^*$. Solving $\mathrm{g}'(\bar{M}) = 0$, we obtain

$$\bar{M}_g^* = \sqrt[3]{Q(P+1)\sigma_P^2/(2\alpha P)}.$$

A.10 Deriving (9.31)

Taking the expectation of $f_r(\mathbf{P}, \mathbf{b})$ leads to

$$\mu_f = \mathbb{E}\left\{f_r(\mathbf{P}, \mathbf{b})\right\} = \Re\left\{ \sum_{m=0}^{M-1} \underbrace{\mathbb{E}\left\{b_m\right\}}_{\mu_{f1}} \underbrace{\mathbb{E}\left\{e^{j\Delta_{hm}}\right\}}_{\mu_{f2}} \right\}, \tag{A.20}$$

where the randomness in $e^{j\Delta_{hm}}$ is caused by $\mathbf{P}$; see (9.30). Based on (9.21), we can calculate μ_{f1} as

$$\mu_{f1} = \left(1 \times \frac{M-Q}{M} + (-1) \times \frac{Q}{M}\right) = \frac{M-2Q}{M}. \tag{A.21}$$

Given a large sample set of $\mathbf{P}$, the second term on the RHS of (9.30) takes $\arg\{\lfloor \mathbf{b}_h \odot \mathbf{a}_{\theta\phi} \rfloor_m\}$ $(\forall m)$ in the probability of $\frac{1}{M}$. Thus, μ_{f2} can be calculated as follows:

$$\mu_{f2} = e^{jm\left(u_{\theta\phi} + \frac{2\pi}{M}\right)} \mathbb{E}\left\{ \frac{\mathbf{1}^{\mathrm{T}}(\mathbf{b}_h \odot \mathbf{a}_{\theta\phi})}{M} \right\}, \tag{A.22}$$

where the expectation still exists due to the presence of $\mathbf{b}_h$. The numerator enclosed in the expectation can be written into $\sum_{m'=0}^{M-1} b_{hm} e^{-jm'\left(u_{\theta\phi} + \frac{2\pi}{M}\right)}$. This further leads to

$$\begin{aligned}
\mathbb{E}\left\{ \frac{\mathbf{1}^{\mathrm{T}}(\mathbf{b}_h \odot \mathbf{a}_{\theta\phi})}{M} \right\} &= \frac{1}{M} \mathbb{E}\left\{ \sum_{m'=0}^{M-1} b_{hm} e^{-jm'\left(u_{\theta\phi} + \frac{2\pi}{M}\right)} \right\} \\
&= \mu_{f1} \frac{S_M(u_{\theta\phi})}{M} e^{-j\frac{M-1}{2}\left(u_{\theta\phi} + \frac{2\pi}{M}\right)},
\end{aligned}$$

where $\mathbb{E}\{b_{hm}\} = \mu_{f1}$ based on (9.21) and (A.21), and $\sum_{m'=0}^{M-1} e^{-jm'\left(u_{\theta\phi} + \frac{2\pi}{M}\right)} = S_M(u_{\theta\phi}) e^{-j\frac{M-1}{2}\left(u_{\theta\phi} + \frac{2\pi}{M}\right)}$ is applied. The expression of $S_M(u_{\theta\phi})$ is given in (9.31). Substituting μ_{f1} and μ_{f2} into (A.20) and after some basic manipulations, we obtain the final μ_f, as given in (9.31).

The variance of $f_r(\mathbf{P}, \mathbf{b})$, as denoted by σ_f^2, can be calculated as follows:

$$\begin{aligned}
\sigma_f^2 &= \frac{1}{2} \operatorname{var}\left\{ \sum_{m=0}^{M-1} b_m e^{-j\Delta_{hm}} \right\} \overset{(a)}{=} \frac{1}{2} \sum_{m=0}^{M-1} \operatorname{var}\left\{ b_m e^{-j\Delta_{hm}} \right\} \\
&\overset{(b)}{=} \frac{1}{2} \sum_{m=0}^{M-1} \left(\mathbb{E}\{b_m^2\} - \mu_{f1}^2 |\mu_{f2}|^2 \right) \\
&\overset{(c)}{=} \frac{1}{2} \sum_{m=0}^{M-1} \left(1 - \mu_{f1}^2 \left(\mu_{f1} \frac{S_M(u_{\theta\phi})}{M} \right)^2 \right) \overset{(d)}{=} \frac{(M - \mu_{f1}^2 \mu_f)}{2},
\end{aligned} \tag{A.23}$$

where the results $\overset{(a)}{=}$ and $\overset{(b)}{=}$ are obtained based on the independence between b_m and $e^{-j\Delta_{hm}}$; the result $\overset{(c)}{=}$ is due to $\mathbb{E}\{b_m^2\} = 1^2 \times \frac{M-Q}{M} + (-1)^2 \times \frac{Q}{M} = 1$ and the substitution of μ_{f2} derived in (A.22), and μ_f given in (9.31) is used to yield the final result $\overset{(d)}{=}$. Replacing μ_{f1} with its expression given in (A.21), we obtain the final result of σ_f^2, as given in (9.31).

References

1 K. Wu, J. A. Zhang, X. Huang, and Y. J. Guo. OTFS-based joint communication and sensing for future industrial IoT. *IEEE Internet of Things Journal*, 1, early access, 2021. doi: 10.1109/JIOT.2021.3139683.

2 H. Seltman. Approximations for mean and variance of a ratio. *unpublished note*, 2012.

3 R. J. Baxley, B. T. Walkenhorst, and G. Acosta-Marum. Complex Gaussian Ratio distribution with applications for error rate calculation in fading channels with imperfect CSI. In *2010 IEEE Global Telecommunications Conference GLOBECOM 2010*, pages 1–5, 2010. doi: 10.1109/GLOCOM.2010.5683407.

4 K. H. Rosen. *Handbook of Discrete and Combinatorial Mathematics*. CRC Press, 2017.

Index

Joint Communications and Sensing: From Fundamentals to Advanced Techniques, First Edition.
Kai Wu, J. Andrew Zhang, and Y. Jay Guo.

Printed and bound by CPI Group (UK) Ltd, Croydon, CR0 4YY
16/04/2025
14658583-0001